PHOTOCHEMISTRY IN ORGANIC SYNTHESIS

Special Publication No 57

Photochemistry in Organic Synthesis

Edited by

J.D. Coyle
Chemistry Department, The Open University

Contributors

H.A.J. Carless, Birkbeck College, University of London
A. Gilbert, University of Reading
W.M. Horspool, University of Dundee
J. Hutchison, The Electricity Council Research Centre, Chester
R.F. Newton, Glaxo Group Research Ltd., Ware
K.H. Pfoertner, F.Hoffmann-LaRoche and Co. Ltd, Basel
T.W. von Geldern, Stanford University
P.A. Wender, Stanford University

The Royal Society of Chemistry
Burlington House, London W1V 0BN

British Library Cataloguing in Publication Data

Photochemistry in organic synthesis.—
(Special publication; no. 57)
1. Chemistry, Organic—Synthesis
2. Photochemistry
I. Coyle, J.D. II. Carless, H.A.J.
III. Royal Society of Chemistry IV. Series
547.2 QD262

ISBN 0-85186-656-5

Printed in Great Britain by
Whitstable Litho Ltd., Whitstable, Kent

Preface

The idea for this book grew out of two beliefs: first, that there are a large number of synthetically useful organic transformations that can only be carried out, or can best be carried out, by means of a photochemical reaction; secondly, that although synthetic chemists should be aware of these transformations, the graduate education of many does not provide a sufficient basis in synthetic photochemistry.

The intention is not an attempt to "sell" photochemistry as a solution to problems that remain in organic synthesis, but rather to equip the chemist engaged in synthesis with as good a working knowledge of photochemical reactions as he has of thermal reactions. For example, a chemist dealing with substituted aromatic systems needs to know about the wide range of substitution reactions that can be carried out readily and efficiently by ultraviolet irradiation, many of which are impossible by a direct thermal route. Often photochemical and thermal methods are complementary tools in the synthetic chemist's armoury.

The chapters are written on the assumption that the reader has a familiarity with organic structures, terminology, compound types and synthetic reactions such as might be acquired in a three-year honours degree course with chemistry as a major subject, or the equivalent gained by experience of working in synthesis. A broad range of reactions and compound types is covered, but there are three strands to the overall theme. First, there are chapters surveying photochemical reactions based on major functional classes of substrate, such as carbonyl compounds or alkenes. Interspersed between these chapters are selections of transformations with particular relevance to the synthesis or interconversion of certain major product groups (steroids, terpenoids, alkaloids), chosen in each case to illustrate reaction types introduced in the immediately preceding chapters. Finally there are chapters dealing with practical aspects of conducting photochemical reactions in the laboratory or on a larger scale.

The aim of the book is to introduce or strengthen elements in the strategies employed by the synthetic chemist that consist of reaction types which can best be accomplished photochemically. In the longer term this may bear fruit in the increased use of such reactions both by laboratory chemists and in the chemical manufacturing industry.

John Coyle Milton Keynes, May 1986

Contents

1
Basic Principles

By J.D. Coyle
CHEMISTRY DEPARTMENT, THE OPEN UNIVERSITY, MILTON KEYNES MK7 6AA, UK

For a balanced or comprehensive account of the principles of photochemistry or of photophysical aspects of the subject, the reader is referred to other sources.[1] This chapter is written from the point of view of an organic chemist with an interest in synthesis, and it covers those properties of the excited states of organic molecules and those photochemical reactions whose appreciation will help the chemist to make the most of the large amount of information available that is of potential use in organic synthesis.

Absorption of Light by the Substrate

Photochemistry is concerned with chemical change that occurs through the electronically excited states of molecules. By far the most widely used method of generating such excited states is through the absorption by a molecule of one photon of visible or ultraviolet light. Molecular electronic spectroscopy is therefore of interest to the photochemist, because the ultraviolet-visible absorption spectrum of a substrate provides information about the wavelengths absorbed. Consider the spectrum of benzophenone shown in Figure 1.

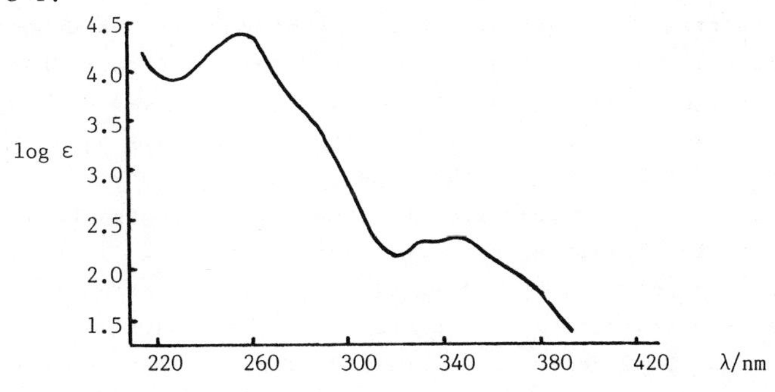

Figure 1. Absorption spectrum of benzophenone

It is clear from the spectrum that visible light (with the exception of far-violet wavelengths) will be ineffective in promoting photochemical reaction, because light that is not absorbed by the molecule cannot induce chemical change. However, near-ultraviolet radiation of about 350 nm wavelength would be well suited to the requirements of this substrate. Shorter-wavelength ultraviolet light (e.g. 254 nm) may also be effective, but it should be noted that a higher-energy excited state will be produced initially (a different absorption band is involved) which might lead to different products. The absorption spectrum provides this indication of the wavelengths required to generate different electronically excited states, though, as we shall see later, most photochemical reactions occur anyway from one of the lowest-energy states. In some contexts the formation of the same products using higher-energy radiation may be more convenient in terms of the light sources that are available, but it does represent a significant waste of energy. Absorption spectra for organic molecules in fluid solution generally consist of broad overlapping bands, and this represents another limit on the amount of information that is available to the photochemist, because some bands may be completely hidden beneath other, more intense bands.[2]

The most commonly used description of electronically excited states in organic chemistry is based on valence-shell molecular orbitals. Normally the molecular orbitals of the unexcited molecule (the ground state) are employed to describe the electronic transitions that can occur and the excited states to which they lead. Figure 2 shows a stylised orbital energy-level diagram for an unsaturated hydrocarbon, indicating some of the valence-shell orbitals. In the lowest-energy, ground-state molecule (Figure 2a) the lower (bonding) orbitals are fully occupied by pairs of electrons and the upper (anti-bonding) orbitals are all unoccupied. Excitation can lead to a higher-energy state (Figure 2b) in which an electron has been promoted from the highest-energy bonding orbital to the lowest-energy anti-bonding orbital. This excited state is designated a (π,π^*) state, and the transition leading to it is a $\pi \rightarrow \pi^*$ transition. In particular, this state is a (π,π^*) singlet state, since the two unpaired electrons have opposed spin; Figure 2c shows the electronic configuration of the related (π,π^*) triplet state, in which the electron spins are parallel.

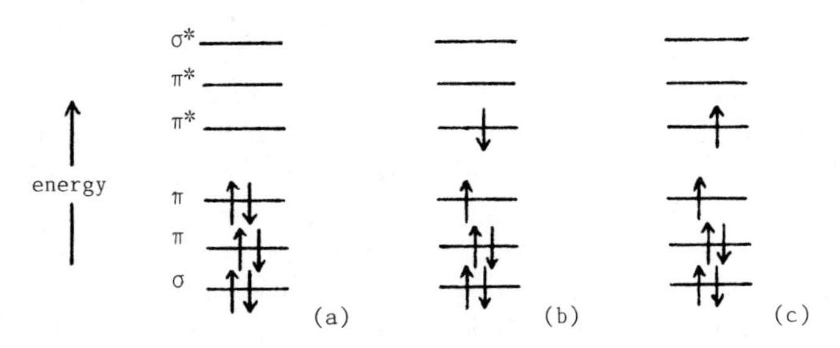

Figure 2. An orbital energy level diagram

For any molecule there is a series of excited states obtained by different electronic transitions, and for convenience they are designated S_1, S_2, S_3 and so on for the singlet states in order of increasing energy, and T_1, T_2, T_3 and so on for the triplet states in order of increasing energy. The ground state is S_0; very few stable organic molecules have a ground state that is not a singlet. Molecules that contain carbon-carbon multiple bonds but no hetero-atom normally have a lowest excited state (S_1) that is (π,π^*) in character. In molecules with heteroatoms but without extensive conjugation the highest filled orbital in the ground state is often non-bonding (n), and so S_1 may be (n,π^*) or (n,σ^*) in nature. Most photochemical reactions can be rationalised by the involvement of excited states described in this way. However, such a simple molecular orbital description is sometimes inappropriate, especially for very small molecules. In particular there are Rydberg excited states that are more appropriately visualised as arising from promotion of an electron to an orbital that extends substantially beyond the core of nuclei and inner-shell electrons.[3] Rydberg states can therefore be regarded as having a large, positively charged core and an outer region of electron density.

The particular interest in the excited states of lowest energy (S_1 or T_1) is a result of the observation that photochemical reaction normally occurs from these states regardless of the state that is produced initially. There are exceptions to this generalisation, but it is a good starting point in considering any particular system. The nature of S_1 can often be deduced from a knowledge of the molecular orbital pattern of the molecule and from the appearance (particularly the relative intensity) of the longest-wavelength band in the compound's absorption spectrum. The situation is not so straightforward for T_1. Triplet states are rarely obtained directly by absorption of a photon, since singlet → triplet absorption is strongly forbidden, but they are obtained indirectly via a singlet excited state or by energy transfer. A triplet state always has a lower energy than the singlet state with the same electronic configuration, but the singlet-triplet energy difference can vary considerably. This energy gap is much larger for (π,π*) states than for (n,π*) states, for example, and it is not uncommon for a molecule to have a lowest (n,π*) singlet excited state but a lowest (π,π*) triplet state. This situation is depicted on a state energy diagram in Figure 3.

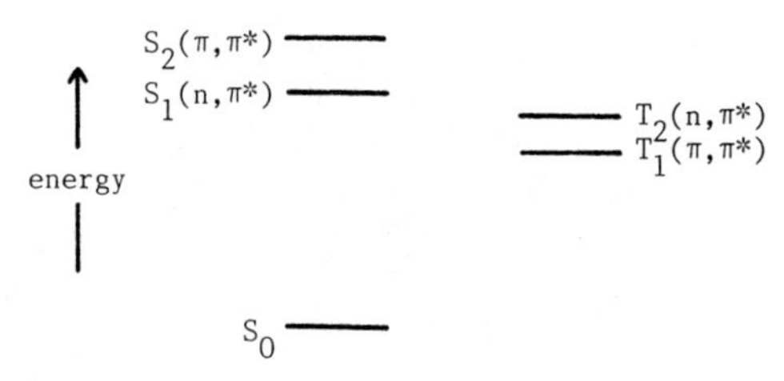

Figure 3. A state energy level diagram.

Absorption of a photon of light is an extremely rapid event, and an excited state is formed initially with the same geometry, that is with the same nuclear coordinates, as the ground state. As well as being electronically excited, this initial state is a vibrationally excited species, and by rapid vibrational deactivation it relaxes to the excited state with "equilibrium" geometry. The relaxed excited state is a distinct species with characteristic properties such as bond lengths, angles, vibrational frequencies, and dipole moment, that can in principle be measured and are generally different from those of the ground-state molecule.[4] The excited state also has its own characteristic range of chemical reactions, and such photochemical change is the issue of especial relevance to organic synthesis.

The Route to Products

A feature of excited states that imposes limitations on observable photochemical change is the very short lifetime of such states. In fluid solution at room temperature the lifetime of the lowest excited states of organic molecules ranges from 10^{-3} to 10^{-12}

seconds, and it is helpful to look into the origin of this property. The lifetime can be expressed as the reciprocal of the sum of first-order rate constants corresponding to all the routes by which the excited state reacts or decays. As well as chemical reactions that may have very high rate constants, there are rapid physical processes by which the excited state regenerates the ground state from which it was derived. For many excited states such fast physical processes dominate the lifetime, that is they are very much faster than chemical reaction. This provides the chief justification for the generalisation that photochemical reaction generally occurs through S_1 or T_1, since most excited states of higher energy than these decay very rapidly to give the S_1 or T_1 state. Such processes are examples of radiationless decay (no light is emitted), and they are known as internal conversion if they occur without a change in spin (e.g. $S_2 \rightarrow S_1$ or $T_2 \rightarrow T_1$), or intersystem crossing if there is a change in spin (e.g. $S_1 \rightarrow T_1$). Decay of the lowest excited states is much slower, because the energy gap between S_1 or T_1 and the ground state (S_0) is quite large. However, chemical reaction still has to compete with radiationless decay to the ground state, and also with luminescent decay, namely fluorescence if there is no change of spin ($S_1 \rightarrow S_0 + h\nu$), or phosphorescence if there is a spin change ($T_1 \rightarrow S_0 + h\nu$).[6] This means that photochemical reactions which are to be efficient in terms of the initial absorption need to have very high rate constants.

This is an appropriate point to emphasise that for photochemical reactions there are two aspects to the term efficiency or yield - the normal "chemical" consideration and the "quantum" consideration. Chemical yield is a measure of the efficiency with which substrate is converted into isolated product, and yields of less than 100% reflect the occurrence of side-reactions and of losses in the isolation procedure. These considerations apply equally to thermal and photochemical processes. The inefficiency just described that arises in photochemical reactions because of the physical decay processes is not a chemical inefficiency, because the processes regenerate the substrate, but rather an inefficiency in the use of the absorbed light. The efficiency with which absorbed radiation causes a compound to undergo a specified chemical change can be expressed in terms of the quantum yield, the ratio of the number of molecules of product formed to the number of photons of light absorbed.[7] High quantum yields are normally desirable for the application of a photochemical reaction in organic synthesis, although a reaction with low, or even very low, quantum yield can be employed if the chemical yield is high by extending the time of the irradiation. In many literature reports of photochemical synthesis quantum yields have not been measured, but the experienced photochemist can often make an order-of-magnitude estimate from the experimental details relating to light source, sample size and concentration, and irradiation time.

The physical processes that contribute to the quantum yield for a particular substrate can be displayed on a state energy diagram of the type shown in Figure 4. This is known as a Jablonski diagram, on which absorption, radiationless decay and luminescence are shown using appropriate arrows. For convenience and clarity the singlet and triplet manifolds are displaced from each other horizontally. Absorption and emission processes are depicted using solid arrows, and radiationless processes are shown with wavy-line arrows. The involvement of vibrationally excited states is indicated by the incorporation of a few vibrational energy levels for each electronically excited state.

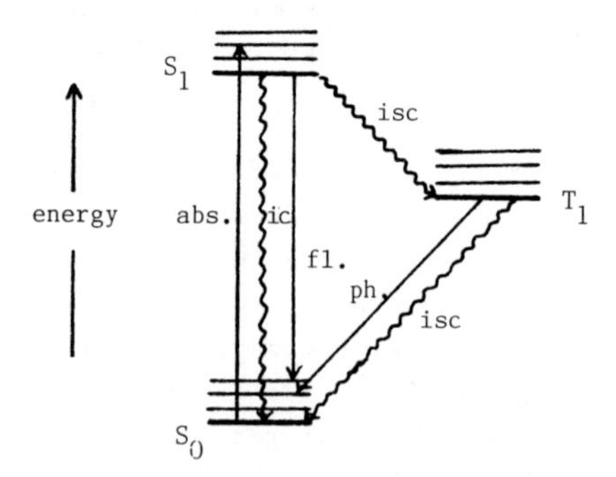

Figure 4. A Jablonski diagram; abs. = absorption, fl. = fluorescence; ph. = phosphorescence; ic = internal conversion; isc = intersystem crossing.

Such a diagram can be constructed for a particular compound on the basis of data from absorption spectra, from luminescence studies and from other photophysical experiments. Often it is not easy to acquire all the detailed quantitative information, but to an organic photochemist there is value in the qualitative information about the nature and relative energy of low-lying excited states and about the photophysical processes with which any observable photochemical conversion of substrate to product has to compete.

Thermodynamic and Kinetic Considerations (Feasibility)

Thermal reactions may operate under conditions where there is effective reversibility, and the substrates and products are in thermodynamic equilibrium. The extent of reaction and the ratios of products depend on differences in standard free energy between the components, and the process is said to be operating under conditions of thermodynamic control. This applies to some thermal reactions, but many are carried out under conditions where thermodynamic equilibrium is not achieved, and in many instances practical reversibility is not feasible. Such reactions are said to operate under conditions of kinetic control, and product ratios are determined instead by relative rates of reaction. In a few instances the balance between kinetic and thermodynamic control is quite fine, and the reaction time can influence the final outcome. Although the general kinetic model used in relation to ground-state chemistry can be applied also to excited-state chemistry, photochemical reactions operate, almost without exception, under conditions of kinetic control. A major reason for this is that the pathway (1) from an excited state of the substrate (R*) to the observed product (P) does not go through an excited state of the product.[8]

$$R \xrightarrow{h\nu} R^* \longrightarrow P \qquad (1)$$

There is usually a large energy difference between R* and P, and product does not revert to the excited state of the substrate. If reversion does take place, especially if P is a reactive intermediate rather than the finally observed product, it yields the ground state of the substrate rather than the excited state, and this does not represent thermodynamic reversibility. The lack of reversibility in photochemical reactions means that product ratios are determined by kinetic factors such as relative rates of competing steps at a branching point in the mechanism, whether this involves alternative routes for an excited state to follow or different fates for an intermediate species.

There are very few exceptions to the generalisation that the product of a photochemical reaction is formed directly in its ground state rather than through one of its excited states, but one group of reactions to which it does not apply are rapid proton transfers. For example, the excited state of a base and the excited state of its conjugate acid can exist in thermodynamic equilibrium (2). It is possible to obtain an indirect measure of pK_a values for such an equilibrium, and they can differ substantially from those for the ground-state equilibrium.[9]

$$B \underset{}{\overset{h\nu}{\rightleftharpoons}} B^* + H^+ \rightleftharpoons (BH^+)^* \overset{h\nu}{\rightleftharpoons} BH^+ \tag{2}$$

Practical reversibility of a different kind arises in some photochemical reactions when the product is capable of absorbing the irradiating light. If the product undergoes a photochemical reaction to give the substrate, then a two-way photochemical reaction can be set up (3). This is not an equilibrium in the thermodynamic sense, because the detailed pathways are different: the forward reaction proceeds by way of an excited state of substrate (4), but the reverse reaction proceeds through an excited state of the product (5).

$$R \underset{h\nu}{\overset{h\nu}{\rightleftharpoons}} P \tag{3}$$

$$R \xrightarrow{h\nu} R^* \longrightarrow P \tag{4}$$

$$P \xrightarrow{h\nu} P^* \longrightarrow R \tag{5}$$

In a situation like this a "photostationary state" can be achieved in which the rate of the forward reaction (4) equals the rate of the reverse reaction (5), and no change in overall composition occurs on further irradiation. The composition at the photostationary state is influenced by the choice of irradiation wavelength, since the relative rates of reaction depend on the relative intensities of light absorbed by the two ground-state species. Relative absorption is measured by absorption coefficients, and these vary with wavelength, so that a change in composition with wavelength is inevitable unless the spectra of substrate and product have exactly the same form in the spectral region considered. For efficient conversion of subtrate to product an irradiating wavelength is required at which substrate absorption is high and product absorption is low, which is a desirable situation in any case because of the possibility that the product may absorb and undergo secondary photochemical reaction with the result that the chemical yield is lower. As an example, consider the spectra of the trans and cis isomers of stilbene shown in Figure 5. Photochemical interconversion of these isomers leads to a photostationary state, and for efficient conversion of trans to cis compound a wavelength around 310-320 nm would be suitable. If the reverse reaction were desired (which is much less likely!), a wavelength around 240-250 nm could be used to maximise the yield, though the differences in absorption coefficients are less marked, and the proportion of trans isomer at the photostationary state correspondingly lower, than for the longer-wavelength irradiation selected to promote the trans → cis conversion.

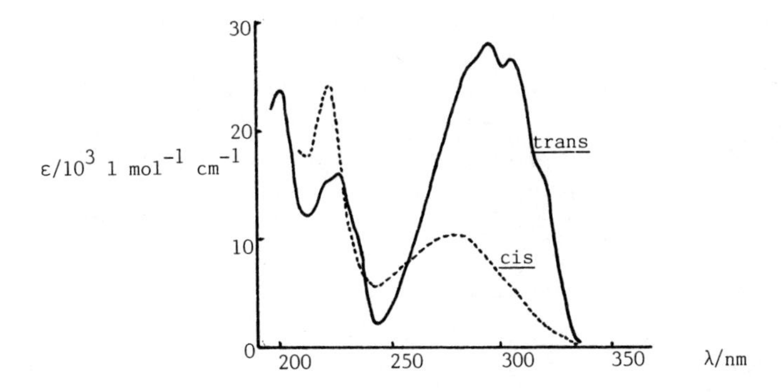

Figure 5. Absorption spectra of cis and trans stilbene.

Thermodynamic considerations do play a part in an understanding of photochemical reactions, because any chemical reaction proceeding to near completion must be thermodynamically favourable, that is it must involve an overall decrease in free energy as it proceeds. The lowest excited states of organic compounds are quite high-energy species, commonly 200 to 500 kJ mol^{-1} higher in energy than the corresponding ground states, and so the range of reactions that is in principle open to an excited state is far greater than that open to the ground state. Figure 6 shows this in a very general way: products that are thermodynamically accessible to the excited state include those that could, in principle at least, be obtained thermally from the ground state, and also those whose standard free energy falls in the (large) energy gap between ground state and excited state.

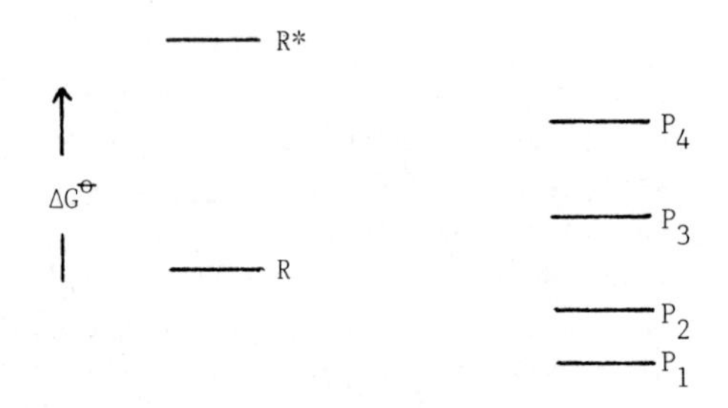

Figure 6. Free-energy diagram showing a wider range of potential photochemical products (P_1-P_4) than of thermal products (P_1-P_2).

Limitations do arise on account of the short lifetime of the excited state, but the very large contribution of excitation to the standard free energy of the reactant (the excited state) in the key elementary reaction step nevertheless underlies the success in applying photochemical reactions to the synthesis of many small-ring compounds, multicyclic compounds, or compounds that are otherwise strained or of high energy.

Electronic Considerations (Mechanism)

An organic chemist normally discusses thermal reactions that operate under conditions of kinetic control in mechanistic terms based on electron distribution within a molecule, and either the preferred sites of attack as a result of this, or the relative stabilities of alternative intermediate species (which are taken to reflect the relative stabilities of activated complexes and hence to be a guide to relative activation energies and rate constants). With certain limitations it is possible to rationalise photochemical reactions in a similar way, bearing in mind that the electron distribution in an excited state may be very different from that in the corresponding ground state. For example, much of the thermal chemistry of ketones can be understood on the basis of either initial attack by a nucleophile at the partly positively charged carbon atom of the carbonyl group, or initial attack by an electrophile at the partly negatively charged oxygen atom. The starting point for this rationale is the polarisation of electrons in the carbon-oxygen bonds, with a higher electron density near the more electronegative oxygen. The lowest excited state of an aliphatic ketone is (n,π^*) in electronic character and is not so strongly polarised, because electron density has been transferred from an orbital (n) located largely on oxygen to an orbital (π^*) covering the carbon and oxygen nuclei more equally. The main electronic feature of the (n,π^*) excited state is that the oxygen atom had odd-electron, radical-like character as a result of the loss of an electron from a non-bonding orbital. Much of the photochemistry of ketones occurring through these excited states can be successfully rationalised on the basis of this odd-electron nature, and the major reactions involve subsequent radical species rather than the electron-paired charged species associated with thermal reactions of ketones.[10]

Approaches to thermal and photochemical reactions of the same substrates can differ more substantially than this, for example in a consideration of aromatic substitution reactions. Ground-state reactions are most successfully rationalised on the basis of the relative stabilities of sigma-bonded intermediates formed by initial attack on the aromatic ring, provided the reaction operates under kinetic control. Such a rationalisation is not valid for excited-state reactions, because it takes no account of the different electron distribution and much higher energy of the excited state as compared with the ground state. For photochemical substitution reactions, arguments based on preferred sites of attack are often successful, with the charge distributions in the excited state determined by calculation or by analogy with compounds for which calculated charge densities are available.

A further dramatic example of the effects on the course of chemical reaction arising from electronic differences between ground states and excited states is seen in the complete reversal of the Woodward-Hoffmann rules for the preferred course of photochemical, as opposed to thermal, pericyclic reactions.[11]

The application of "ground-state" electronic considerations to excited-state reactions helps to account for many of the distinctive features of photochemical processes in organic chemistry, but there is another property of excited states that enhances the differences further. This is the ability of a molecule to donate, or to accept, an electron. Ground-state molecules do take part in electron-transfer reactions, but this is most usually in the context of the supply or removal of electrons through such agencies as electrode processes in electrochemistry, or the use of reactive metals like sodium. It is unusual for a thermal electron-transfer reaction to involve two electron-paired molecules. However, electronic excitation of any molecule makes it both a better electron donor and a better electron acceptor. Figure 7 illustrates that the enhanced donor properties arise because less energy is required to remove an electron completely from the molecule if there is already one in a higher energy orbital.

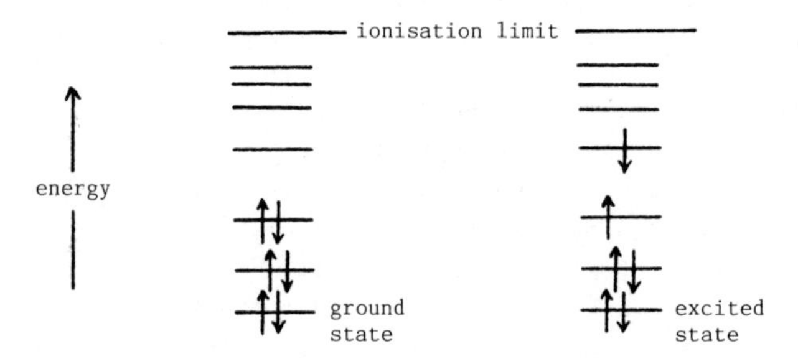

Figure 7. An orbital energy level diagram showing the more ready loss of an electron from an excited state.

The enhanced electron-acceptor properties arise similarly because there is a greater release of energy on addition of an electron to an excited state, in which a half-filled orbital of lower energy is available. The result is that electron transfer (6) is a common first step in photochemical mechanisms,[12] and the immediately following chemistry is that of radical cations and radical anions.

$$R^* + Q \longrightarrow R^{+\cdot} + Q^{-\cdot} \quad \text{or} \quad R^{-\cdot} + Q^{+\cdot} \qquad (6)$$

Electron-transfer interaction of this kind is a common first step in the photochemistry of ketones in the presence of electron donors such as tertiary amines, giving ketone radical anions and amine radical cations. The subsequent release of protons from the radical cations forms the basis for the use of ketone-amine pairs as photo-initiators for the cationic polymerisation of epoxides and related compounds. Electron transfer also occurs in the first stage of many photobiological processes such as photosynthesis. In synthetic photochemistry it is often possible to promote reaction through a radical cation or radical anion by the use of a sensitiser known to be very strongly electron-accepting or electron-donating in its excited state. Cyano-substituted aromatic compounds, such as 1,4-dicyanobenzene (7) or 9,10-dicyanoanthracene (8), are widely used as electron-accepting sensitisers in photochemical reactions, and methoxy-substituted aromatics, such as 1,4-dimethoxybenzene (9), as electron-donating sensitisers. In some systems it has been shown that irradiation of methoxy-substituted aromatic compounds in the absence of a suitable acceptor leads to complete ejection of an electron from the molecule (10), so that further reaction may involve the solvated electron.[13]

CN CN CN CN OMe OMe

(7) 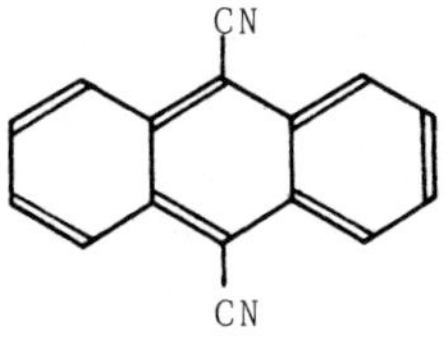 (8)  (9)

$$R \xrightarrow{h\nu} R^{+\cdot} + e^{-} \qquad (10)$$

The factors discussed in this section provide the synthetic chemist with the extra considerations that need to be applied to a mechanistic understanding of organic photochemical reactions. Together with the relevant factors that relate to all reactions, whether photochemical or thermal, they form a basis for appreciating and optimising existing photoreactions that are of synthetic potential, and for developing new reactions.

Quenching of Excited States - Boon and Bane

Interaction of an electronically excited state with a second molecular species may lead to chemical reaction, but it can also lead to deactivation of the excited state to regenerate the original substrate molecule in its ground state. This general process is called quenching of the excited state, and it may lead to the formation of an electronically excited state of the quencher (11), or it may not involve excited quencher (12).

$$R^* + Q \longrightarrow R + Q^* \tag{11}$$

$$R^* + Q \longrightarrow R + Q \tag{12}$$

Numerous mechanisms exist for the quenching process.[14] Energy transfer may occur over relatively large distances without the need for close approach of the two species; quenching may occur in a collisional complex, or in an intermediate excited complex (an exciplex); rapid reversible photochemical reaction may also account for quenching, such as electron transfer (6) followed by reverse electron transfer from the radical anion to the radical cation. Whatever the mechanism, the outcome is a very useful tool in the hands of the synthetic photochemist. Selective quenching used qualitatively can remove an unwanted component from a photochemical reaction which occurs by way of two different excited states, and this is particularly useful in inhibiting a triplet-state reaction in the presence of an accompanying singlet-state process. A very important use of quenching is in the generation of excited states (Q*) that are not readily accessible by the normal process of light absorption. This use of quenching is often called sensitisation, and it can be employed as a route to triplet states of molecules for which the indirect route of absorption followed by intersystem crossing from singlet to triplet is very inefficient, that is, very slow by comparison with other decay processes open to the singlet state.[15] For example, aromatic ketones such as benzophenone ($Ph_2C{=}O$) undergo intersystem crossing very rapidly, and there is no problem in studying their triplet-state chemistry. However, intersystem crossing for conjugated dienes such as penta-1,3-diene ($CH_2{=}CH{-}CH{=}CH{-}CH_3$) is very inefficient, and their triplet-state chemistry can only be carried out with the help of a sensitiser. Sensitisation can also introduce an element of photochemical selectivity, in that a triplet state can be generated in the absence of a related singlet excited state whose chemical reactions may be desirable.

A general requirement for quenching processes that involve transfer of excitation energy is that the reaction should be thermodynamically feasible; this means that the energy of the substrate excited state should be higher than, or at least not much lower than, the energy of the quencher excited state. The energy-level diagram in Figure 8 illustrates this point, showing why conjugated dienes are good quenchers for the triplet states of aromatic ketones, but they are not efficient quenchers of the singlet excited states because such quenching would be required on the basis of spin selection rules to generate the singlet excited state of the diene.

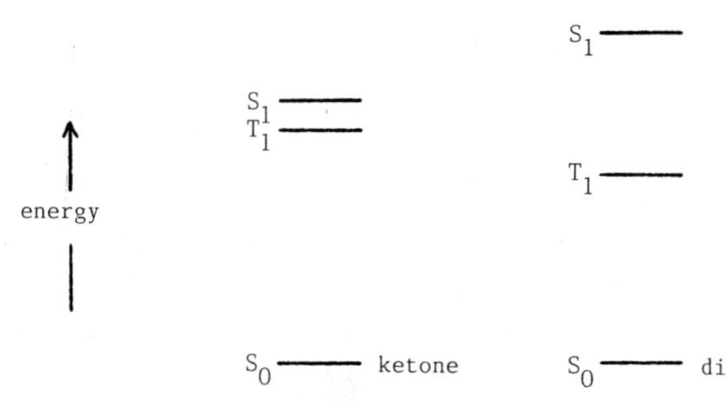

Figure 8. State energy level diagrams for an aromatic ketone and a conjugated diene.

The efficiency of quenching in a particular system depends on the concentration of the quencher, the rate constant for the quenching interaction, and the lifetime of the excited state in the absence of quencher. Quantitative quenching studies allow the properties of excited states taking part in photochemical reactions to be investigated. For example, it may be possible to determine the lifetime of an excited state, which in turn provides a starting point for estimating the rate constant for the primary photochemical step. Such studies may not be of direct relevance to photochemical synthesis, but they are a major source of data that can be used to construct a Jablonski diagram for a substrate.

Triplet states are, in general, much longer lived than singlet states, and so they are quenched more readily. This is the main reason why phosphorescence (luminescence from the lowest triplet state to the ground state) is seldom observed in fluid solution at room temperature, because very small amounts of quenching impurities inhibit the process. One such impurity is molecular oxygen, and oxygen is an important, though often unwanted, triplet quencher in organic photochemistry. The ground state of molecular oxygen (O_2) is a triplet state. It can interact with many excited triplet states of other molecules,[16] either by a transfer of energy (13), which generates a low-energy excited singlet state of molecular oxygen that is usually called singlet oxygen, or by a transfer of an electron (14), which generates the dioxygen radical anion and the radical cation of the organic compound, or by a quenching mechanism that does not produce a different oxygen species (15).

$$R\ (T_1) + O_2\ (T_0) \longrightarrow R\ (S_0) + O_2\ (S_1) \qquad (13)$$

$$R\ (T_1) + O_2\ (T_0) \longrightarrow R^{+\cdot} + O_2^{-\cdot} \qquad (14)$$

$$R\ (T_1) + O_2\ (T_0) \longrightarrow R\ (S_0) + O_2\ (T_0) \qquad (15)$$

The general outcome is a quenching of the excited triplet state of the substrate and, in the first two processes, production of a very reactive oxygen species that may react chemically with any organic compound present. The chemical reactivity of singlet oxygen is employed usefully in synthesis with a variety of unsaturated compounds such as alkenes and dienes. However, if this is not the intended reaction, oxygen has to be regarded as an undesirable component of an organic photochemical reaction. It is common practice to remove as much oxygen as possible by carrying out reactions with a steam of oxygen-free nitrogen or some other inert gas bubbling through the irradiated solution. Whether such a practice is necessary can only be determined for a particular system by trial-and-error or from a knowledge of related systems. If the relevant excited state is quenched only very inefficiently by oxygen, either because of its electronic nature or, more likely, because its lifetime is extremely short on account of a fast primary photochemical step, degassing may not be essential. However, it is a wise precaution to take in laboratory photoreactions, to avoid unwanted oxidation products and unnecessarily long irradiation times.

With quenching by oxygen or by deliberately added compounds, as with other mechanistic aspects of organic photochemistry, the synthetic chemist is in a better position to influence the outcome of a particular reaction if he appreciates the mechanistic features sufficiently to understand how they affect the conversion of substrate into the desired product. The purpose of this chapter has been to facilitate such an understanding without going into a level of detail that, whilst interesting, is not necessary for the specific aim.

References

1 J.A. Barltrop and J.D. Coyle, "Principles of Photochemistry", Wiley, Chichester, 1978.

2 H.H. Jaffé and M. Orchin, "Theory and Applications of Ultraviolet Spectroscopy", Wiley, New York, 1962.

3 M.B. Robin, "Higher Excited States of Polyatomic Molecules", Volume 3, Academic Press, New York, 1985, Chapter 1.

4 H. Okabe, "Photochemistry of Small Molecules", Wiley, New York, 1978.

5 S.L. Murov, "Handbook of Photochemistry", Marcel Dekker, New York, 1973.

6 R.S. Becker, "Theory and Interpretation of Fluorescence and Phosphorescence", Wiley, New York, 1969.

7 W.M. Horspool, "Synthetic Organic Photochemistry", Plenum, New York, 1984, Chapter 9.

8 N.J. Turro, "Modern Molecular Photochemistry", Benjamin/ Cummings, Menlo Park, 1978, Chapter 4.

9 J.F. Ireland and P.A.H. Wyatt, Adv. Phys. Org. Chem., 1976, 12, 131.

10 N.J. Turro, "Modern Molecular Photochemistry", Benjamin/ Cummings, Menlo Park, 1978, Chapter 7.

11 T.L. Gilchrist and R.C. Storr, "Organic Reactions and Orbital Symmetry", Cambridge University Press, Second edition, 1979.

12 R.S. Davidson, Adv. Phys. Org. Chem., 1983, 19, 1.

13 G. Grabner, W. Rauscher, J. Zechner and N. Getoff, Chem. Commun., 1980, 222.

14 D.O. Cowan and R.L. Drisko, "Elements of Organic Photochemistry", Plenum, New York, 1976, Chapter 6.

15 P.S. Engel and B.M. Monroe, Adv. Photochem., 1971, 8, 245.

16 H.H. Wasserman and R.W. Murray, "Singlet Oxygen", Academic Press, New York, 1979.

2
Practical Photochemistry: General Considerations

By J. Hutchison
THE ELECTRICITY COUNCIL RESEARCH CENTRE, CAPENHURST, CHESTER CHI 6ES, UK

Lamps

The lamps most commonly used for preparative photochemistry fall into two main categories:

medium pressure mercury discharge tubes
(with or without a metal halide dopant);
low pressure mercury discharge tubes
(with or without an internal phosphor coating).

In the medium pressure discharge, the pressure of mercury covers the range 1 to 10 atmospheres, being determined by the amount of mercury introduced into the tube during manufacture, and the operating temperature of the lamp, which is in the region of 600 to 800°C. The spectral distribution of the radiation emitted by an undoped medium pressure mercury lamp is mainly in the form of sharp, discreet lines (Figure 1). Metal halide doped lamps produce additional lines characteristic of the metal dopant (e.g. gallium, magnesium, iron, thallium).

Medium pressure lamps used for laboratory scale synthesis will normally have a power rating of between 100 watts and 500 watts, an arc length of between 5cm and 15cm and a tube diameter of between 1cm and 2.5cm. For industrial scale synthesis, lamps can be obtained with power ratings up to 60 kilowatts, a 60 kW lamp having an arc length of 2m and a tube diameter of 6cm.

Low pressure lamps operate at a lamp tube temperature just above ambient (40-50°C usually being the optimum), at which the vapour pressure of mercury, present in excess in the tube, is in the region of 10^{-5} atmospheres. Power ratings for low pressure lamps, in terms of power input per unit length of arc, are a factor of roughly 100 lower than the power ratings of medium pressure lamps. For example a low pressure lamp used for laboratory scale synthesis may, typically, have an arc length of 40cm and a power rating of 16 watts (0.4 watts/cm) compared with a typical medium pressure lamp having an arc length of 10cm and a power rating of 400 watts (40 watts/cm).

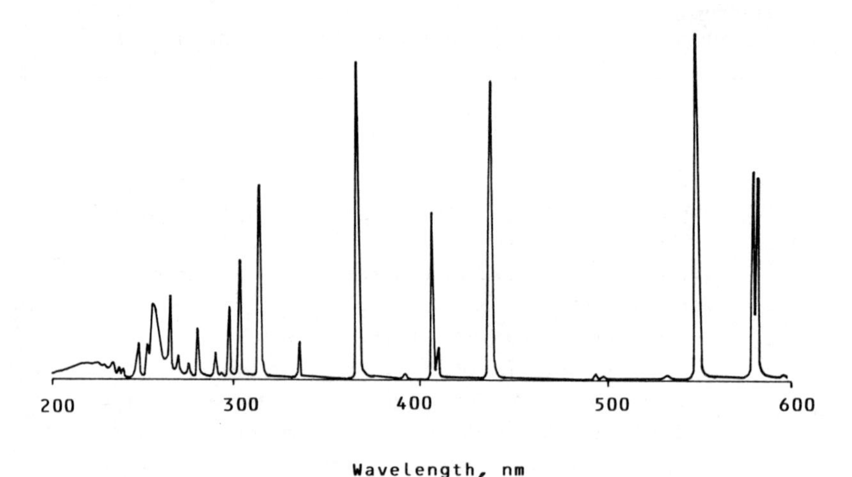

Figure 1. Spectral output of undoped medium pressure mercury lamp.

Low pressure lamps can be obtained in the form of U-shaped tubes, spirals and grids, as well as straight tubes.

Around 90% of the primary radiation emitted by a low pressure discharge consists of the mercury resonance lines at 254nm and 185nm, produced in the ratio of approximately 6:1. The 185nm line dissociates atmospheric oxygen, giving rise to the formation of ozone, which is toxic, and strongly absorbing in the 200-300nm region. Ozone production can be avoided by operating the lamp in an oxygen-free atmosphere, or by constructing the lamp tube from a type of glass which absorbs the 185nm line; the glass may also, however, absorb a significant proportion of the 254nm line.

By means of a phosphor coating on the internal surface of a low pressure lamp tube, the 254nm line can be converted into longer wavelength radiation having a spectral distribution characteristic of the phosphor, or mixture of phosphor materials used (Figure 2). The everyday fluorescent lighting tube is a familiar example of this. Phosphor emission bandwidths range from over 100nm down to about 20nm (the bandwidth being the spread of wavelengths over which the output is greater than 50% of the peak output).

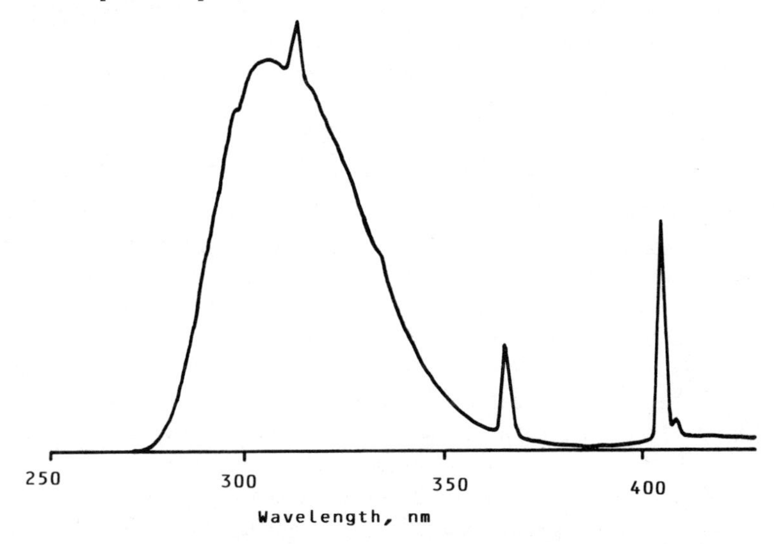

Figure 2. Spectral output of a low pressure phosphor coated lamp. Bandwidth 45nm (290-335nm).

High pressure lamps (a third category) are particularly suitable for quantitative photochemical studies, and may also be used for preparative work. The lamp fill can be mercury, or a rare gas such as xenon, or a mixture of mercury and rare gas. There are two types of construction, capillary arc and compact arc. The latter have arc lengths of only a few millimeters, and can be regarded as essentially point sources. Power ratings range from tens of watts to several kilowatts, and operating pressures from 10 atmospheres to several hundred atmospheres. A stout housing needs to be provided (usually supplied by the lamp manufacturer) for protection in case the lamp fractures, explosively, during operation, and to shield personnel from the intense ultraviolet and visible radiation. The main feature of the spectral distribution emitted by high pressure lamps is the presence of a strong continuum, which becomes more dominant for the higher operating pressures.

From this full range of available wavelengths, any particular region which may be required can be selected using a monochromator or a suitable combination of filters.

All types of discharge lamps must be operated with the appropriate power supplies, obtained from or recommended by the lamp manufacturer. The power supply will incorporate an impedance, usually an inductance, in series with the lamp in order to overcome the otherwise unstable voltage-current characteristic of the discharge. Overall power consumption by the lamp circuit exceeds the nominal lamp power rating, usually by between 10% and 30%.

Lamp output data. Given in Table 1 are some representative figures for the outputs in different spectral regions emitted by a typical medium pressure lamp used for laboratory scale synthesis. The reason for dividing the spectrum into 30nm wide segments is that this corresponds to the minimum wavelength spread that will normally be required for preparative work, although it may turn out to be an individual narrow line which is in fact utilised. An individual absorption band of the photoactive component in the reaction mixture will usually be considerably broader than 30nm, and absorption anywhere within the band will usually lead to the same photochemical reaction, proceeding with the same quantum efficiency (in the absence of possible complications to be discussed later).

Lamp manufacturers usually quote outputs in watts, whereas einstein per hour (an einstein being 6.023×10^{23} quanta) is a more useful unit for photochemists because it can be related more readily to the photochemical reaction rate.

$$\text{einstein } h^{-1} = \text{watts} \times \text{wavelength (nm)} \times 3 \times 10^{-5}$$

Reaction rate (mol h^{-1})
= einstein h^{-1} absorbed by the photoactive component in the reaction mixture
x quantum yield for the reaction.

Table 1. Typical figures for spectral output of 400 watt undoped medium pressure mercury lamp, arc length 10cm.

Spectral region, nm	Major lines	Output	
		watts	einstein h^{-1}
210 - 240	(Continuum)	10	0.07
240 - 270	248, 254-7, 265	20	0.15
270 - 300	280, 289, 297	10	0.09
300 - 330	302, 313	20	0.19
330 - 360	334	2	0.02
360 - 390	365	25	0.28
390 - 420	405-8	10	0.12
420 - 450	436	15	0.20
450 - 480	-	0	0
480 - 510	-	0	0
510 - 540	-	0	0
540 - 570	546	20	0.33
570 - 600	577-9	17	0.30

Returning to Table 1, if the precise spectral region which is to be utilised to bring about a photochemical reaction is known (say the 270 - 330nm region); and if it can be assumed that all of the lamp output in this region will be transmitted to the reaction mixture, and absorbed by the desired photoactive component; and if the quantum yield for the formation of the desired product is known (say this is 0.5); then it could be predicted that the product would be formed at a rate of (0.19 + 0.09) x 0.5 = 0.14 mol h^{-1}. If the molecular weight of the product is, say, 200, then to produce 100g would take about 3½ hours. Very much longer irradiation periods, stretching to several days, may often be required, if the reaction proceeds with a much lower quantum yield, or if practical considerations to be discussed later preclude the above assumptions from being realised or maintained.

Consider now a typical low pressure lamp used for laboratory scale synthesis, with a power rating of 16 watts and an arc length of 40cm. Figures given for the efficiencies with which various different types of low pressure lamp produce the 254nm line cover a fairly wide range, from less than 20% to more than 40%. Taking 30% as a representative figure, this would give

$$16 \times 0.3 \times 254 \times 3 \times 10^{-5} = 0.037 \text{ einstein h}^{-1}$$

which is about a quarter of the 240 - 270 nm output from the 400 watt medium pressure lamp. For a phosphor coated low pressure lamp, the quantum efficiency of the phosphor would appear to be typically around 60%, giving an output of 0.022 einstein h^{-1} over a spectral region characteristic of the particular phosphor. If this were such that essentially all of the output covered the 270 - 330nm region, then the 0.022 einstein h^{-1} from the 16 watt low pressure lamp would compare with 0.28 einstein h^{-1} in this region from the 400 watt medium pressure lamp (see Table 1). Although rather contrived, this comparison does serve to illustrate the relative outputs and efficiencies of medium pressure and low pressure lamps. Medium pressure lamps produce much high intensities of radiation, but low pressure lamps can be significantly more efficient at converting power input into radiated output over the desired wavelength range. In the above cases, the efficiencies for producing 270 - 330nm radiation work out at 7.7% for the medium pressure lamp and 15% for the low pressure lamp, bearing in mind that all of the low pressure lamp output has been assumed to lie in the 270 - 330nm region, whereas this represents only a portion of the medium pressure lamp output.

It is worth noting here that the rate at which a photochemical reaction proceeds is not always directly proportional to the absorbed light intensity (einstein $l^{-1}h^{-1}$), although the rate for the primary photochemical step generally is. In the case of photoinitiated free radical chain reactions, the overall reaction rate is usually proportional to the square root of absorbed light intensity, and for a given reaction mixture to the square root of incident light intensity (einstein $cm^{-2}h^{-1}$). If this square root dependence is incorporated into the above comparison, and it is assumed that the intensity of useful radiation incident on the reaction mixture is proportional to the 270 - 330nm output per unit length of arc (that is considering the lamps being used in reactors having the same geometry), then

$$\frac{\text{Rate from unit length of medium pressure lamp}}{\text{Rate from unit length of low pressure lamp}} = (0.28/10 \div 0.022/40)^{\frac{1}{2}} = 7.1$$

Since the power input per unit length of arc for the medium pressure lamp is 100 times that for the low pressure lamp, this means that the low pressure lamp will be 14 times more efficient than the medium pressure lamp in terms of the reaction rate obtained per unit power input to the lamps. In other words, for the same electricity consumption, 14 times more product would be expected from the low pressure lamp.

The Reaction Mixture

The first step in any photochemical reaction is the absorption of ultraviolet or visible radiation by (usually only) one particular component of the reaction mixture, each quantum of radiation (photon) absorbed photoexciting (promoting to an electronically excited state) a single molecule of this "photoactive" component. Absorption by other components will not usually bring about the desired reaction, and may result in the occurrence of unwanted side reactions.

The photoactive component may itself be a reactant, being consumed as a result of the photoexcited state undergoing a unimolecular reaction, or reacting chemically with another component of the mixture; or it may act as a photosensitiser, transferring its excitation energy to another component which goes on to react, the photosensitiser reverting (unconsumed) to its ground state. The photoactive component could also be the photoinitiator of a chain reaction. Photoinitiators are consumed, but each reactive intermediate (radical or ion) it produces in the photochemical step may lead to the reaction of tens, hundreds or thousands of reactant molecules, giving correspondingly high quantum yields for the overall reaction. The consumption of the photoinitiator may therefore be insignificant compared with the consumption of the reactant.

The starting point for determining the wavelength region required to carry out a particular photochemical reaction is the absorption spectrum of the photoactive component in the solvent to be used, or in a number of possible solvents. It is always advisable to record several spectra, covering a wide range of concentrations, so that extinction coefficients for all absorption bands, from very strong to very weak, can be calculated.

Absorbance = $\log_{10}(I_o/I_t) = \epsilon cd$ (Beer-Lambert Law)

where I_o = incident intensity (of parallel beam)

I_t = transmitted intensity

c = concentration of absorbing component, mol l^{-1}

d = path length, cm

ϵ = molar extinction coefficient, l mol^{-1} cm^{-1}

Consider, for illustration, the spectra shown in Figure 3.

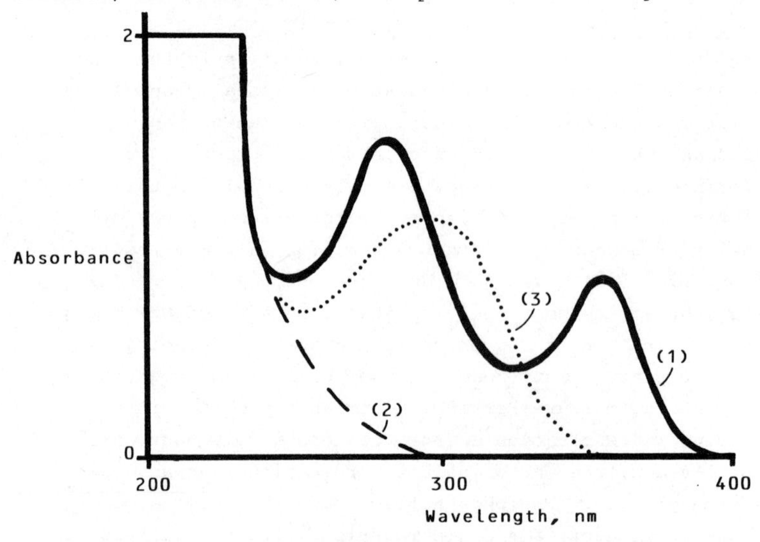

Figure 3. Typical absorption spectra of photoactive component (1), solvent (2) and product (3).

Absorption by the solvent presents the first constraint on the range of wavelengths which will be effective in bringing about the reaction. Fortunately there is a fairly wide choice of solvents which transmit down to quite short wavelengths including many alcohols, ethers, saturated hydrocarbons and water (Table 2).

Table 2. Cut-off wavelengths for a selection of solvents. (Shorter wavelengths strongly absorbed).

Solvent	Cut-off wavelength, nm
hexane	195
1,2-dichloroethane	225
di-isopropyl ether	220
toluene	285
acetone	330
ethyl acetate	255
acetonitrile	190
pyridine	305
methanol	205
water	<190

Other requirements of the solvent are that it is free from impurities, especially those which absorb outside the solvent cut-off, and that it should not (unless desired) react with the photoexcited states or with other intermediates involved in the course of the reaction. For example if free radicals are involved, solvents labile to hydrogen abstraction by the free radicals need to be avoided. This can exclude alcohols, ethers and saturated hydrocarbons from consideration. Benzene, acetonitrile and tert-butanol are relatively, but not completely, unreactive towards free radicals. Dissolved oxygen is an efficient quencher of photoexcited (specifically triplet) states, and also reacts with free radicals. In preparative work, oxygen is usually removed by bubbling nitrogen through the reaction mixture prior to and during irradiation.

Absorption by reactants other than the photoactive component, and by the product(s), provide further contraints on the wavelength region that can be utilised. As the product builds up in solution, it will compete more and more effectively with the photoactive component for absorption of incident wavelengths which lie within the region of overlap of their absorption spectra. This will reduce the efficiency with which these wavelengths bring about the reaction and, if all the incident radiation lies within the region of overlap, may lead to premature cessation of the reaction. The same of course applies to absorption by side products, with the added complication that this will generally be less predictable.

If the desired product is decomposed by the radiation it absorbs, and the lamp eventually selected to carry out the reaction emits in this region, then it will be essential to incorporate some sort of filter which will efficiently absorb these unwanted wavelengths, whilst efficiently transmitting the wavelength region that is to be utilised. As illustrated in Figure 3, it is quite often the shorter wavelengths that need to be removed. This can be done by means of an appropriate glass selected from a range of different glass types available having different cut-off wavelengths (Table 3). The glass filter can be used in the form of a tubular sleeve surrounding the lamp. Alternatively the appropriate glass can be used in the construction of the reaction vessel, most obviously if it turns out that Chemical Pyrex fulfils the transmission requirements. A general purpose reaction vessel would normally be constructed from high purity quartz, to allow the transmission of wavelengths down to 200nm if required.

Table 3. Cut-off regions* for some different glass types.

Quartz, e.g. Suprasil	<200
Vycor 7910	210 - 240
Corex D	270 - 310
(Chemical) Pyrex 7740	290 - 330
Crown (optical) glass	300 - 335
Nonex 7720	320 - 360
Wood's Glass	235 - 285, 405 - 380**

*Approx. region (nm) over which transmission increases from 10% to 80% for a 2mm thickness of glass.
**Long wavelength cut-off, extending to 680nm.

Filter solutions can be used instead of glass filters to remove unwanted short wavelength radiation, or in combination with glass filters as a means of isolating the required wavelength region, if the filter solution transmits this and absorbs longer wavelengths. A wide range of filter solutions are described in the literature. Many of the recipes given necessitate the radiation passing through several filter solutions, which can be readily arranged if small scale quantitative studies are being carried out, but is generally not feasible for preparative work.

<u>Concentration of the photoactive component.</u> Having established the wavelength region which is to be utilised, further consideration needs to be given to the concentration of photoactive component that should be used. It is worthwhile carrying out some simple calculations based on the Beer-Lambert Law, taking into account the extinction coefficient of the photoactive component over the selected wavelength region, and the spectral distribution of incident radiation within this region. The most straightforward case would be utilisation of a single wavelength (one of the mercury lines, for example), which is absorbed only by the photoactive component. The concentration required for absorption of 99% of the incident radiation (<u>i.e.</u> for absorbance = 2) over a particular path length d cm, would then be given by $c_{99\%} = 2/\epsilon d$ mol l^{-1} where ϵ is the molar extinction coefficient of the absorbing component at this wavelength. Put the other way, for a particular concentration c mol l^{-1}, the distance over which 99% absorption occurs will be $d_{99\%} = 2/\epsilon c$ cm. Taking $\epsilon = 10$ as an example of a fairly low extinction coefficient, $c_{99\%} = 0.2$ mol l^{-1} for d = 1cm, and $d_{99\%}$ = 2mm for c = 1 mol l^{-1}. Taking $\epsilon = 10^4$ as an example of a fairly high extinction coefficient, $c_{99\%} = 2 \times 10^{-4}$ mol l^{-1} for d = 1cm, and $d_{99\%}$ = 2 microns for c = 1 mol l^{-1}.

It is obviously more complicated when, within the selected wavelength region, there are a number of spectral lines and/or a continuum and the extinction coefficient varies, possibly by a factor of ten or more. In general though, given a reaction vessel path length of a cm or more, there will be a wide range of concentrations of the photoactive component for which essentially total absorption of the selected wavelength region will occur. The particular concentration chosen can influence the course of the reaction, and the optimum concentration, to give the maximum yield of the desired product, may need to be found by trial and error. Relatively low concentrations/long absorption path lengths will be desirable when it is important to discourage excited states or other reaction intermediates (such as free radicals, or biradicals) from reacting with each other or with the ground state of the photoactive component, and thereby encourage an intramolecular reaction of the excited state, or a reaction involving another component of the reaction mixture (the concentration of which will also have to be optimised). If a low concentration is used, and the photoactive component is consumed by the reaction, then it may need to be added throughout the irradiation period to maintain the optimum concentration, and of course to use sufficient in total to prepare the desired amount of product.

The presence or likelihood of any unavoidable absorption by other components within the selected wavelength region may necessitate the use of a relatively high concentration of the photoactive component in order to ensure that it competes successfully for the absorbed radiation.

$$\frac{\text{Absorption by component 1}}{\text{Total absorption}} = \frac{\epsilon_1 c_1}{\epsilon_1 c_1 + \epsilon_2 c_2 + \epsilon_3 c_3 \cdots\cdots\cdots}$$

The correspondingly short path length for absorption ($d_{99\%}$ of maybe only a few microns) will result in relatively high concentrations of excited states and other intermediates being produced close to the wall of the reaction vessel, and to the accumulation of product(s) in this region. These conditions will tend to favour unwanted secondary reactions, which can lead to the deposition of tarry residues or highly absorbing polymeric films on the vessel wall. Although these problems can be alleviated to some extent by vigorous stirring, or by other features of reactor design, they may be symptomatic of the need to select more carefully the wavelengths which are allowed to be incident on the reaction mixture, or of (failure!) the impracticality of the chosen synthesis.

Preparative Photoreactors

The photoreactor design most widely adopted for preparative work brings lamp and reaction mixture into close proximity by the use of an immersion well (Figure 4). The lamp is located in the central portion of this double walled vessel which is constructed from quartz or Pyrex, depending on the wavelength transmission requirements. When a medium pressure lamp is used, cooling water is circulated between the two walls in order to shield the reaction mixture from the heat of the lamp

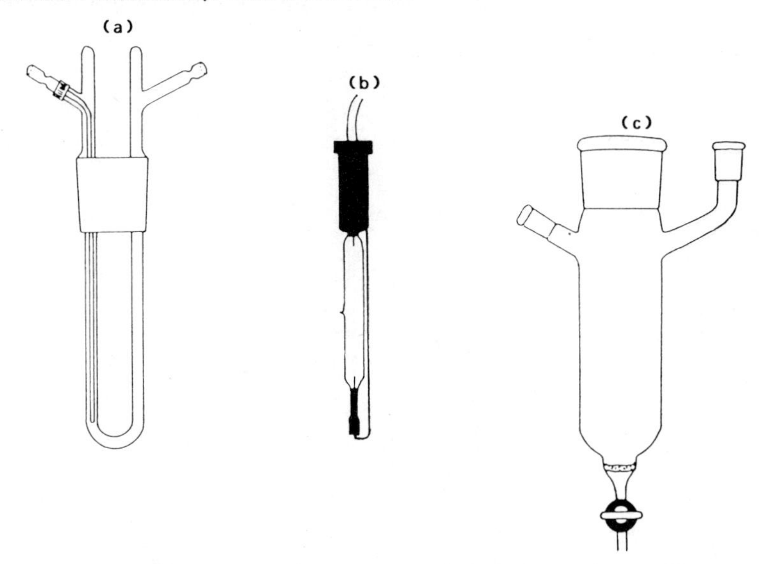

Figure 4. Immersion well photoreactor, (a) immersion well, (b) lamp, (c) outer vessel.
(Reproduced with the permission of Applied Photophysics Limited).

(not to cool the lamp as this would impair its performance). If necessary, a filter solution can be circulated via an external heat exchanger, and the lamp can be surrounded by a glass filter sleeve. The assembly is immersed in the reaction mixture contained in an outer Pyrex vessel. Nitrogen will normally be bubbled through the reaction mixture. The central portion of the immersion well should also be purged with nitrogen to prevent ozone formation and, in the case of medium pressure lamps, to protect the lamp electrode seals and other electrical connections from oxidation.

For strongly absorbing reaction mixtures, a falling film reactor may be more suitable. In the design shown in Figure 5, the reaction mixture is pumped from a reservoir to a nozzle positioned so that the emerging jet of liquid forms a thin film which covers the outer surface of a quartz or glass tube, and is irradiated by a lamp located inside the tube. By continuously replenishing the layer of liquid closest to the lamp, the potential problem of wall deposits building up may be overcome.

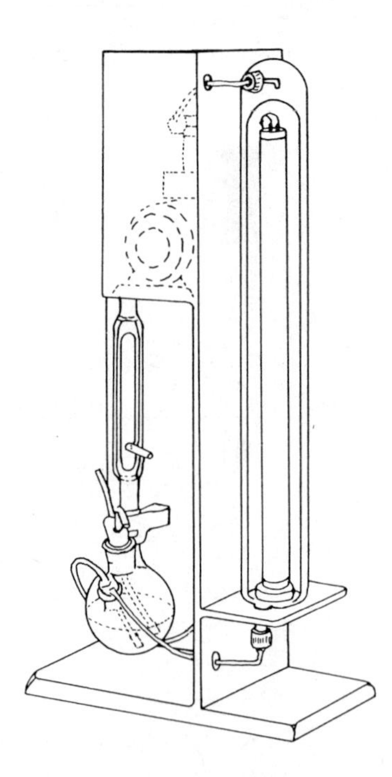

Figure 5. Falling film photoreactor.
(Reproduced with the permission of Applied Photophysics Limited).

The alternative configuration to the lamp being surrounded by the reaction mixture is the reaction mixture being surrounded by lamps. For example a reaction vessel may be positioned in the middle of a circular array of maybe 12 low pressure lamps. If medium pressure lamps are used, two lamps will normally suffice, placed on opposite sides of a double walled, water cooled reaction vessel. Too many medium pressure lamps in a confined space would lead to serious heat dissipation problems, and even with two lamps a fan would normally be incorporated to avoid overheating of the equipment. Light capture efficiency can be improved by placing curved reflectors behind the lamps. The reflectors must be made from aluminium, as this is the only material having high specular reflectivity down to 200nm.

In the elliptical reactor (Figure 6), the aluminium reflector forms a full ellipse. A lamp tube is positioned along one focal line of the ellipse, and the reaction mixture is circulated through a tube positioned along the other focal line. This geometry maximises the amount of radiation which is directed towards the reaction tube. The fact that the radiation converges towards the centre of the reaction tube can be advantageous, since this tends to compensate for the diminution in intensity resulting from absorption, producing a more uniform reaction rate throughout the tube.

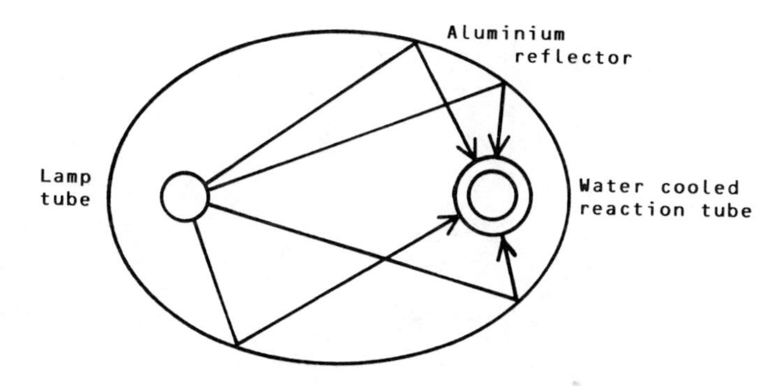

Figure 6. Elliptical photoreactor.

Apparatus for Quantitative Work

Carousel or merry-go-round arrangements (Figure 7) are designed to enable the simultaneous irradiation of a number of reaction tubes, and to ensure that the tubes receive equal exposure. This is a convenient way of carrying out comparative studies. The quartz or glass reaction tubes are held in a rack which rotates either inside a circular bank of lamps, or around a single lamp in an immersion well. It is possible with the latter configuration to contain the whole assembly in a water bath, to keep the temperature constant, and perhaps investigate a range of temperatures. The reaction tubes can be fitted with cones and taps for attachment to a vacuum line, so that dissolved oxygen can be removed from the reaction mixtures (by freeze-pump-thaw degassing cycles) prior to irradiation.

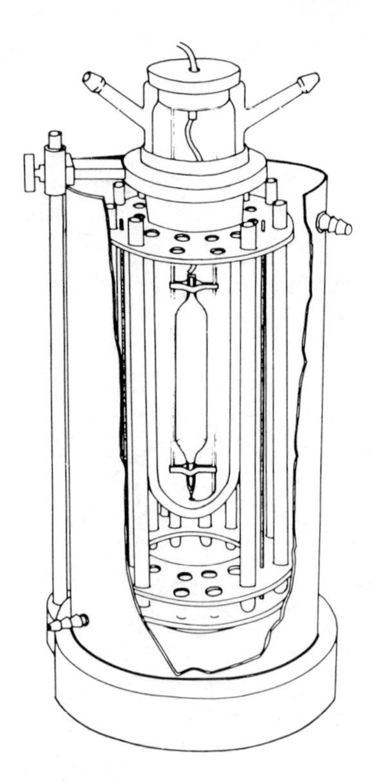

Figure 7. Carousel Assembly
(Reproduced with the permission of Applied Photophysics Limited).

Examples of comparative studies which may be required are the investigation of a series of related compounds, or a range of solvents, or a range of concentrations of one of the components of the reaction mixture. Quantum yields can be determined by carrying out a reaction of known quantum yield (*i.e.* using an "actinometer" solution) in one of the tubes. In all these cases, it is essential to take into account the proportion of incident light which is absorbed by each reaction mixture, and usually to try and arrange for essentially total absorption, by the photoactive component, to occur.

The experimental arrangement illustrated in Figure 8 is used when precise control and accurate monitoring of irradiation conditions are required. By means of a compact arc high pressure lamp and a "train" of components mounted on an optical bench, a parallel beam of narrow bandwidth radiation is directed on to a flat surfaced reaction vessel.

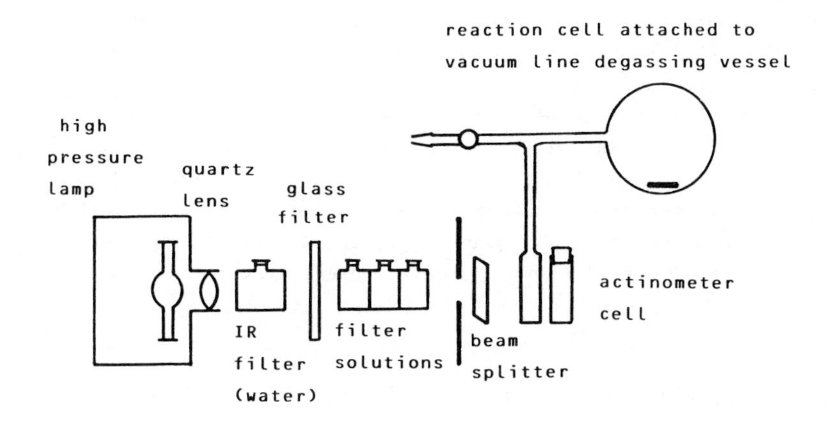

Figure 8. Optical bench arrangement.

Spectrophotometer cells serve well as reaction vessels, being readily obtainable in a wide range of shapes and sizes contructed from quartz or glass. Cells are available with Pyrex tubes attached, enabling the construction of vacuum line degassing vessels which incorporate the cell as a side-arm. A small PTFE covered magnet may be placed in the reaction cell if stirring during irradiation is considered necessary.

In place of glass cut-off filters and filter solutions, a diffraction grating monochromator can be used to rapidly select the required wavelength and bandwidth of radiation. Another alternative is the use of interference filters which pass only a narrow band of radiation, and can be readily obtained for selection of the main mercury lines, for example. A gauze, or quartz/glass neutral density filter having known transmission may also be included in the optical train if the dependence of reaction rate on light intensity is to be investigated (by carrying out a series of experiments using neutral density filters having different transmissions). The light intensity incident on the reaction cell can be monitored by means of a beam splitter, in the form of a quartz plate held at an angle of 45° and placed immediately in front of the reaction cell in order to reflect a small, known portion of the radiation towards a photodetector, or a cell containing actinometer solution. (The most widely used actinometer solution utilises the photo-reduction of potassium ferrioxalate. The resulting ferrous ions are complexed with o-phenanthroline and the concentration of the complex is determined spectrophotometrically from its 510nm absorption maximum). By placing another actinometer cell behind the reaction cell, the intensity of transmitted light can be measured, enabling light absorption by the reaction mixture to be calculated. Quantum yields can then be determined following analysis of the reaction mixture at the end of the irradiation period.

Hazards

Before carrying out any practical photochemistry, it is essential to be aware of the potential hazards involved, and to implement the necessary safety precautions.

The eyes and the skin can be severely affected by exposure to ultraviolet radiation. Wavelengths shorter than 300nm are particularly dangerous to the eye, causing photokeratitis (inflammation of the cornea) and conjuctivitis. The painful symptoms which result, characteristic of welders' arc eye and snow blindness, resemble sand being rubbed into the eyes, and do not usually appear until 6 to 12 hours after exposure (in the middle of the night for instance!), so there is no immediate indication of the damage being done. This is especially so for lamps having only low visible output, as is the case with low pressure, germicidal lamps, producing almost all of their output at 254nm. The symptoms usually recede after a few distressing hours, and disappear after 24 to 36 hours, normally leaving no permanent damage. Repetition of the experience tends to be avoided.

It is again the shorter ultraviolet wavelengths which are most damaging to the skin. The range 280nm to 315nm is referred to as the "erythemal" region becuase of the skin reddening and in extreme cases blistering effects it produces, although shorter wavelengths and, to a lesser degree, longer wavelengths, can also give rise to these sunburn symptoms. The time taken for the symptoms to reach maximum severity varies from 1 hour to 48 hours. Individuals vary in their susceptibility. Increased sensitivity, especially to longer wavelength exposure, can be brought about by the presence of photosensitisers in the skin, which can be derived from certain drugs and antiseptics, or, in the case of photochemists, the materials they may handle in the course of the experiments they are carrying out!

The most obvious precaution is routinely to wear special UV absorbing safety spectacles or goggles (obtainable from Ultra-Violet Products Limited, Cambridge). Standard-issue face shields will usually be sufficiently UV absorbing to provide protection for facial skin. Protection of the neck and forearms with normal clothing should not be overlooked, and of course suitable gloves should be available. It is good practice to locate photochemical apparatus in a screened off area (Embalex Limited, Hayling Island, Hants, supply UV absorbing plastic curtaining and screens) with clear notices describing the hazards, and warning lights to indicate when the lamps are operating.

Apart from exposure of you and your colleagues to ultraviolet radiation, a number of other hazards, against which precautions must be taken, require attention when carrying out photochemical experiments:-

Production of ozone (threshold limit value 0.1ppm) by interaction of short wavelength UV with atmospheric oxygen. (Similarly phosgene from chlorinated hydrocarbon vapours.) Adequate ventilation required.

High voltages associated with operation of discharge lamps. Proper standards of electrical safety necessary.

High internal pressures and/or fragility of lamps. Guard against breakage when handling and operating.

Lamps operating at high temperatures. Beware of handling until cooled sufficiently after switch off.

Fire hazard. Especially stringent precautions necessary if equipment to be left operating whilst unattended.

References

J.G. Calvert and J.N. Pitts, "Photochemistry", Wiley, New York, 1966.

A. Schonberg, "Preparative Organic Photochemistry", Springer Verlag, New York, 1968.

W.M. Horspool, "Aspects of Organic Photochemistry", Academic Press, London, 1976.
S.L Murov, "Handbook of Photochemistry", Marcel Dekker, New York, 1973.
R. Phillips, "Sources and Applications of Ultraviolet Radiation", Academic Press, London, 1983.
National Radiological Protection Board, "Protection against ultraviolet radiation in the workplace", HMSO, 1977.
D. Hughes, "Hazards of Occupational Exposure to Ultraviolet Radiation", University of Leeds Industrial Services Ltd, 1978.

3
Carbonyl Compounds: α-Cleavage

By R.F. Newton
GLAXO GROUP RESEARCH LTD., WARE, HERTS. SG12 0DJ, UK

Simple aliphatic aldehydes and ketones have a characteristic weak absorption band in the 3400-2300 Å region of their ultra-violet spectra. This band is due to an $n \rightarrow \pi^*$ transition involving the promotion of a non-bonding 2p-electron of the oxygen atom to the anti-bonding π-orbital of the carbonyl group. Photochemical excitation of aliphatic carbonyl compounds in the 3400-2300 Å region leads to four general types of reaction.

(i) Norrish Type I cleavage (α-cleavage)
(ii) Norrish Type II cleavage (intramolecular photoelimination)
(iii) Photoreduction
(iv) Cycloaddition to olefines.

This chapter will only be concerned with the Norrish Type I α-cleavage process and, since the general theme is synthetic chemistry, it will concentrate on solution rather than gas phase photochemistry.

When an aliphatic aldehyde undergoes photodecomposition via a Type I cleavage (Scheme 1) the major mode of free radical decomposition is by

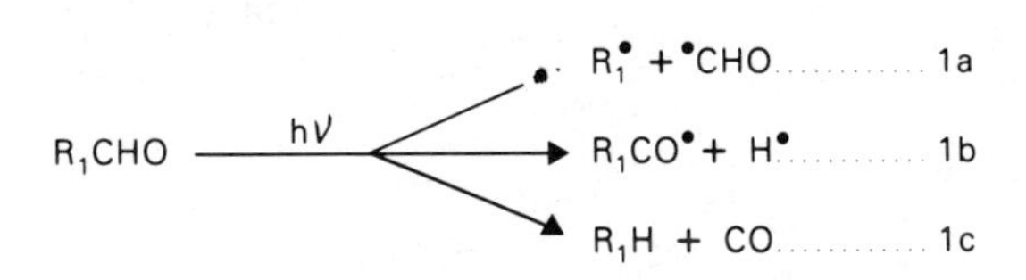

Scheme 1. Photochemical decomposition of aliphatic aldehydes

process **1a** and process **1b** is not important. At long wavelengths (3130 Å with acetaldehyde) process **1c** is also of little importance, however at shorter wavelengths (2537 Å with acetaldehyde) it becomes a major reaction pathway.[1]

Norrish Type I photodecomposition of aliphatic ketones in which R_1 and R_2 are different alkyl groups (Scheme 2) occurs via both process **2a** and process **2b** with a preference for the products that arise from cleavage of

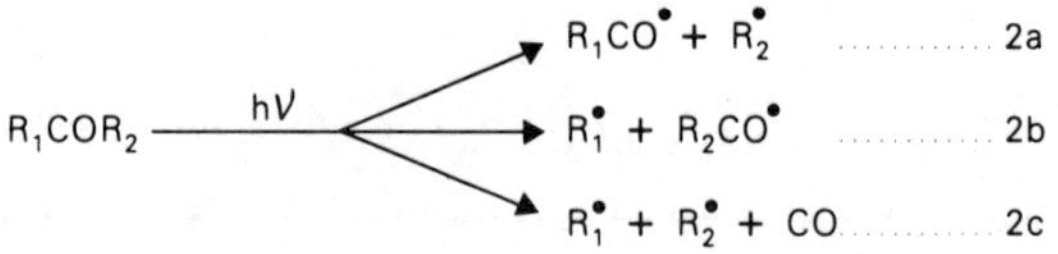

$$R_1COR_2 \xrightarrow{h\nu} \begin{cases} R_1CO^\bullet + R_2^\bullet & \text{2a} \\ R_1^\bullet + R_2CO^\bullet & \text{2b} \\ R_1^\bullet + R_2^\bullet + CO & \text{2c} \end{cases}$$

Scheme 2. **Photochemical decomposition of aliphatic ketones**

the weakest bond. This is particularly true when the absorbed light is of long wavelength. At shorter wavelengths the radicals which are formed may carry considerable excess energy. These "hot" radicals can then undergo further decomposition by process **2c**. Intramolecular decompositions of the type **1c** seen in aldehydes are unknown in simple aliphatic ketones.

Cyclic aliphatic ketones present further possibilities (Scheme 3), several of which have important synthetic utility. The particular reactions that occur depend on the size of the ring, the substitution pattern on the ring and the conditions under which the photolysis is carried out.

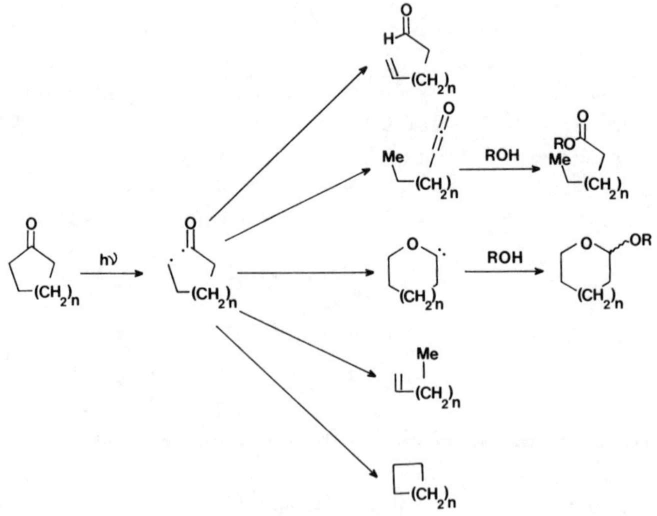

Scheme 3. **Photochemical decomposition of cyclic aliphatic ketones**

In the case of cyclopropanones the major process is decarbonylation and gives rise to carbon monoxide and a simple alkene.

Cyclobutanones undergo three types of photoreactions (Scheme 4) virtually to the exclusion of all others.[2] These processes reflect the inherent ring strain of this system.

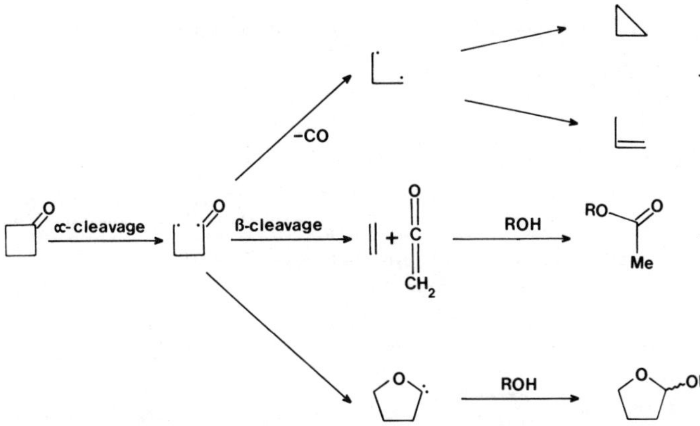

Scheme 4. Photochemical decomposition of cyclobutanones

After the initial α-cleavage the diradical may lose carbon monoxide and the new diradical can either cyclise to cyclopropane or undergo hydrogen abstraction to yield propene (**4a**).

Alternatively (Scheme 4b) the initial product of α-cleavage can

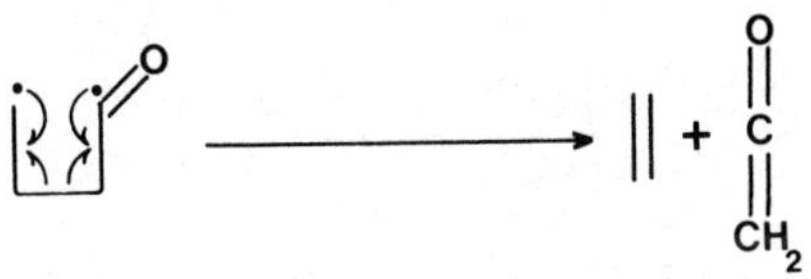

Scheme 4b. β-Cleavage mechanism

undergo a subsequent β-cleavage to give ethylene and ketene. In the presence of alcoholic solvents the ketene forms the corresponding ester.

The third type of reaction that the diradical can undergo is perhaps the most useful synthetically (Scheme 4c). Electronic rearrangement of the acyl-alkyl diradical leads to an oxacarbene[3] which in the presence of an

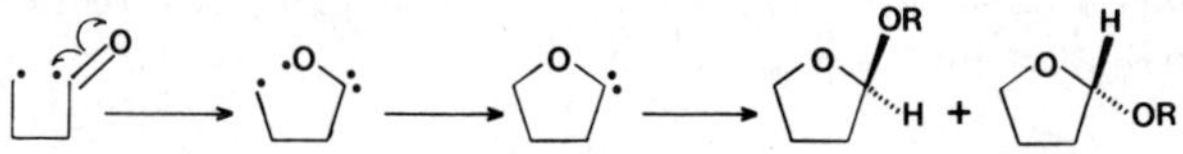

Scheme 4c. Ring expansion mechanism for cyclobutanones

alcohol can insert into the oxygen-hydrogen bond to produce a pair of acetals.[4,5]

In order to use these reactions to synthesise target molecules the factors that influence both the regiochemistry of the initial α-cleavage

Table 1. Photolysis of various cyclobutanones in methanol

Process	Yield
	8%
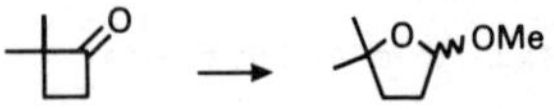	41%
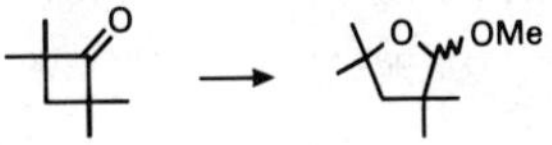	68%
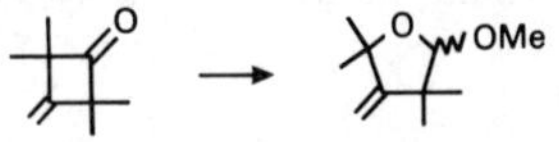	89%
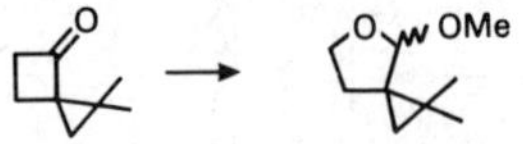	88%
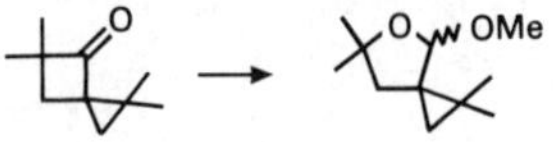	100%
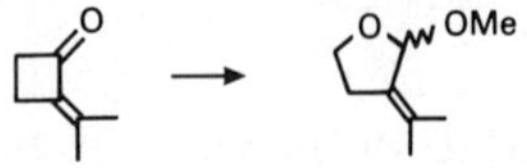	100%

reaction and the subsequent chemistry of the diradical must be understood. Turro and others have shown that the α-cleavage generally occurs at the most substituted bond adjacent to the carbonyl group thus producing the most stable alkyl radical. Furthermore it was found (Table 1) that the yield of the product arising from the ring expansion reaction improved, relative to the other processes, with increasing α-substitution.[4-6]

The incorporation of an exo-methylene group β- to the carbonyl group increases the ring strain and leads to a further improved yield of the ring expanded product.

When either a spirocyclopropyl or an exo-methylene group are introduced onto the α-position with respect to the carbonyl function the expected yield enhancement of ring expanded product is observed.[5] However the regiochemistry of the α-cleavage is reversed[7] and bond breaking occurs in these cases at the least substituted α-position. This has been interpreted (Scheme 5) in terms of the resonance stabilisation available to the oxacarbene thus formed.

Scheme 5. Mechanism of oxacarbene stabilisation

With this information it is now possible to apply the photochemical transformations of cyclobutanones synthetically. The examples are all taken from our own work on the synthesis of prostaglandins.

The Corey Lactone 1 is undoubtably one of the most useful synthetic intermediates for making prostaglandins (Scheme 6). Thus improved

DIBAH

Wittig

1

COREY LACTONE

2

Scheme 6. The Corey lactone in prostaglandin synthesis

procedures for making either the lactone 1 or the corresponding lactol 2 are important synthetic goals.

A retrosynthetic analysis of the lactol (Scheme 7) suggested that reaction of the epoxy-ketal **3** with a lithium organo cuprate reagent to incorporate the prostaglandin sidechain followed by removal of the ketal protecting group should afford the cyclobutanone intermediate **4**.

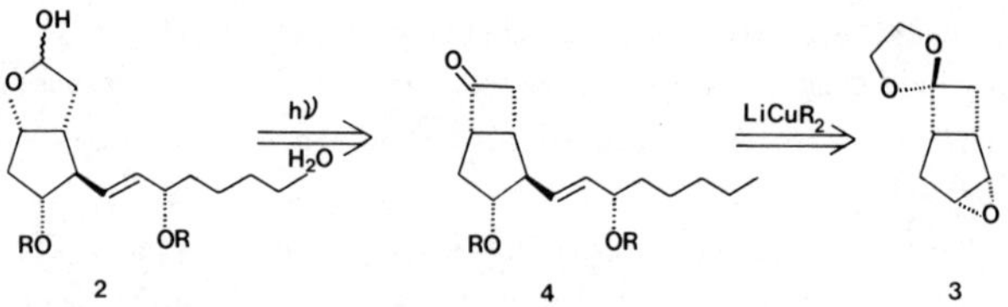

Scheme 7. Retrosynthetic analysis of lactol

A photochemical ring expansion reaction in the presence of an alcohol could then give the target lactol **2**.

In order to test the feasibility of this approach we first synthesised the corresponding epoxy-ketone **5**[8] (Scheme 8).

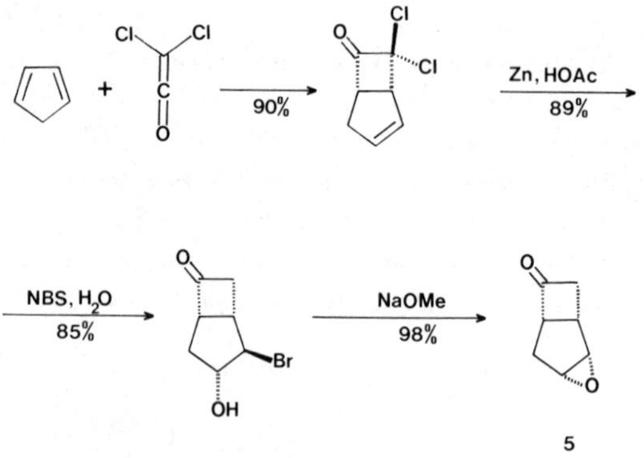

Scheme 8. Synthesis of the epoxy-ketone

To determine the optimum conditions for photolysis of the bicyclo [3.2.0] heptanone ring system a series of reactions were carried out on the epoxy-ketone **5** using methanol as the solvent[9] (Table 2).

Table 2. Optimisation of photochemical conditions

hν, MeOH

conditions	time (h)	yield of acetals % exo-	yield of acetals % endo-	yield of starting epoxide %
PYREX	52.0	25	4	7
QUARTZ	0.5	20	3	10
QUARTZ, $NaHCO_3$,	2.5	33	10	15
QUARTZ, $NaHCO_3$,	2.5	37	10	21
QUARTZ, $NaHCO_3$,	3.5	42	11	6

In Pyrex apparatus, which is opaque below 2800 Å, the reaction was extremely slow when compared with equivalent experiments in quartz apparatus. Previous workers[10] had shown that acidic products, presumably formic acid, could be produced during the photochemical irradiation of methanol and that under the photochemical conditions they further reacted with epoxides. We found that the addition of sodium hydrogen carbonate lead to an improved yield of the acetals **6** and **7** although no products due to acid catalysed ring opening of the epoxide were detected in the absence of base. We also deoxygenated the reaction solvent since cyclobutanones are known to form the corresponding lactones during photolysis in the presence of oxygen (Scheme 9).

hν, O_2

Scheme 9. Lactone formation

A further small improvement in the yield of the acetals was obtained by adding the triplet quencher 2,5-dimethyl-2,4-hexadiene. The cyclobutanone ring expansion reaction is thought to occur via a singlet or an unquenchable triplet state[11,12] so the quenching agent is presumably suppressing a side reaction that proceeds via the triplet state such as decarboxylation.[6,7]

During a large scale reaction a 6% yield of the unsaturated epoxy-ester **8** (Scheme 10) was isolated. This was formed by an unexpected

hν, MeOH

MeO

8

Scheme 10. Formation of epoxy-ester

β-cleavage reaction which lead to rupture of the 5-membered ring, ketene formation and thus the product.

Table 3. Effect of changing alloy alcohol

hν, QUARTZ, ROH, $NaHCO_3$, 2.5h

R	yield of acetals % exo-	endo-	yield of starting epoxide %
Me	37	10	21
Et	21	0	20
Pr^i	16	0	20
Bu^t	11	0	22

When photolyses were carried out using the preferred conditions and the alcohol was changed to ethanol, isopropanol or t-butanol the exo-acetal was the only product observed (Table 3). This is presumably simply a reflection of the steric factor.

The epoxy methoxy acetal **6** is a known prostaglandin intermediate.[13] Corey had shown that upon reaction with the 1,3-bis (methylthio)

Scheme 11

allyl lithium reagent (Scheme 11) a mixture of the two alcohols **9** and **10** was obtained. The minor regioisomer **10** was converted into prostaglandin $F_{2\alpha}$.

Our initial studies had indicated that epoxy ring opening reactions of the corresponding bicyclo[3.2.0] system should exhibit considerable regioselectivity in the desired sense.[14] This is because unlike the rather flexible five-membered acetal ring of Corey's intermediate **6** the more rigid four-membered ring of the epoxy-ketal **3** exerts marked steric control over the chemistry of the cyclopentane ring. As shown in Table 4 a variety of organometallic reagents reacted regioselectively with the epoxide to give mainly the required regioisomer.[15,16]

Table 4. Regioselectivity of epoxy ring opening reactions

reagents	conditions	yield %	isomer ratio
$Li^+MeS\text{-allyl-}SMe$	4h −78°	72	83 : 17
BuMgCl,CuI	2h −30°	98	80 : 20
$LiCu-C{\equiv}C-C_3H_7$ / $CH=CH-CH(C_5H_{11})OSiMe_2Bu^t$	16h −78°→ −30°	86	80 : 20

Separation of the regioisomers and hydrolysis of the ketal protecting group afforded the target photochemical substrates.

Table 5. Photochemistry of cyclobutanones

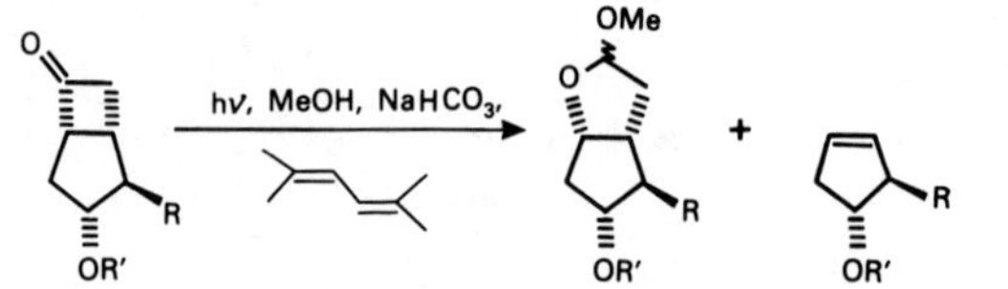

hν, MeOH, $NaHCO_3$

R′	R	yield of acetal %	yield of cyclopentene %
H	 SMe / SMe	33	18 (PYREX)
H	Bu	40	0
$SiMe_2Bu^t$	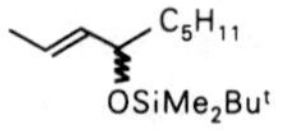 C_5H_{11} / $OSiMe_2Bu^t$	43	24
H	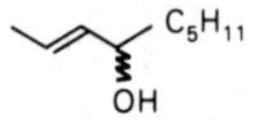 C_5H_{11} / OH	31	37
$COCH_3$	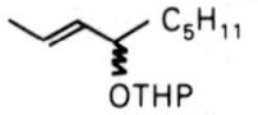 C_5H_{11} / OTHP	35	28

Photolyses were carried out under the previously optimised conditions. The 1,3-bis(methylthio)allyl analogue gave a complex mixture of products; however in pyrex apparatus a slow reaction occurred to generate a mixture of the required acetal **10** and the corresponding alkene (Table 5). The latter compound is the product of an initial α-cleavage followed by a β-cleavage (Scheme 12).

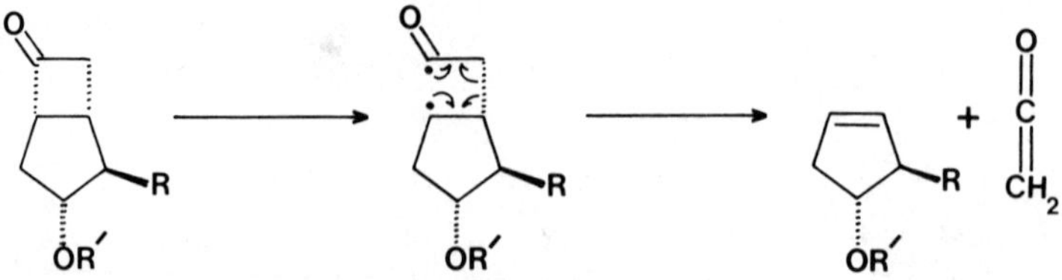

Scheme 12. Formation of cyclopentenes

The other substrates all reacted satisfactorily in quartz apparatus. Efforts to modify the ratio of products in favour of the acetal included the use of triplet sensitisers, low energy light sources and varying temperatures. None of these measures affected the product ratio suggesting that the products derived from a singlet or an unquenchable triplet excited state.

In order to utilise the derived acetals as prostaglandin intermediates it is first necessary to generate the corresponding lactols by acid hydrolysis. This process removes all of the acid sensitive protecting groups and is most suitable therefore for the synthesis of the deprotected lactol **11** which after the appropriate Wittig reaction yielded prostaglandin $F_{2\alpha}$ (Scheme 13).

OMe O RO OR H^+ OH O HO OH Wittig OH CO_2H HO OH PG $F_{2\alpha}$

11

Scheme 13

In order to generate prostaglandin E_2 and prostaglandin D_2 it is necessary to differentially protect the C15 sidechain hydroxyl group and the C11 or the C9 hydroxyl group respectively.

O CO_2H HO OH PG E_2 HO CO_2H O OH PG D_2

There was one report in the literature of the use of an aqueous solvent in the photolysis of a cyclobutanone leading directly to a lactol[5] (Scheme 14).

O hν, H_2O O OH OH CHO

Scheme 14. Photolytic synthesis of a lactol

If this reaction was generally applicable it would allow the direct synthesis of lactols in which the C15 and C11 or C9 hydroxyl groups were protected with acid sensitive protecting groups. The results of our investigation of this process are summarised in Table 6.

The low yield of lactol obtained from the bis-silylated cyclobutanone probably reflects the low concentration of water and high volume of co-solvent necessary to solubulise the reactant. In pure THF the cyclopentene is the only identifiable product.

Table 6. Aqueous photolyses of cyclobutanones

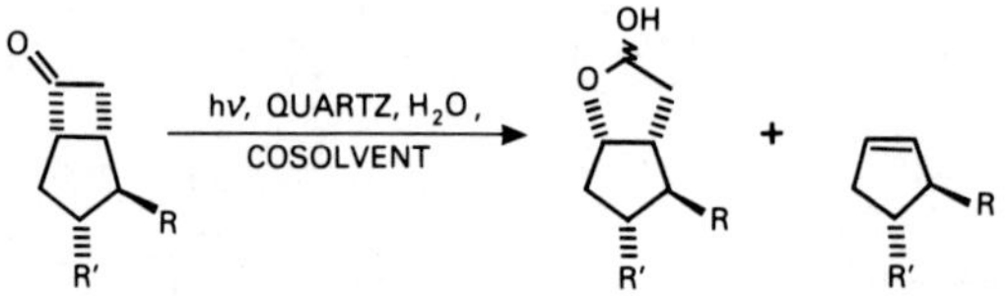

hν, QUARTZ, H_2O, COSOLVENT

R′	R	cosolvent	yield of lactol %	yield of cyclo-pentene %
H	H	THF	75	5
OH	Bu	THF	77	5
OH	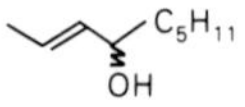 C_5H_{11} / OH	THF	41	10
OH	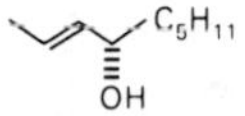 C_5H_{11} / OH	CH_3CN	43	15
$OSiMe_2Bu^t$	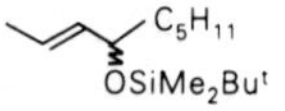 C_5H_{11} / $OSiMe_2Bu^t$	THF	18	40
$OCOCH_3$	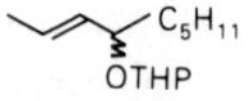 C_5H_{11} / OTHP	Acetone	50	4
$OCOCH_3$	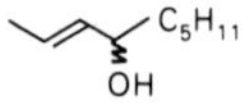 C_5H_{11} / OH	THF	36	8

The 11,15-bis-silylated lactol **12** has been converted into prostaglandin E_2 methyl ester[17] as shown in Scheme 15. The 11-acetoxy-15-tetrahydropyranyl protected lactol is a known[18] precursor for prostaglandin D_2.

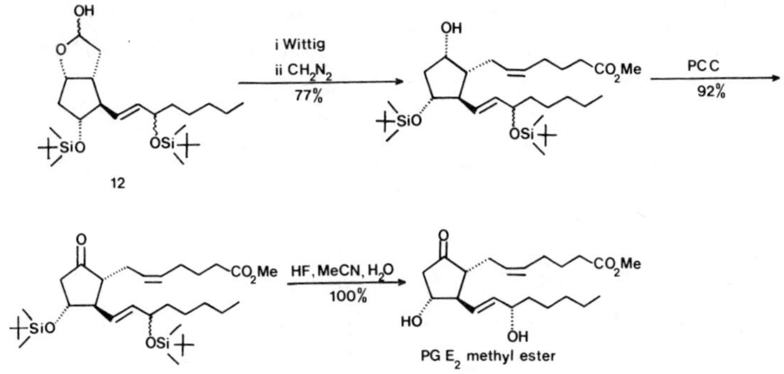

Scheme 15. Synthesis of prostaglandin E_2 methyl ester

Although the dihydroxyl lactol **11** may be obtained pure in 43% yield and subsequently converted into prostaglandin $F_{2\alpha}$ in 70% yield (29% overall) by a Wittig reaction a more effective sequence involves azeotropic removal of water from the photolysis using benzene and a Wittig reaction on the crude lactol. In this way prostaglandin $F_{2\alpha}$ was obtained in 49% overall yield[19] (Scheme 16). This suggests that the photolysis proceeds in about 70% yield and that isolation and purification of the lactol causes a considerable loss of yield.

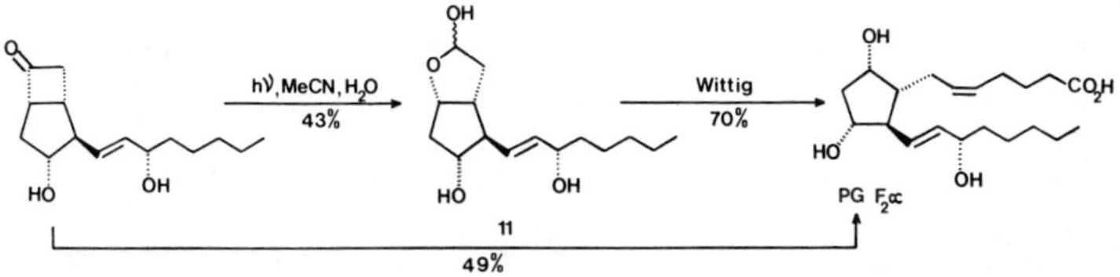

Scheme 16. Synthesis of prostaglandin $F_{2\alpha}$

The photochemical behaviour of simple cyclopentanones reflects the decreased ring strain in these systems compared with the cyclobutanones. Thus the acyl-alkyl diradical generated by α-cleavage undergoes a hydrogen abstraction to give either the unsaturated aldehyde **13** or the ketene **14** (Scheme 17). In the presence of an alcoholic solvent the latter generates the corresponding ester **15**.

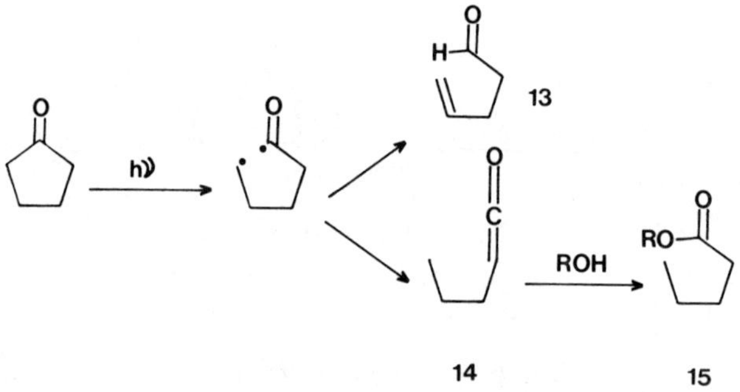

hν ROH RO 13 14 15

Scheme 17. Photolysis of cyclopentanone

However, when the cyclopentanone ring is part of a strained polycyclic molecule or where structural features inhibit aldehyde and ketene formation ring expansion reactions may become important.

In a simple unstrained molecule such as 1-methyl-norbornanone (**16**) photochemical behaviour is normal and α-cleavage leads to abstraction of the syn-C7 hydrogen to give the unsaturated aldehyde **17** as the sole product.[20] (Scheme 18). With Camphor (**18**) however in which the syn-C7 hydrogen is no longer available the diradical transfers the endo-C6 hydrogen

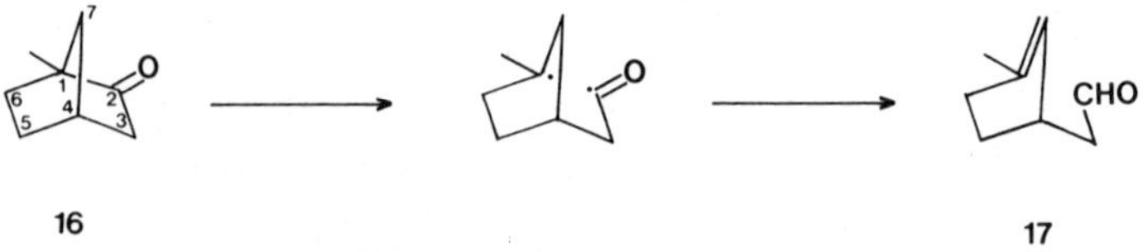

CHO 16 17

Scheme 18. Photolyses of 1-methyl-norbornanone

atom to give the unsaturated aldehyde **19**. A small amount of the acetal **20** is also formed[21] (Scheme 19). When the endo-C6 hydrogen atom is

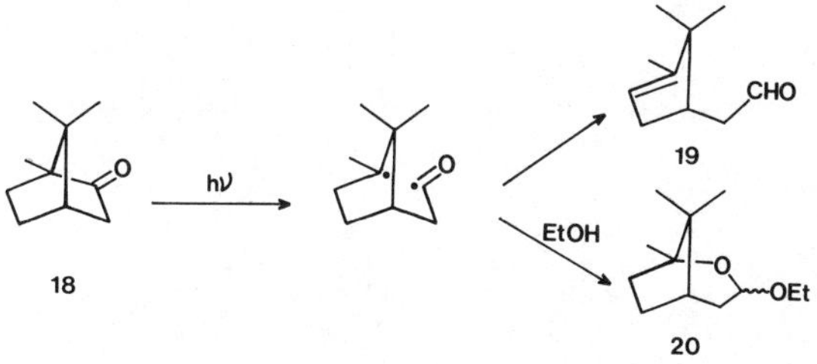

Scheme 19. Photolysis of Camphor

unavailable as is the case with the corresponding ester **21** (Scheme 20) the acetal **22** is the major photolytic product along with a small amount of the diester **23** derived from the corresponding ketene.[22]

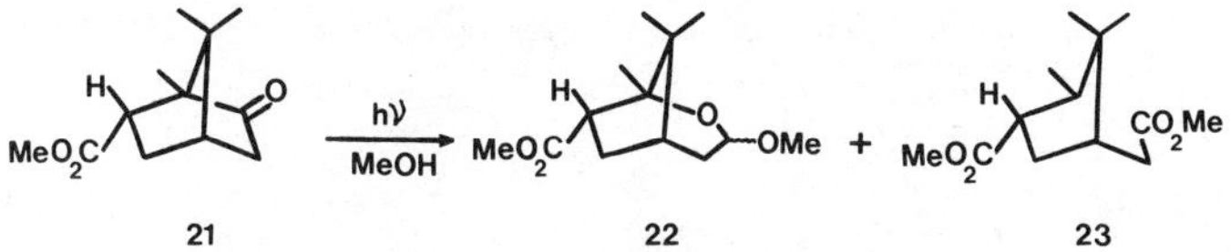

Scheme 20. Photolysis of 6-endo-substituted Camphor

Photolysis of the more strained nortricyclanone **24** (Scheme 21) gave a single lactol ring expansion product **25** in good yield even though the C7 and C6 hydrogen atoms are available.[23] It appears therefore that the

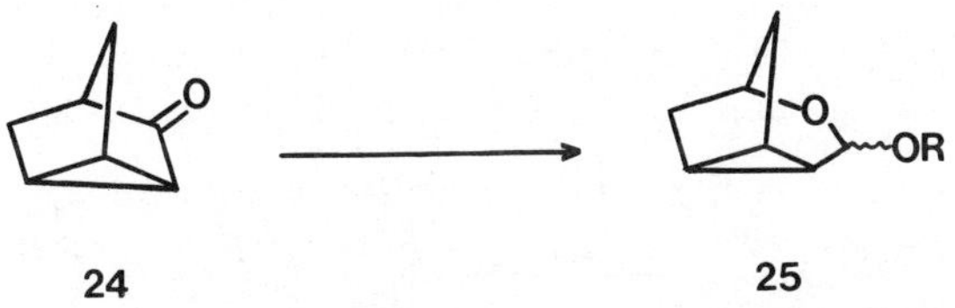

Scheme 21. Photolysis of nortricyclanone

fused cyclopropyl ring facilitates oxacarbene formation and controls the regiochemistry of the initial α-cleavage reaction. This was confirmed by the fact that the spirocyclopropyl cyclopentanone **26** (Scheme 22) gave the acetal **27** as the major product on photolysis in methanol[5] and the bicyclo[3.1.0]hexanone **28** reacted similarly when photolysed in

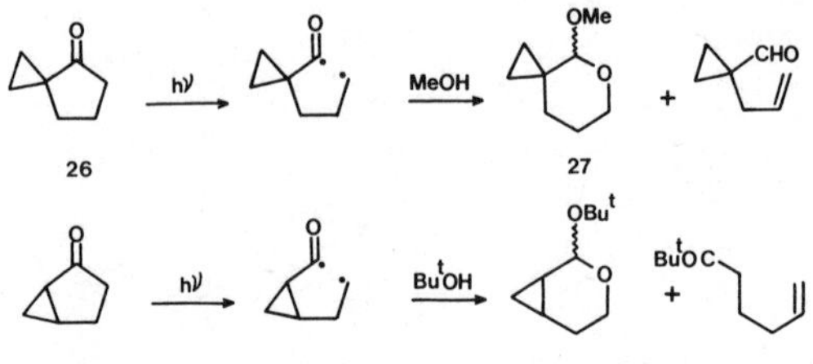

Scheme 22. Photolysis of cyclopropyl substituted cyclopentanones

t-butanol.[24] The precise explanation for these observations is unclear but the resonance stabilisation argument invoked to explain the regiochemistry of the α-cleavage in the analogous cyclobutanones could be important.[6] It has also been pointed out that for the spiro-substituted compounds the angle between the geminal substituents of the cyclopropane ring is 116° (Scheme 23), an increase of 7° over the normal tetrahedral situation.[7]

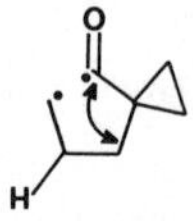

Scheme 23

As a result the distance between the acyl radical and the γ-hydrogen atom will be increased and this might decrease the efficiency of aldehyde formation.

We have utilised the photochemical behaviour of cyclopentanones in a synthesis of prostaglandin C_2. Previous workers had shown that the

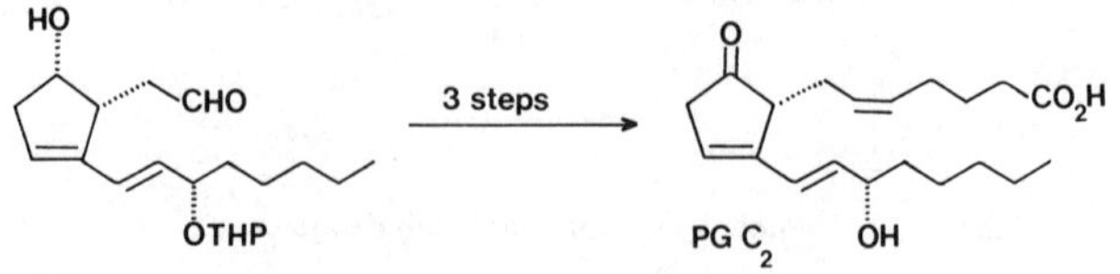

Scheme 24. Synthesis of prostaglandin C_2

unsaturated aldehyde **29** (Scheme 24) could be converted into prostaglandin C_2 by a process of chain extension, oxidation and deprotection.[25] We reasoned that the intermediate aldehyde should be available photochemically from the corresponding norbornanone (Scheme 25).

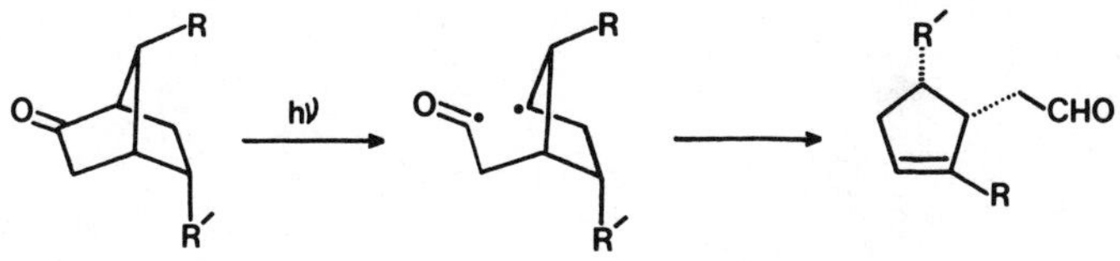

Scheme 25. Strategy for the synthesis of aldehyde 29

In order to synthesise this molecule we adapted the chemistry that we had already evolved for the synthesis of prostaglandins $F_{2\alpha}$,[26] E_2[27] and D_2.[28]

The bromoacetate **30** (Scheme 26) was treated with potassium t-butoxide to generate the key tricyclic ketone intermediate **31** in very high yield. This type of highly strained molecule can be isolated; however they react very rapidly with nucleophilic reagents.[29] Thus on treatment of the tricycle with a lithium organocuprate reagent and work up with potassium carbonate the required norbornanone **32** was obtained. Photolysis of this molecule in deoxygenated methanol with sodium hydrogen carbonate buffer, pyrex apparatus and a medium pressure Hanovia lamp afforded the

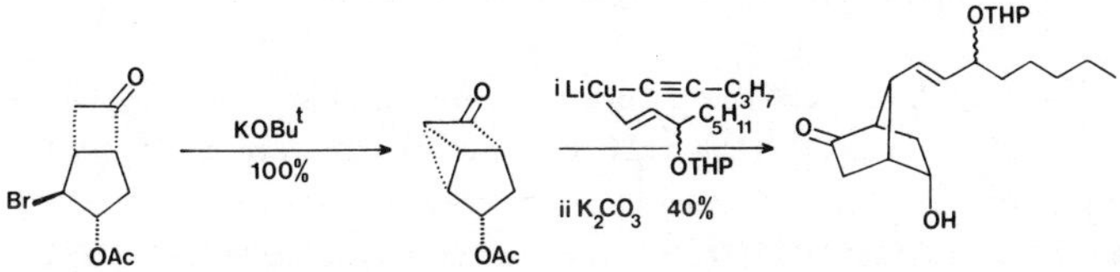

Scheme 26. Synthesis of substituted norbornanone

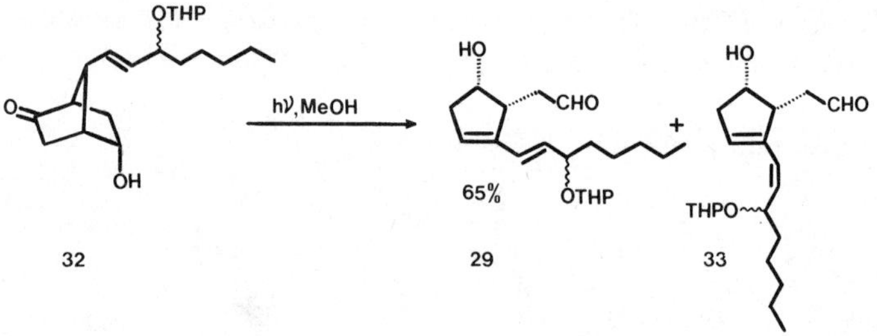

Scheme 27. Photolysis of norbornanone

required hydroxy-aldehyde **29** (Scheme 27) in 65% yield along with a small amount of the double bond isomer **33**.[30] The intermediate **29** was converted into prostaglandin C_2 using the literature procedure.

One of the most challenging synthetic targets in the field of prostaglandins is Thromboxane A_2. The difficulties arise because of the

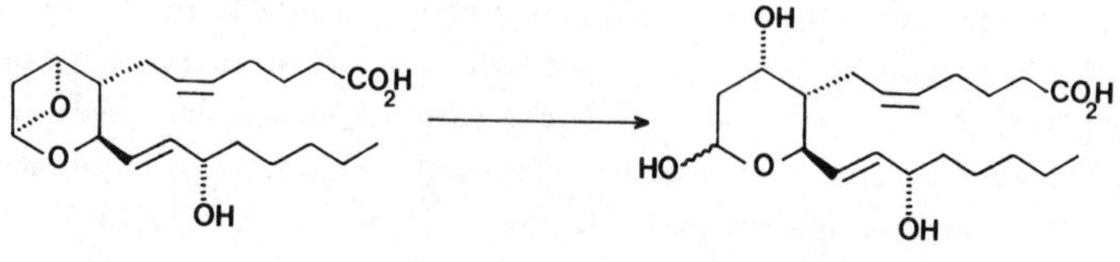

Scheme 28. Formation of thromboxane B_2

extremely acid labile 2,6-dioxabicyclo[3.1.1] heptane portion of the molecule. At pH 7.4 and 37°C the half life for the conversion of TxA_2 into TxB_2 (Scheme 28) is about 30 seconds. The recent syntheses of this ring system and of TxA_2 itself[31,32] have overcome this problem by using the more stable 10-heterosubstituted dioxabicycloheptanes and removing the heteroatoms at a late state.

Our strategy was to trap intramolecularly a photochemically generated oxacarbene (Scheme 29) by an hydroxyl group. Since unsubstituted cyclopentanones do not undergo photochemically induced ring expansion reactions we

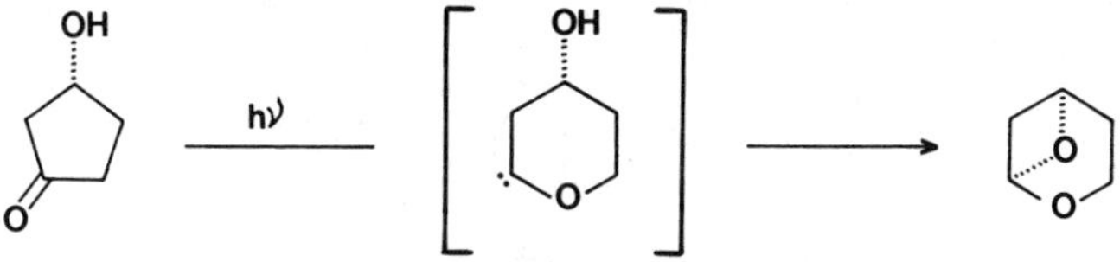

Scheme 29. Synthetic strategy

planned to utilise the unusual photochemical behaviour of the spiro-substituted cyclopentanone system.

When a model compound **34** (Scheme 30) was photolysed at room temperature in dry methanol and pyrex apparatus for 47 hours the methoxy

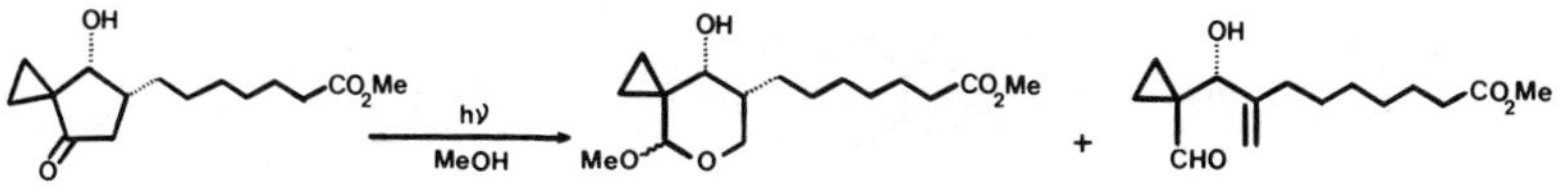

Scheme 30. Formation of a thromboxane-B_2 analogue

acetal **35** (28%) and the unsaturated aldehyde **36** (26%) were the major products.[33] A similar reaction in quartz apparatus at -78^0 was complete in 4.5 hours.

When the photolysis was repeated in quartz apparatus and using dry acetonitrile as the solvent, analysis of the products by g.l.c. showed that the unsaturated aldehyde **36** was present (32%) and n.m.r. spectroscopy of the reaction mixture indicated the presence of the required TxA_2 analogue (**37**). Further proof for the presence of the 2,6-dioxabicyclo[3.1.1]heptane system was obtained by adding methanol to the solution containing the photolysis products. Re-analysis by g.l.c. showed that the methoxy acetal **35** had been formed in 25% yield. The presence of the unsaturated aldehyde **36** and the methoxyacetal **35** was confirmed by separation and purification.

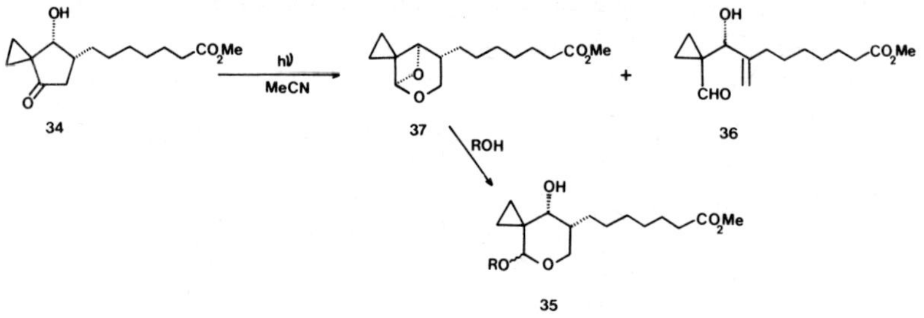

Scheme 31. Formation of a thromboxane-A_2 analogue

The photochemical behaviour of cyclohexanones and larger ring ketones is for the most part restricted to the formation of the corresponding unsaturated aldehyde or the ketene. An interesting exception is provided by the pentulose analogue **38** (Scheme 32). This was shown[34] to give the corresponding acetals (**39**) in 72% yield after photolysis in ethanol.

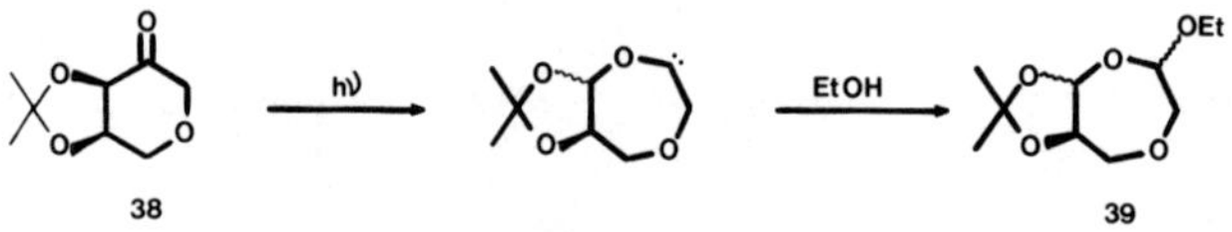

Scheme 32.

Yates[35] has suggested that this reflects the ability of the annular oxygen atom to stabilise and thus facilitate the formation of the corresponding oxacarbene.

REFERENCES

1. C.S. Parmenter and W.A. Noyes, J.Amer.Chem.Soc., 1963, 85, 416.
2. H.V. Hostettler, Helv.Chim.Acta, 1966, 49, 2417.
3. P. Yates and L. Kilmurray, J.Amer.Chem.Soc., 1966, 88, 1563.
4. N.J. Turro and R.M. Southam, Tetrahedron Letters, 1967, 545.
5. D.R. Morton, E. Lee-Ruff, R.M. Southam and N.J. Turro, J.Amer.Chem.Soc., 1970, 92, 4349.

6. D.R. Morton and N.J. Turro, J.Amer.Chem.Soc., 1973, 95, 3947.
7. N.J. Turro and D.R. Morton, J.Amer.Chem.Soc., 1971, 93, 2569.
8. S.M. Ali, N.M. Crossland, T.V. Lee, S.M. Roberts and R.F. Newton, J.C.S. Perkin 1, 1979, 122.
9. N.M. Crossland, S.M. Roberts and R.F. Newton, J.C.S. Chem. Comm., 1978, 661.
10. G. Roussi and R. Beugelmans, Tetrahedron Letters, 1972, 1333.
11. R.F. Klemm, Can.J.Chem., 1970, 48, 3320.
12. N.J. Turro and D.M. McDaniel, Mol.Photochem., 1970, 2, 39.
13. E.J. Corey and R. Noyori, Tetrahedron Letters, 1970, 311.
14. S.M. Ali, N.M. Crossland, T.V. Lee, S.M. Roberts and R.F. Newton, J.C.S. Perkin 1, 1979, 122.
15. R.F. Newton, C.C. Howard, D.P. Reynolds, A.H. Wadsworth, N.M. Crossland and S.M. Roberts, J.C.S. Chem. Comm., 1978, 662.
16. R.J. Cave, C.C. Howard, G. Klinkert, R.F. Newton, D.P. Reynolds, A.H. Wadsworth and S.M. Roberts, J.C.S. Perkin 1, 1979, 2954.
17. C.C. Howard, R.F. Newton, D.P. Reynolds and S.M. Roberts, J.C.S. Perkin 1, 1981, 2049.
18. E.F. Jenny, P. Schaublin, H. Fritz and H. Fuhrer, Tetrahedron Letters, 1974, 2235.
19. C.C. Howard, R.F. Newton, D.P. Reynolds, A.H. Wadsworth, D.R. Kelly and R.F. Newton, J.C.S. Perkin 1, 1980, 852.
20. P. Yates and R.O. Loutfy, Acc. Chem. Res., 1975, 8, 209.
21. W.C. Agosta and D.K. Herron, J.Amer.Chem.Soc., 1968, 90, 7025.
22. J.C. Tam, Ph.D. Thesis, University of Toronto, 1970.
23. P. Yates, Pure Appl. Chem., 1968, 16, 93.
24. W.G. Dauben, L. Schutte, G.W. Schaffer and R.B. Gagosian, J.Amer.Chem.Soc., 1973, 95, 468.
25. E.J. Corey and G. Moinet, J.Amer.Chem.Soc., 1973, 95, 7185.
26. T.V. Lee, S.M. Roberts, M.J. Dimsdale, R.F. Newton and K. Rainey, J.C.S. Perkin 1, 1978, 1176.
27. R.F. Newton, D.P. Reynolds, C.F. Webb, S.N. Young, Z. Grundzinski and S.M. Roberts, J.C.S. Perkin 1, 1979, 2789.

28. R.F. Newton, D.P. Reynolds, C.F. Webb and S.M. Roberts, J.C.S. Perkin 1, 1981, 2055.
29. T.V. Lee, S.M. Roberts and R.F. Newton, J.C.S. Perkin 1, 1978, 1179.
30. N.M. Crossland, S.M. Roberts and R.F. Newton, J.C.S. Perkin 1, 1979, 2397.
31. S.S. Bhagwat, P.R. Hamann and W.C. Still, Tetrahedron Letters, 1985, 26, 1955.
32. S.S. Bhagwat, P.R. Hamann and W.C. Still, J.Amer.Chem.Soc., 1985, 107, 6372.
33. S. Jones, F. Scheinmann, B. Wakefield, D. Middlemiss and R.F. Newton, unpublished work.
34. P.M. Collins, N.N. Oparaeche and B.R. Whitton, J.C.S. Chem. Comm., 1974, 292.
35. P. Yates, A.K. Verma and J.C.L. Tam, J.C.S. Chem. Comm., 1976, 933.

4
Carbonyl Compounds: Hydrogen Abstraction

By W.M. Horspool
DEPARTMENT OF CHEMISTRY, THE UNIVERSITY, DUNDEE DD1 4HN, UK

The last twenty or so years has seen striking changes in the methods available to organic chemists for the synthesis of compounds old and new. One of the advances which has been of great interest is the availability of methods which enable the carrying out of reactions in a selective manner. As you will be aware from this set of lectures it is clear that the electronic excited state of molecules has contributed to this and has provided a new dimension to the synthetic routes available for achieving a specific step in a reaction sequence. Not the least of these is the photochemical behaviour of the humble carbonyl group.

Hydrogen abstraction is one of the oldest and most extensively studied photochemical reactions of the carbonyl group.[1] It can occur both intra- and intermolecularly and can lead either to reduction of the carbonyl function or to the synthesis of new compounds by cyclization or fragmentation. This article is directed towards an understanding of these reactions.

Firstly, it is an advantage to have a knowledge of the excited state or states involved in the process and the first section deals with this subject. Later sections are devoted to the synthetic utility of hydrogen abstraction reactions.

1. Model for The Excited State.

Photochemical excitation of a simple carbonyl compound like formaldehyde[2], which has a weak (ϵ ca 10-30) band in the 280-320nm region, involves the promotion of an electron from a lower orbital into a higher one. The electron undergoing excitation comes from the non-bonding *n*-orbital and ends up in the anti-bonding π^*-orbital. This is referred to as an *n*-π^* transition. In the excited state the electrons which were spin paired (antiparallel spin) in the ground state(S_0) can have either antiparallel spin, a singlet (S_1) state, or parallel spin, a triplet(T_1) state. The singlet and triplet states are designated S_1 and T_1 since Kasha's rule[3] permits the assumption that only vibrationally equilibrated lowest excited states will be involved in primary photochemical processes of organic molecules in solution. The population of the triplet state on direct irradiation is dependent on the intersystem crossing efficiency which for aryl ketones is unity but is variable for alkyl ketones e.g. acetone ϕ_{ISC} = 0.9–1.0, hexan-2-one ϕ_{ISC}= 0.27 and 5-methylhexan-2-one ϕ_{ISC}= 0.11.[4] The actual position of the n-π^* absorption maximum in alkyl ketones and aldehydes is subject to substituent effects. Thus acetone has a maximum at 278.5 nm, 3-methylbutan-2-one a maximum at 284.5 nm and 2,2,4,4-tetramethylpentan-3-one a maximum at 296.8 nm.[5] With aryl ketones the maximum is pushed towards longer wavelengths and benzophenone shows the n-π^* absorption maximising at 348 nm.[5] There is mixing of the π-π^* with the n-π^* and the extinction coefficient is enhanced for these carbonyl compounds.

The reactivity of the n-π^* state becomes obvious when we examine a simple model for the excited state. The outcome of the electronic excitation changes the spatial electronic distribution around the carbonyl group. Thus there is a stereo-electronic factor intrinsic in the n-π^* state of such a group. This is illustrated in figure 1 where it can be seen that the excitation yields a half filled electrophilic n-orbital in the plane of the molecular structure and a nucleophilic π^*-orbital perpendicular to the molecular plane. These changes make nucleophilic attack at the carbon of the carbonyl group an unlikely event in the excited state and the outstanding feature is the unpaired electron in a p-type orbital on oxygen. The similarity between this and an alkoxy radical, which also has an odd electron on oxygen, should be obvious. As a result of this it is not surprising that many of the reactions of a carbonyl group in its n-π^* state will

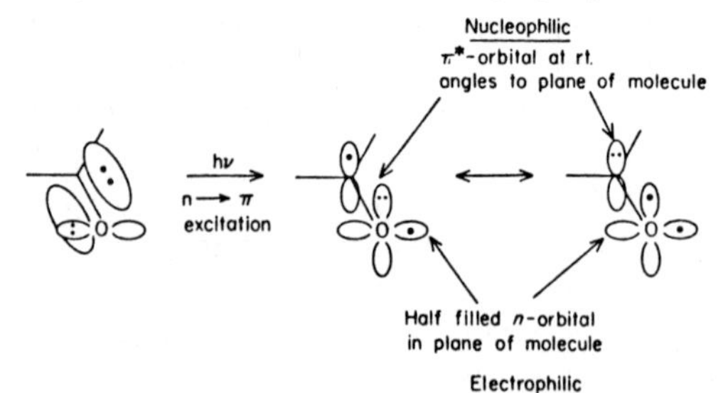

Figure 1

be similar to reactions of alkoxy radicals. Indeed this simple model gives a reasonable account of all the processes which arise from the carbonyl n-π^* excited state such as inter- and intra-molecular hydrogen abstraction, photoaddition, and photofragmentation.

2. Intermolecular Hydrogen Abstraction

2a. Ketones

The irradiation of a carbonyl compound such as acetone in a solvent can result in the abstraction of hydrogen from the solvent. This reaction can occur with a variety of substrates such as alkenes, alkanes, alcohols, and ethers. The result of this is the production of free radicals which can undergo disproportionation or coupling. The irradiation of acetone in *n*-hexane, for example, yields propan-2-ol efficiently.[6] The formation of the product is quenched by added diene and thus a triplet state is implicated. Indeed the reduction arises solely from the triplet state. This ties in with the fact that the intersystem crossing efficiency of acetone is almost unity. Other alkyl ketones such as hexan-2-one also undergo reduction from the triplet state but do so less efficiently than acetone. Indeed the intersystem crossing efficiency of hexan-2-one is only 0.27. Cyclic ketones fit into this general pattern, having a low efficiency of intersystem crossing, and undergo reduction solely from the triplet state.[7]

Aryl ketones also undergo reduction. Benzophenone is particularly unique and has been studied almost for as long as modern photochemistry has excited interest.[1] The reduction reaction of benzophenone is highly efficient and in a hydrocarbon solvent such as toluene affords the pinacol with a quantum yield of 0.39. This reaction scheme is shown in figure 2. The solvent in which the reaction

$$Ph_2CO \xrightarrow{h\nu} Ph_2CO(S_1) \rightarrow Ph_2CO(T_1)$$
$$Ph_2CO(T_1) \xrightarrow{RH} Ph_2\dot{C}OH + R\cdot$$
$$2Ph_2\dot{C}OH \rightarrow Ph-C(Ph)(OH)-C(Ph)(OH)-Ph$$
$$2R\cdot \rightarrow R-R$$
$$Ph_2\dot{C}OH + R\cdot \rightarrow Ph_2C(R)OH$$

Figure 2

is carried out often plays a large part in determining the overall efficiency of the process. Thus in propan-2-ol the quantum yield for the disappearance of benzophenone has a limiting value of 2. This remarkably efficient process is outlined in figure 3. The excitation of the benzophenone yields the triplet state of the carbonyl group which abstracts a hydrogen from C-2 of the alcohol yielding the two acyl radicals (1) and (2). The latter of these can transfer hydrogen to another molecule of benzophenone thus giving a molecule of acetone and another partially reduced benzophenone.

Substantiation of this mechanism has been obtained from an e.s.r. study which identified both ketyl radicals. The reaction described above is only efficient when low light intensities are employed. With high light intensity highly coloured transients e.g. (3) and (4) are produced which act as filters and thus prevent the photoreduction.[8] As was pointed out earlier other substrates, such as thiols, sulphides, ethers and amines, can act as donors of hydrogen to the excited

$$
\begin{gathered}
Ph_2CO \xrightarrow{h\nu} Ph_2CO(S_1) \rightarrow Ph_2CO(T_1) \\
Ph_2CO(T_1) + Me_2CHOH \rightarrow \underset{\mathbf{1}}{Ph_2\dot{C}OH} + \underset{\mathbf{2}}{Me_2\dot{C}OH} \\
Me_2\dot{C}OH + Ph_2CO \rightarrow Me_2CO + Ph_2\dot{C}OH \\
2Ph_2\dot{C}OH \rightarrow Ph_2C(OH){-}C(OH)Ph_2
\end{gathered}
$$

Figure 3

$$
\begin{gathered}
Ph_2CO \xrightarrow{h\nu} Ph_2CO(S_1) \rightarrow Ph_2CO(T_1) \\
Ph_2CO(T_1) + RCH_2NHR' \rightarrow Ph_2\dot{C}\bar{O} + RCH_2\dot{N}^{+}HR \\
Ph_2\dot{C}\bar{O} + RCH_2\dot{N}^{+}HR' \rightarrow Ph_2\dot{C}OH + R\dot{C}HNHR' \\
2Ph_2\dot{C}OH \rightarrow Ph_2C(OH){-}C(OH)Ph_2
\end{gathered}
$$

Figure 4

state carbonyl group. The involvement of electron transfer steps have been shown to be important in photoreduction in the presence of amines.[9] This is outlined for pinacol formation from benzophenone and secondary amines(figure 4). Because of the intervention of the electron transfer step the reduction of the ketone cannot be quenched by triplet quenchers even although the reaction does involve the triplet state of the carbonyl group. Sulphides and arylamines behave in a like manner.

Many ketones fail to undergo photoreduction in alcoholic media. The ability of diaryl ketones to abstract hydrogens from a substrate depends to a great extent on the substitution pattern on the molecule. Substituents on the molecule can change the nature of the lowest triplet state from n-π^* to π-π^*. In the n-π^* state the excitation energy is localized on the carbonyl group while in the π-π^* the excitation is associated with the whole system of the aromatic molecule and so tends to increase the electron density on oxygen thus diminishing its electrophilicity. Thus compounds which fail to photoreduce are those where the π-π^* triplet state is lower in energy than the n-π^* state e.g. *o*-hydroxybenzo, *p*-methoxy. and *o*-methyl acetophenone derivatives and *p*-hydroxy and *p*-amino benzophenones.

The last compound mentioned above, *p*-aminobenzophenone(5), exhibits photochemical properties which depend on the polarity of the solvent in which the reaction is being carried out. In propan-2-ol the hydrogen abstraction process is very inefficient due to the presence of a charge transfer state (6) which is stabilized relative to the n-π^* state(figure 5). In non-polar solvents the charge transfer state does not occur and the n-π^* triplet state is then lower in energy. Thus in cyclohexane photoreduction does take place yielding the pinacol(7). The formation of the charge transfer state can be suppressed in propan-2-ol by the addition of HCl. This protonates the nitrogen and so prevents the occurrence of the charge transfer. Under these conditions the photoreduction to afford pinacol does take place. Analogous behaviour is reported for dimethylamino benzophenone.

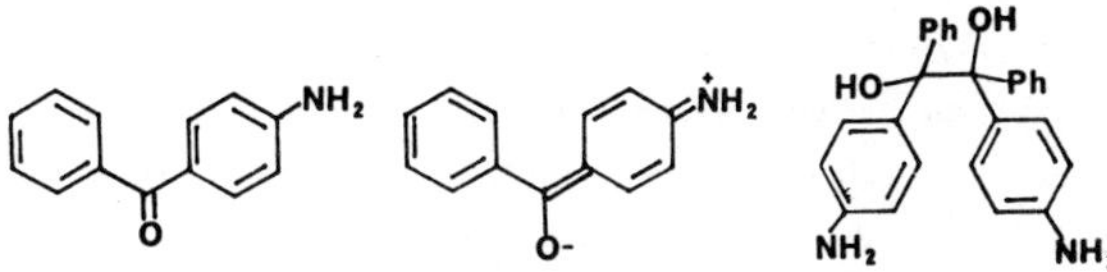

Figure 5

It is possible to bring about the photo-reduction of ketones with low lying π-π^* states using electron transfer from an amine as illustrated for fluorenone in figure 6.

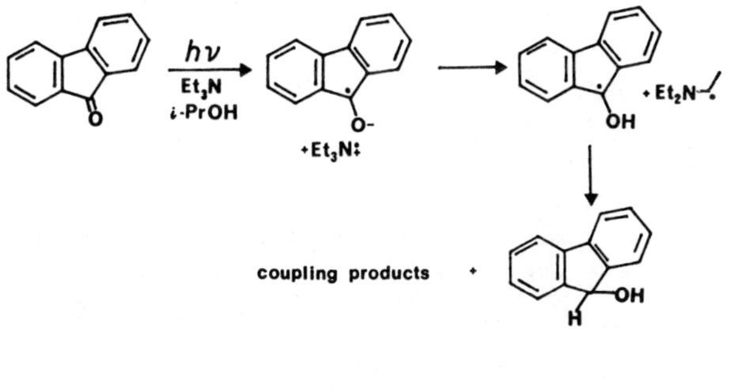

Figure 6

2b. 1,2-Dicarbonyl Compounds

1,2-Diketones and *o*-benzoquinones have long wavelength absorptions, are often highly coloured and also participate in hydrogen abstraction reactions. Thus pinacols and other coupling products can arise from such irradiations. This is shown in general for 1,2-diketones in figure 7. The existence of a semidione radical(8) has been proven by flash photolysis and e.s.r. studies.[10]

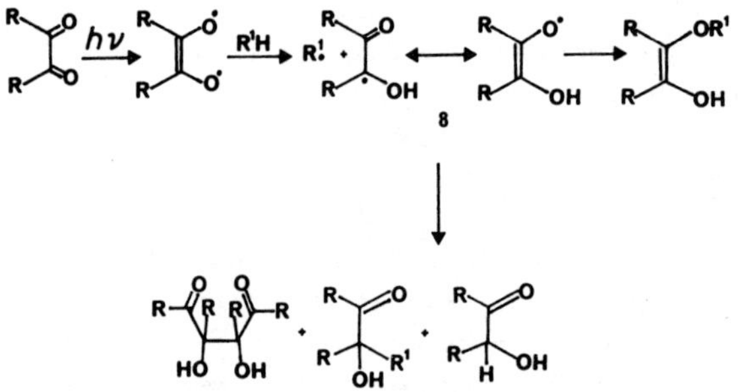

Figure 7

3. Intramolecular Hydrogen Abstraction

3a. γ-Hydrogen Abstraction

On excitation carbonyl compounds with a γ-hydrogen undergo a characteristic 1,5-hydrogen transfer to yield a biradical. This biradical can either undergo cleavage or cyclization or can undergo back hydrogen transfer to yield the starting material. The first two processes are outlined in figure 8. This is a well known reaction referred to as a Norrish Type II process after its discoverer.[11]

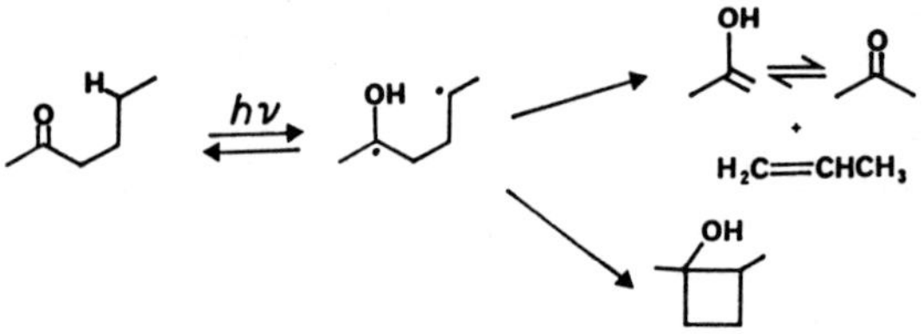

Figure 8

The fragmentation step outlined above in figure 8 has its analogy in mass spectrometry where the McClafferty rearrangement[12] brings about a similar process involving the abstraction of the γ-hydrogen.

From the model for the excited state it is clear that hydrogen abstraction will occur on the half-filled electron deficient (electrophilic) *n*-orbital on the oxygen. Thus the hydrogen to be transferred has to be capable of close approach to this centre. Wagner[13] has studied this problem in considerable detail both for the acyclic and the cyclic situations. For a conventional open chain system involving an excited triplet state γ-hydrogen abstraction, the 1,5-hydrogen transfer is the most rapid. Available data shows that such transfers are 20 times faster than 1,5-transfers in cyclic systems. Thus we see the molecule, because of its flexibility, adopting the most strain free transition state available, a six-membered system, from all the possible transition states open to it. The situation in a cyclic system is quite different and there are only a few conformations available to the molecule within which an hydrogen transfer can take place. This is illustrated in figure 9 for the 2-*n*-propyl-4-*t*-butylcyclohexanones.[14]

Figure 9

Aliphatic ketones undergo Norrish Type II hydrogen abstraction from both the singlet and the triplet $n-\pi^*$ states. The evidence for this is that only part of the reaction can be quenched by the addition of known triplet state quenchers, such as dienes, to the reaction. The percentage of reaction arising from the singlet is dependent on the strength of the γ-hydrogen bond. For example, pentan-2-one with a γ-H bond strength of 410 $kJmol^{-1}$ reacts almost completely from the triplet state, 5-methylhexan-2-one, a γ-H bond strength of 380$kJmol^{-1}$, reacts mostly from the singlet and hexan-2-one with a γ-H bond strength of 395 $kJmol^{-1}$ reacts equally from the singlet and the triplet.[15] Aryl ketones, because of there unit intersystem crossing efficiency, always react from the triplet state. There is a dependency on substitution upon the the rate at which the γ-hydrogen atom is abstracted. In the series of ketones(9) it was demonstrated that the tertiary γ-hydrogens were 165 more reactive than primary γ-hydrogens. There is also a rate dependence on the nature of the substituent on the γ-carbon. Thus the abstraction is faster when the substituents are better capable of stabilizing a free radical than is a methyl group.

The abstraction of the γ-hydrogen yields a biradical. It is obvious, as a result of the concepts of conservation of spin, that a singlet carbonyl group will give rise to a singlet biradical while a triplet carbonyl group will afford a triplet biradical. This influences the lifetime of the biradical and the disappearance of a singlet biradical will be considerably faster than that of a triplet due to the need for spin inversion in the latter case. Substituents have an influence on the lifetime of a triplet biradical as does the solvent in which the reaction is carried out. Scaiano[16] has demonstrated that the triplet lifetime of the biradical is enhanced in polar solvents. Apparently the solvent effect shows itself as a redistribution of the biradical conformational populations. However, it was Wagner[17] who was the first to observe that alcohol solvents could be used to enhance the fragmentation in phenones(10) and that racemization, back H-transfer, could be completely suppressed in such solvents. These two effects can be rationalized in terms of H-bonding by solvent as illustrated in figure 10. Thus the H-bonded biradical is not free to undergo reverse H-transfer on intersystem crossing to the singlet state and so fragmentation can compete favourably.

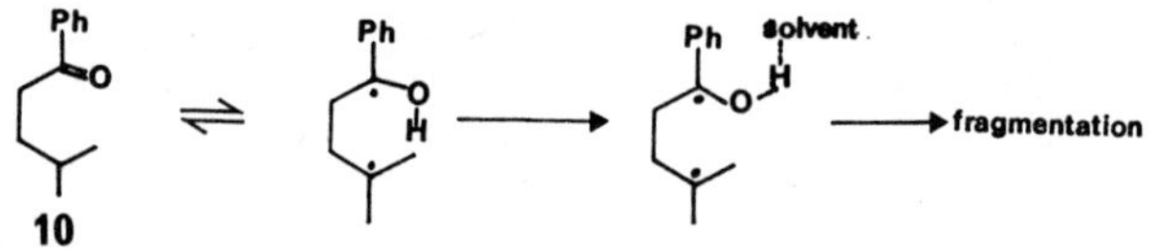

Figure 10

The existence of biradical intermediates in such processes is without doubt and their presence has been substantiated by spectroscopic techniques and by trapping of biradical intermediates by a variety of reagents such as di-*t*-butylselenoketone, alkenes, and oxygen as illustrated in figure 11.

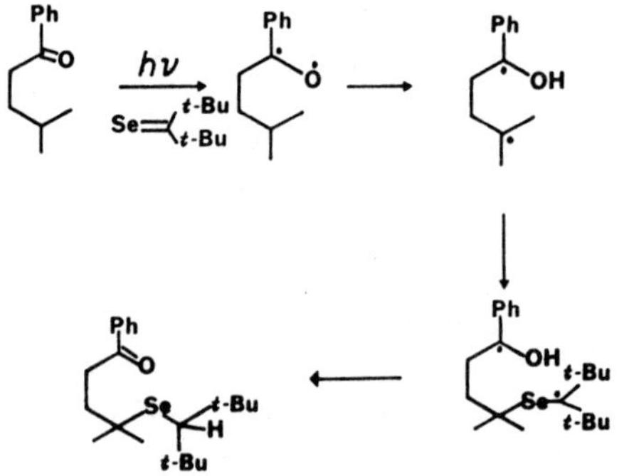

Figure 11

A typical example of the outcome of γ-H abstraction is shown in figure 12. In some ways this has become the archetypal reaction and illustrates all the possible events which can happen to the biradical; racemization, fragmentation and cyclization.[18] The fragmentation reaction does have considerable synthetic potential. Some of the possibilities for its use are shown in figure 13. In this we see the synthesis of isomerically pure alkenes in high yield,[19] and the use of the fragmentation in the photoremoval of a side chain, a protecting group, from some carbohydrate systems.[19] Another example, figure 14, illustrates the photofragmentation of a thiobenzoate in the selective *syn*-elimination in the steroid derivative(11).[21]

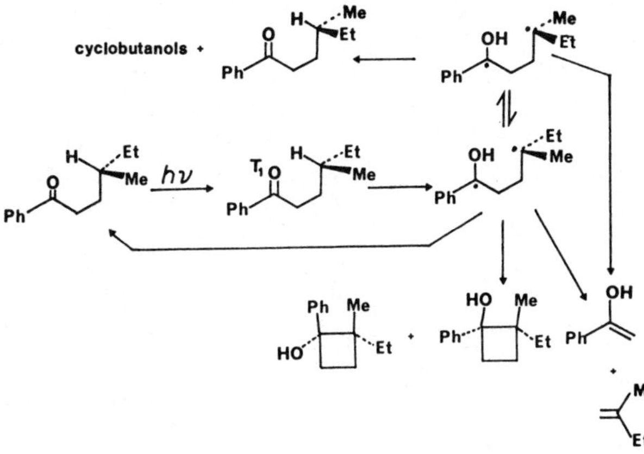

Figure 12

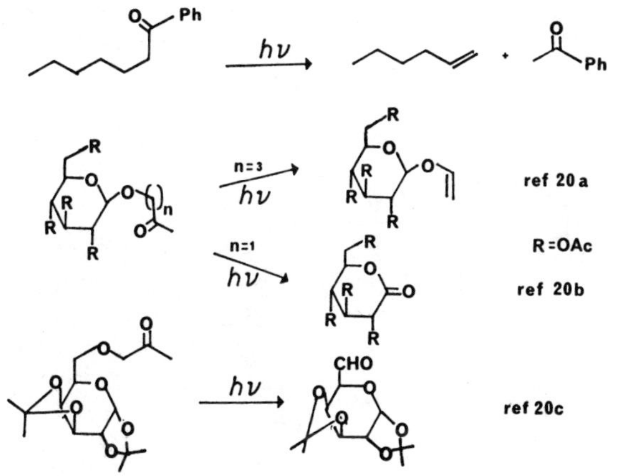

ref 20 a

R = OAc

ref 20b

ref 20c

Figure 13

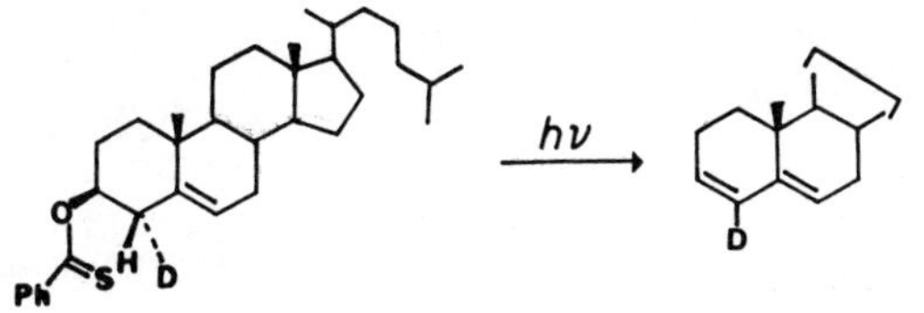

Figure 14

Another application of the Norrish type II fragmentation has been to chain breaking reactions in suitably substituted polymers.[22]

The 1,4-biradical resulting from the γ-H abstraction has also the possibility of undergoing cyclization. Turro and his coworkers[23] have demonstrated that steric factors can be important. These factors manifest themselves in the stereo-electronic requirements necessary for good overlap. In the absence of overlap cyclization occurs in preference to fragmentation. This is seen in figure 15 where hydrogen abstraction yields a biradical which has the β-bond, the bond which usually fragments, orthogonal to the orbitals of the biradical. Thus cyclization arises in preference to fission and yields the cyclobutanols(12).

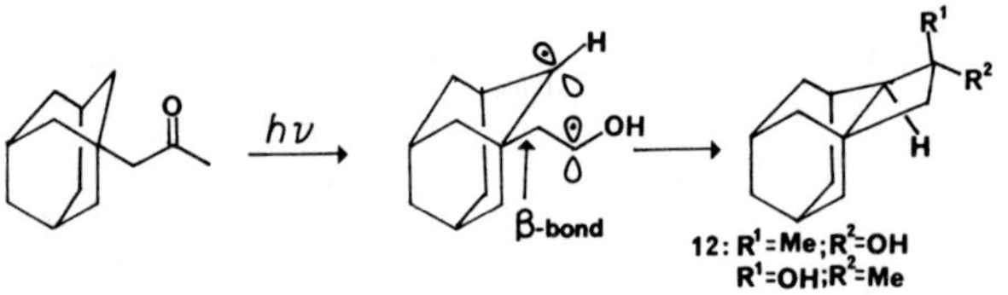

Figure 15

Cyclobutanol synthesis by this route is well-exploited. Much of the effort has been directed towards the synthesis of strained molecules the formation of which would be tedious or impossible by ground state paths. Several examples are outlined in figures 16 and 17.

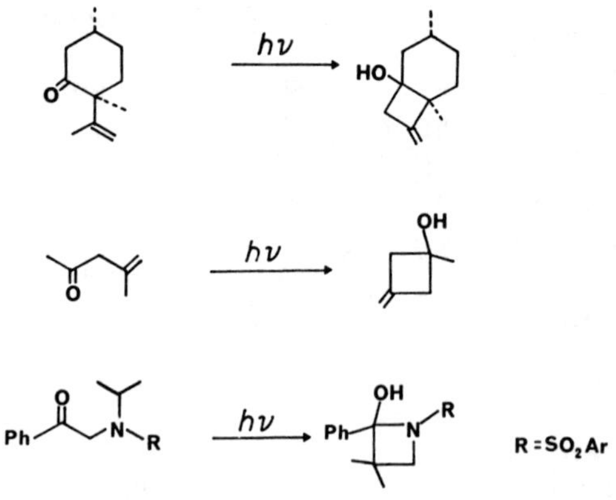

Figure 16

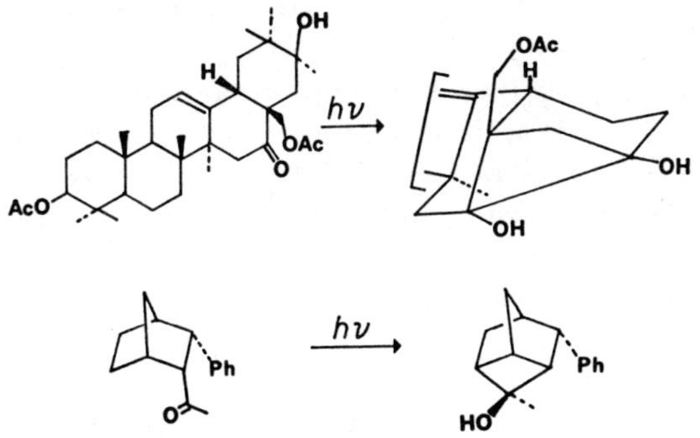
OH
OAc
H
hν
OAc
OH
AcO
O
OH
Ph
hν
Ph
O
HO

Figure 17

While the normal reaction path of the Norrish type II process involves a six-membered transition state alternative reaction paths are fairly common. In these either larger or smaller transition states provide biradicals which cannot fragment. Often the route followed in the hydrogen abstraction can be controlled by the absence of γ-hydrogens or the presence of hetero-atoms. The following examples will illustrate the use that can be made of the reaction for the synthesis of a variety of cyclic compounds.

3b. β-Hydrogen Abstraction

This class of hydrogen abstraction can be used for the synthesis of cyclo-

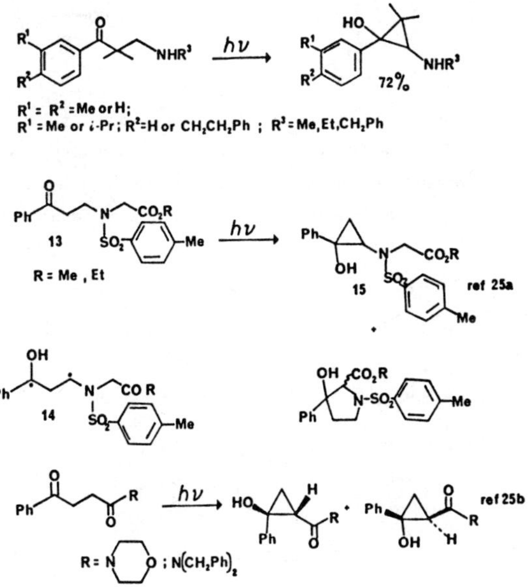
NHR³
hν
HO
NHR³
72%
R¹ = R² = Me or H;
R¹ = Me or i-Pr; R²=H or CH₂CH₂Ph ; R³=Me,Et,CH₂Ph
CO₂R
13
R = Me , Et
hν
15
ref 25a
14
ref 25b
R = N O ; N(CH₂Ph)₂

Figure 18

propanols. The reaction follows the usual path and on excitation the triplet excited carbonyl group abstracts the β-hydrogen affording a biradical which undergoes ring closure. In the first example[24] in figure 18 the yields are high. In the other examples[25] the yields of cyclopropane are lowered because of competition between alternative reaction paths. Thus irradiation of (13) leads to two biradical species one (14) from the β-abstraction yielding the cyclopropanol(15) while the five membered ring is produced by a path involving δ-hydrogen abstraction. In the last example isomeric cyclopropanols are formed in a ratio of 3:1.

3c. δ-Hydrogen Abstraction

δ-Hydrogen abstraction is also fairly common and arises in situations where γ-hydrogens are not available. Some examples of this type of behaviour have been reported[26,27] for carbohydrate derivatives(figure 19).

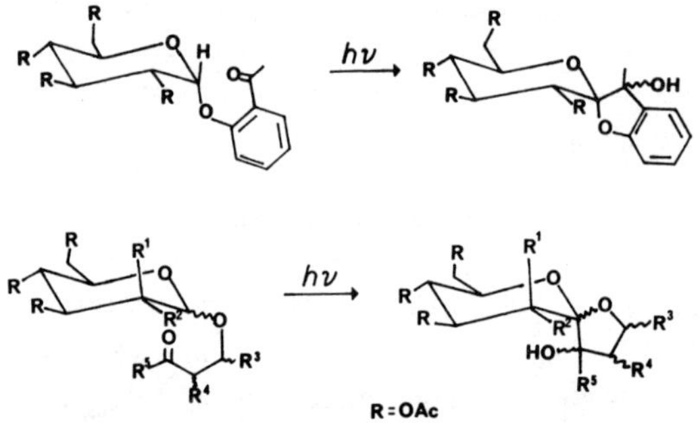

R:OAc

Figure 19

Excitation again results in hydrogen abstraction and cyclization yielding the derivatives shown. Other workers[28] have also demonstrated that cyclizations of this type occur in simpler molecules such as the substituted aryl ketones(16) and (17). Irradiation brings about cyclization to afford the indanols(18). The same

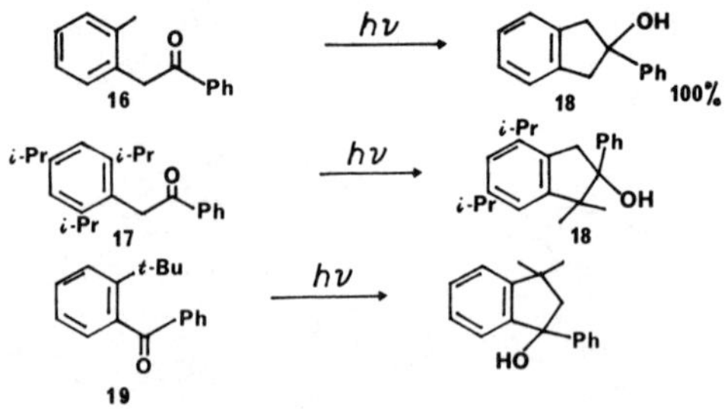

Figure 20

reactivity is also seen[29] in the irradiation of *o*-*t*-butylbenzophenone(19). For many years it had been thought that *t*-butyl groups were reluctant donors of hydrogen in radical reactions. This has been proven to be wrong. The H-abstraction from a methyl of a *t*-butyl group leads to the formation of a primary radical which undergoes cyclization to yield an indanol(figure 20). The reaction is efficient with a quantum yield of unity in methanol. Another example of this type of cyclization is illustrated in figure 21 and involves the abstraction from an *o*-methoxy group on benzophenone and yields the dihydrobenzofuranol(20).[29]

OMe Ph O $h\nu$ O Ph HO 20

Figure 21

Another area where δ-H abstraction is common is in phthalimide photochemistry. Some examples of this, shown in figure 22, are used in the formation of five membered rings. Most of these examples involve the directional influence of a hetero-atom in the side chain. The presence of the hetero-atom can complicate the interpretation of the mechanism of the process since electron transfer reactions could be involved.[30]

O N O OMe $h\nu$ O N HO OMe

O N O $h\nu$ O N HO

Figure 22

3d. ϵ-Hydrogen Abstraction

Hydrogen abstractions in this classification are not as well known. Two examples of the process are shown in figure 23. The first of these illustrates[31] the formation of a dihydropyran(21) from the cyclization of the aryl ketone derivative(22). The second example shows[32] the formation of a tetrahydropyran(23). The excitation of the ketone(24) brings about hydrogen abstraction from the ϵ-position. The attack at this site is controlled to some extent by the presence of the oxygen atom. The biradical produced is an acyl radical/ allyl radical system and this is to be preferred over the other biradical which could have arisen by β-hydrogen abstraction.

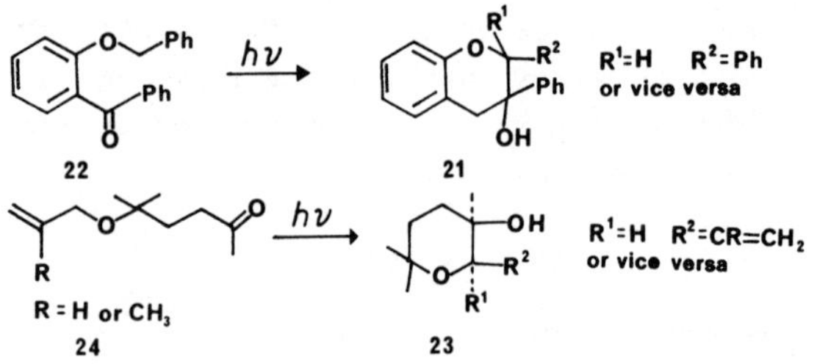

Figure 23

3e. Hydrogen Abstraction from Distant Sites

Considerable interest has been shown in reactions in this classification. The principal area in which they are seen is in the study of phthalimide derivatives

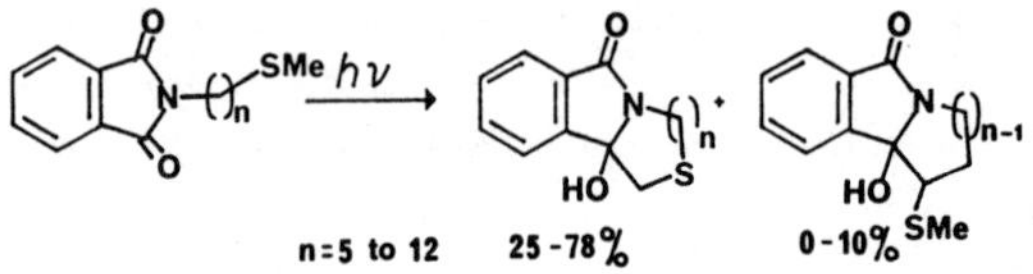

Figure 24

which are known to undergo hydrogen abstraction and cyclization as illustrated previously in figure 22. Moderate to large-ring diaza systems can be prepared by the irradiation of long chain substituted phthalimides as shown in figure 24. The yields are variable although it is possible to achieve high yield cyclizations producing large rings as illustrated in figure 25.[30]

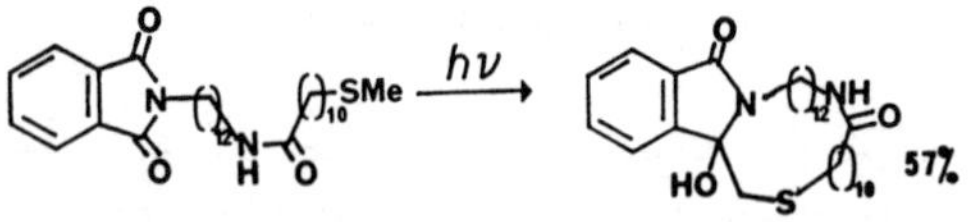

Figure 25

3f. Biomimetic Hydrogen Abstraction

Also within this area interest has been shown in processes which are designed to mimic those which occur in biological or biochemical environments. These reactions have been studied mainly by Breslow and his coworkers[33] and involve hydrogen abstraction from sites far removed from the carbonyl function undergoing the excitation.

The approach to this problem involved the attachment of a benzophenone unit by chains of varying length to a steroidal molecule. In this way they were able to investigate the selectivity of the photochemical hydrogen abstraction from various points on the steroid by the photo-excited carbonyl oxygen. In the example shown in figure 26 the chain holding the two units together permit the carbonyl oxygen to abstract the α-hydrogen at C-14 of the steroid. The hydrogen on C-15 is removed by the ketyl radical resulting in the formation of the single olefinic product(65%). Thus the reaction is regio- and stereo-specific in that only the C-14 α-hydrogen is abstracted. Needless to say this is not the only reaction which occurs during the irradiation. Attack on solvent results in the reduction of the benzophenone moiety and there is also the formation of the two lactones(25) and (26) in a total yield of 25%. These products are the result of the attack at the C-7 and C-12 sites followed by recombination of the developed radical centre with the ketyl radical.

The influence of longer chains was also studied as is shown in figure 27. The extended carbon chain gives the reactive site a longer reach and in this example the C-17 hydrogen comes under attack. However, the chain is still flexible and attack closer to the point of attachment is also encountered, i.e. at C-14. Later studies in this area demonstrated that the benzophenone unit did not need to be directly bonded to the steroid. Thus figure 28 illustrates the use of a hydrogen bonded benzophenone unit with which selective attack at C-16 resulted.

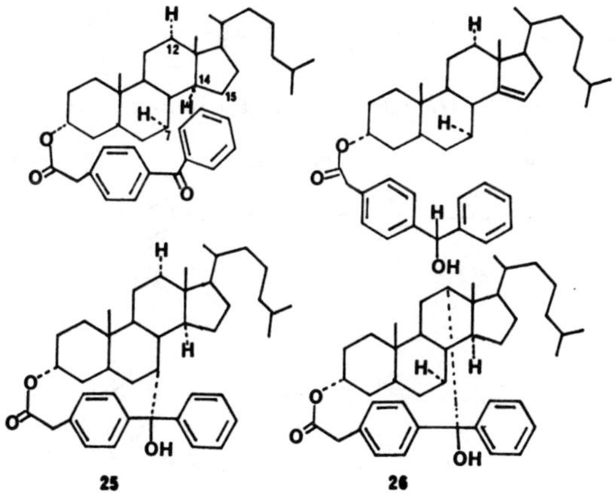

25 26

Figure 26

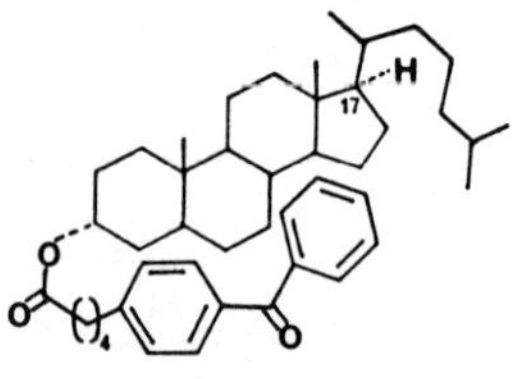

Figure 27

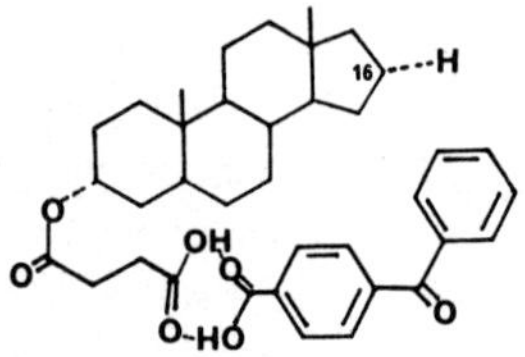

Figure 28

Conformational effects on H-abstraction from distant sites have also been studied by Winnick[34] who observed that in a series of *n*-alkylesters of 4-benzoylbenzoic acid (27) that the rate constant for remote hydrogen abstraction is very small when the chain is relatively short, n<9. The hydrogen abstraction becomes more common with chains longer than 9 with an increase of 0.5 x 10^4 s^{-1} for each methylene unit added into the chain.

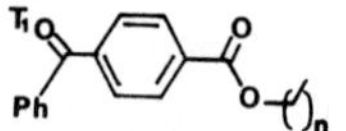

27

4. Formation of Photoenols

These species are formed by the path shown in figure 29 and are obviously a close

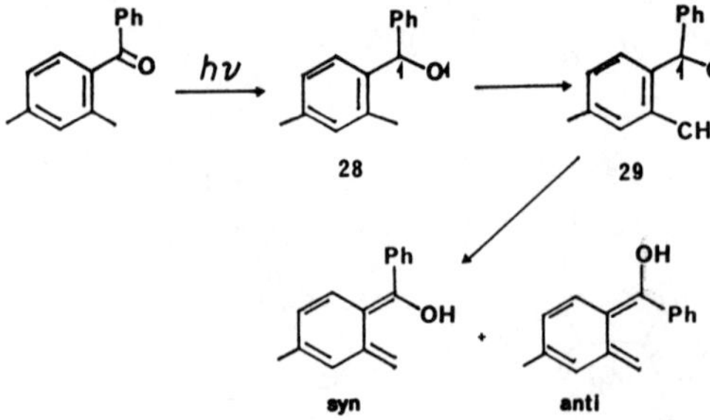

Figure 29

analogue of Norrish Type II processes. The enols are formed from o-alkyl aryl carbonyl compounds and involve a complicated series of excited states.[35] Detailed studies have identified the states and conformations involved. Thus the *syn*-triplet state(28) is the only conformation in which H-abstraction can occur. The enol (29) so formed is also a triplet which can be quenched by oxygen but in the absence of quencher the excited state enol decays to the *syn-anti* ground state.

It should be obvious, because of the ease of formation of photo-enols, that their trapping is of use in synthesis. An example[36] is seen in figure 30 where the addition of a dienophile to the reaction mixture affords the adduct(30).

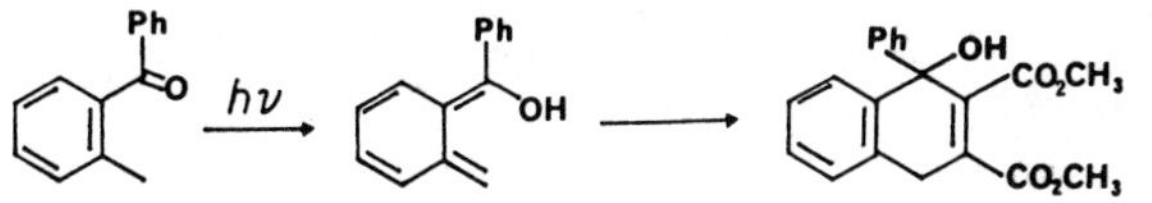

Figure 30

The work of Quinkert and his co-workers[37] is especially notable among the many experiments where trapping has been employed. This example(figure 31) is an approach to the synthesis of estrone(31).

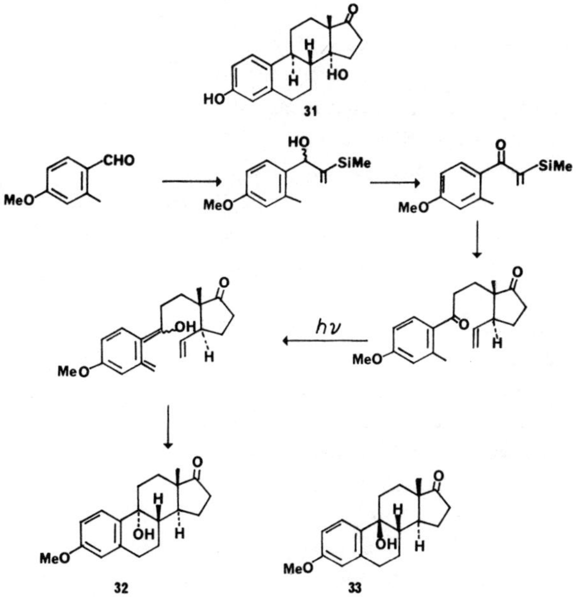

Figure 31

The irradiation affords two isomers the major of which was identified as (32) and the minor as (33). The major product is obtained from the least strained transition state, the *Z*/*exo*/β-conformation. The possible transition state leading to the minor isomer is not as easy to rationalize and it could equally be formed from the two transition states, *Z*/*endo*/α-(34) or *E*/*exo*/β-(35), represented in figure 32. Subsequent development of this approach has sought to synthesize the enantiomerically pure estrone.

Z/exo/β 34 35

Figure 32

References

1. G. Ciamician and P. Silber, Chem. Ber., 1900, 33, 2911; W.D. Cohen, Rec. Trav. Chim., 1920, 39, 243.
2. N.J. Turro, "Modern Molecular Photochemistry", Benjamin Cummings Publishing Co Inc., Menlo Park, California, 1978, p22.
3. M. Kasha, Radiation Research, Supplement 2, 1960, 2, 243.
4. S.L. Murov, Handbook of Photochemistry, Marcel Dekker Inc., N.Y., 1973, p49.
5. C.N.R. Rao, Ultra-violet and Visible Spectroscopy, Butterworth, London, 1967.
6. P.J. Wagner, J. Am. Chem. Soc., 1966, 88, 5672.
7. N.J. Turro and D.M. McDaniel, Mol. Photochem., 1970, 2.
8. S.G. Cohen and J.I. Cohen, Israel J. Chem., 1968, 6, 757; *idem*, Tetrahedron Lett., 1968, 4823.
9. P.S. Mariano and J.L. Stavinoha, in "Synthetic Organic Photochemistry", Ed. W.M. Horspool, Plenum Press, New York and London, 1984, 145.
10. M.B. Rubin, Topics in Current Chemistry. Photochemistry, Springer-Verlag, Berlin, 1968, 13, 251; Top. Curr. Chem., 1985, 129, 1.
11. R.G.W. Norrish, Trans. Faraday Soc., 1939, 33, 1521.
12. H.C. Hill, Introduction to Mass Spectrometry, Heyden and Son Ltd., 1966, p.66.
13. P.J. Wagner, Acc. Chem. Res., 1983, 16, 461.
14. N.J. Turro and D.S. Weiss, J. Am. Chem. Soc., 1968, 90, 2185.
15. P.J. Wagner, Acc. Chem. Res., 1971, 4, 168.
16. R.D. Small jnr. and J.C. Scaiano, Chem. Phys. Lett., 1978, 59, 246.
17. P.J. Wagner, Tetrahedron Lett., 1967, 1753.

18. see reference 15 and R.M. Wilson, Org. Photochem., 1985, 7, 339.
19. D.C. Neckers, R.M. Kellogg, W.L. Prins, and B. Schoustra, J. Org. Chem., 1971, 36, 1838.
20. a. L. Cottier, G. Remy, and G. Descotes, Synthesis, 1979, 711.
    b. G. Bernasconi, L. Cottier, G. Descotes, and G. Remy, Bull. Soc. Chim. Fr., 1979, 322.
    c. R.W. Binkley and H.F. Jarrell, J. Carbohydr. Nulceosides, Nucleotides, 1980, 7, 347.
21. D.H.R. Barton, M. Bolton, P.D. Magnus, K.G. Marathe, G.A. Poulton, and P.J. West, J. C. S. Perkin Trans., I, 1973, 1574.
22. P. Hrdlovic and I. Lukac, Dev. Polym. Degrad., 1982, 4, 101.
23. R.B. Gagosian, J.C. Dalton, and N.J. Turro, J. Am. Chem. Soc., 1975, 97, 5189.
24. A. Abdul-Baki, F. Rotter, T. Schrauth, and H.J. Roth, Arch. Pharm., 1978, 311, 341.
25. a. H.G. Henning, H. Haber, and H. Buchholz, Pharmazie, 1981, 36, 160.
    b. H.G. Henning, R. Berlinghoff, A. Mahlow, H. Koeppl, and K.D. Schleintz, J. Prakt. Chem., 1981, 323, 914.
26. G. Bernasconi, L. Cottier, G. Descotes, J.P. Praly, G. Remy, M.F. Gernier-Loustalot, and F. Metras, Carbohydr. Res., 1983, 115, 105.
27. G. Descotes, Bull. Soc. Chim. Belg., 1982, 91, 973.
28. M.A. Meador and P.J. Wagner, J. Am. Chem. Soc., 1983, 105, 4484.
29. P.J. Wagner, M.A. Meador, B.P. Giri, and J.C. Scaiano, J. Am. Chem. Soc., 1985, 107, 1087.
30. J.D. Coyle, in "Synthetic Organic Photochemistry", Ed. W.M. Horspool, Plenum Press, New York and London, 1984, 259.
31. M.A. Meador and P.J. Wagner, J. Org. Chem., 1985, 50, 419.
32. H.A.J. Carless and G.K. Fekarurhobo, Tetrahedron Lett., 1984, 25, 5943.
33. R. Breslow, Acc. Chem. Res., 1980, 13, 170.
34. M.A. Winnick, C.K. Lee, S. Basu, and D.S. Saunders, J. Am. Chem. Soc., 1974, 96, 6182.
35. R.M. Wilson, Org. Photochem., 1985, 7, p373.
36. N.C. Yang and C. Rivas, J. Am. Chem. Soc., 1961, 83, 2213.
37. G. Quinkert and H. Stark, Angew. Chem. Int. Ed. Engl., 1983, 22, 637.

5
Steroids

By A. Gilbert
CHEMISTRY DEPARTMENT, UNIVERSITY OF READING, WHITEKNIGHTS, PO BOX 224, READING, BERKS. RG6 2AD, UK

The photochemistry of the carbonyl group in various molecular environments is considered in other Chapters of this text. The purpose of the present Chapter is to amplify and illustrate these processes by reference firstly to the use of this chemistry to synthesise the steroid skeleton, and secondly to reactions of the carbonyl chromophore in steroidal systems. The latter treatment will thus outline the chemical changes that can readily be achieved in these naturally occurring compounds and thereby demonstrate the synthetic potential of photochemistry in this series.

1. Synthesis of the Steroid Skeleton

There have been few reports which describe a photochemical route to the steroid skeleton. The three examples outlined below which involve carbonyl compounds, however, do demonstrate the diversity of approaches that may be applied to this problem.

The de Mayo reaction involves the $2\pi+2\pi$ photocycloaddition of an ethylene to the enol form of a β-diketone and in its application to steroid synthesis the diketone used is (1) and the addend is cyclopentene.[1] Both enol forms (2) and (3) of (1) react to give a 1:1 mixture of (4) and (5). The former product is converted into 18,19-bis-nor-$\Delta^{9(11)}$androsten-12-one (6) of non-defined stereochemistry and in 30% yield overall. A second approach to the synthesis of the steroid skeleton involves two photochemical steps, the first of which converts the keto-alkene (7) into the trans isomer and the second initiates intramolecular oxetan formation to give a 1:9 mixture of (8) and (9) respectively.[2] Treatment of the latter with HI in acetic acid results in opening of the oxetan ring and the formation of 1α-hydroxy-$\Delta^{3,5}$-cholestadiene (10) and 1α-hydroxy-3β-acetoxy-Δ^{5}-cholestene (11).

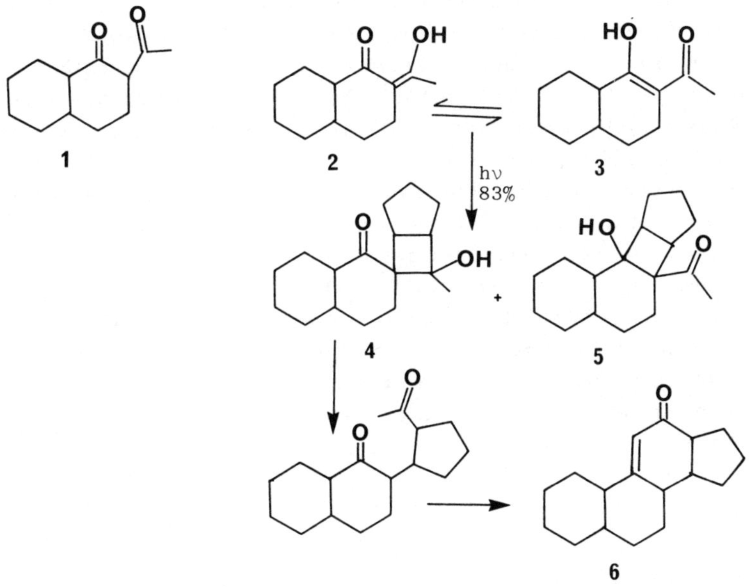

1 2 3

hν 83%

4 + 5

6

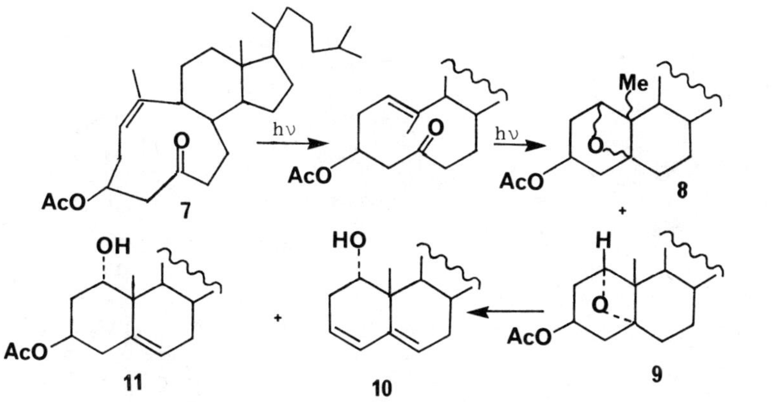

7 hν hν 8

+

11 + 10 9

It is the goal of all synthetic organic chemists to produce enantiomerically pure compounds. This goal is rarely attained and in the steroid series is particularly difficult to achieve. However, by elegant chemistry and the use of a photochemical reaction as the key step, enantiomerically pure 19-norsteroids have been synthesised by a route which neither involves racemate resolution nor depends on the use of protecting groups.[3] The key step is the photo-initiated enolisation of an _o_-tolyl ketone and intramolecular trapping of the _o_-quinodimethanoid intermediate with an ethylene. Thus irradiation of (12) yields a mixture of (13) and (14) _via_ the enol (15). Dehydration of the product mixture gives (16) and (17) in a respective ratio of 19:1 and in a yield of 65% from (12). These ethylenes are suitable starting materials for several biologically active 19-norsteroids [_e.g._ (18)] and these conversions are achieved in yields of 39-71%.

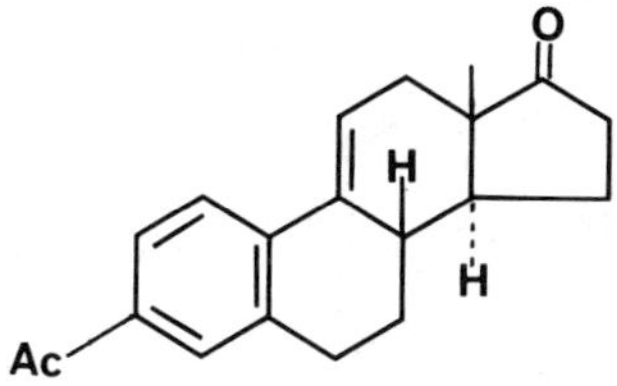

16 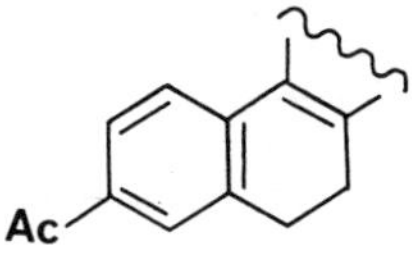 17 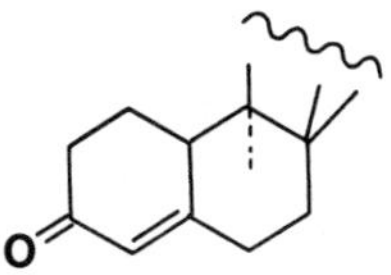 18

2. Reactions of the Carbonyl Chromophore in Steroid Systems

Study of this area of photochemistry has been popular for many years and the published literature is extensive. The examples of the various processes cited here are intended only to illustrate reaction types and the resultant changes that may be readily induced in steroid systems.

2.1 α-Cleavage - Norrish Type I Processes

Photolysis of cyclolaudanone (19) induces ring-opening which in aqueous dioxan leads to the formation of the acid (20) in 35% yield.[4] The efficiency of trapping of the intermediate is improved considerably in methanol or solutions containing cyclohexylamine when the corresponding ester and N-cyclohexylamide

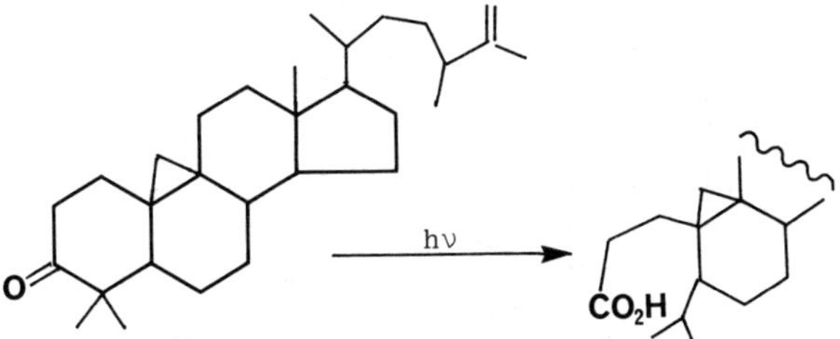

19 hν 20

are obtained in 58 and 73% yields respectively. α-Cleavage of the lactone (21) in methanol solution results in opening of both A and B rings of the steroid to give (22) which is considered to arise by fission of the 5,10-bond to give a trappable ketene.[5] The cleavage takes a different route with the lactam (23) and on irradiation in t-butanol it forms the isocyanate (24) but in methanol the major products are (25) and (26).[6]

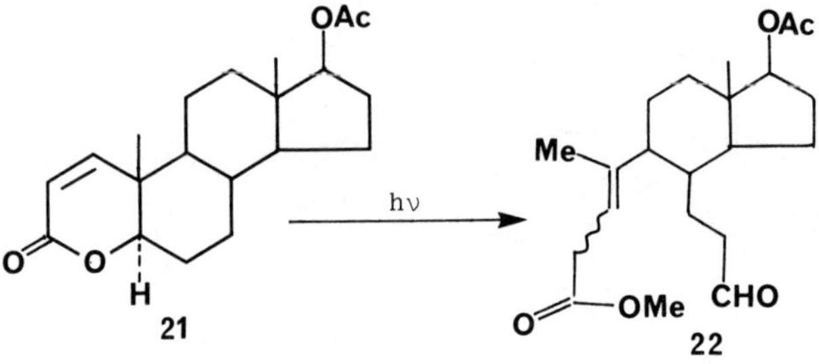

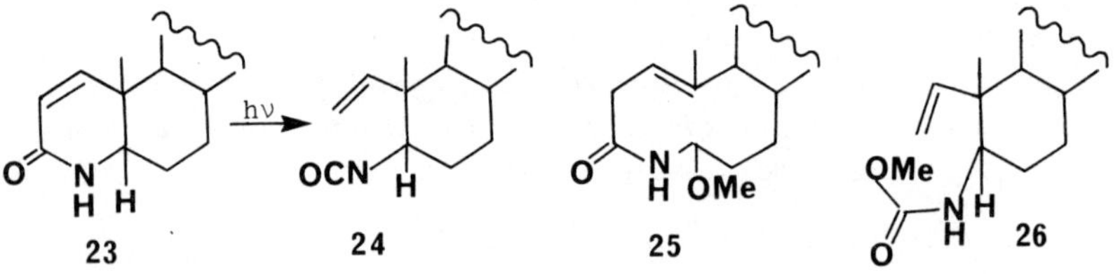

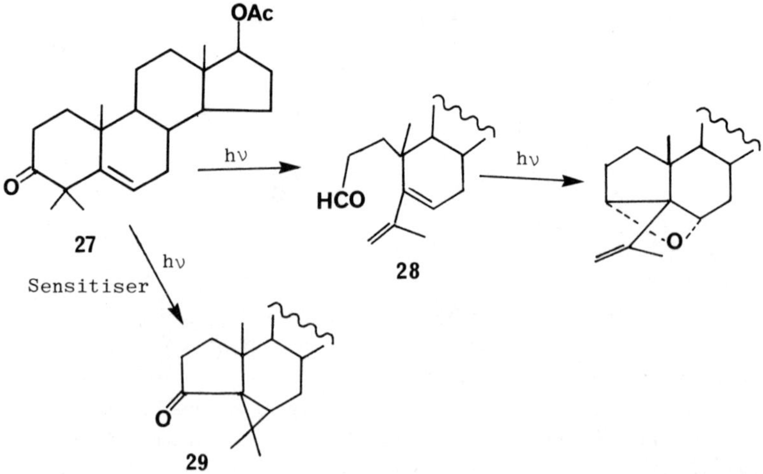

The reaction pathway followed by the steroidal enone (27) is, not surprisingly, dictated by the multiplicity of the excited state.[7] Direct irradiation causes α-cleavage and formation of the aldehyde (28) which subsequently yields an intramolecular oxetan in a second photoprocess, whereas on triplet sensitisation the β,γ-unsaturated ketone undergoes an oxadi-π-methane reaction to give the ring contracted ketone (29). This latter type of process has also been observed with the enone (30) which in t-butanol gives a 50% yield of the bicyclo[3.1.0]hexanone (31).[8]

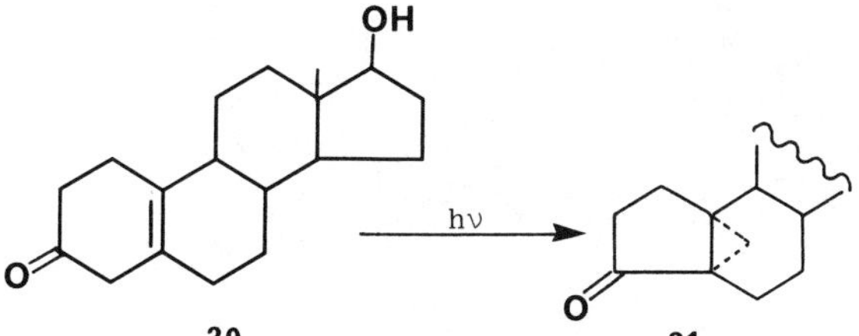

α-Cleavage can also be induced in the B-ring of steroids and although the 9β,10β-epoxide (32) is photostable, the 9α,10α-isomer undergoes photodecarbonylation to yield (33) in 20% and 3% of the other Norrish Type I product (34).[9] In methanol the yield of isolated product from this cleavage is increased considerably and the ester (35) is obtained in 73%.

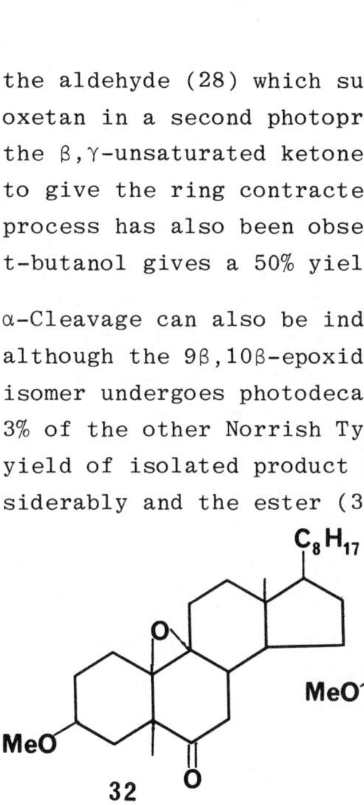

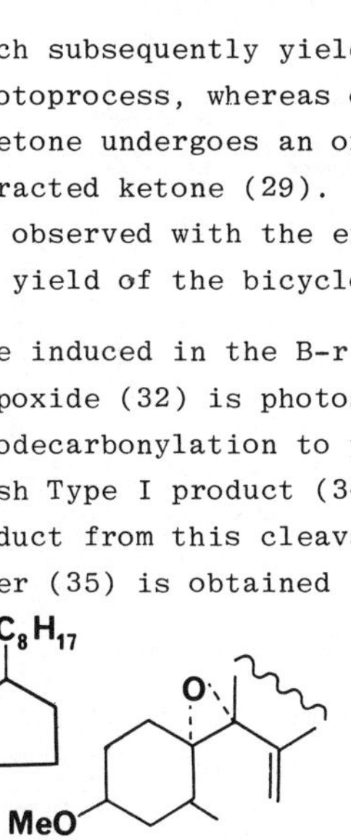

There are several reported examples of this type of reaction which involves the steroid D-ring. Thus the androstene (36) efficiently yields the aldehyde (37),[10] and the process has been used for the epimerisation of the C,D-ring angular methyl group. Examples of the latter type of reaction include the conversions of (38)[11] and (39)[12] to the α-isomers (40) and (41) respectively.

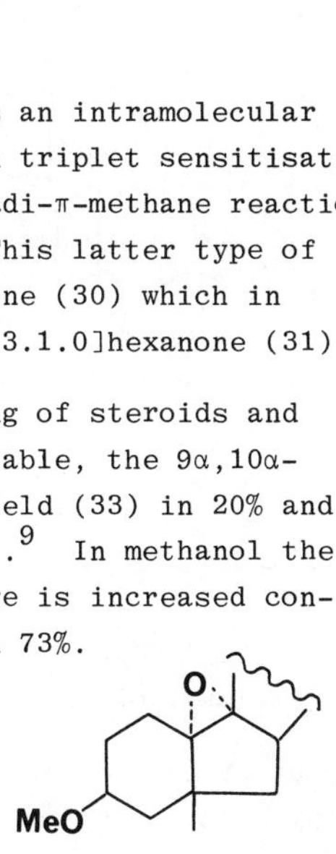

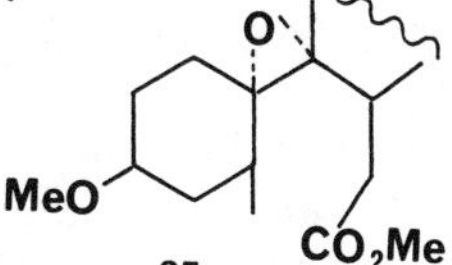

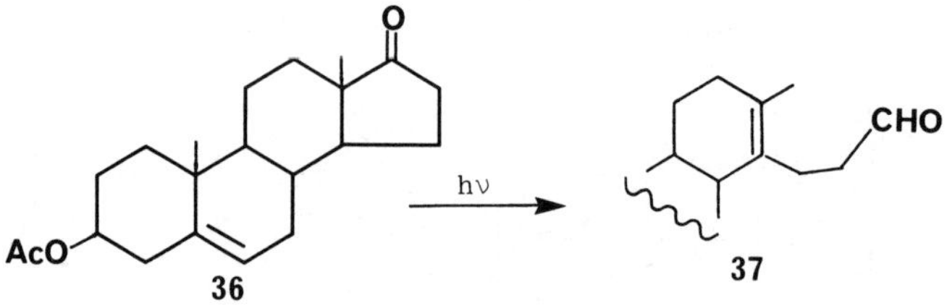

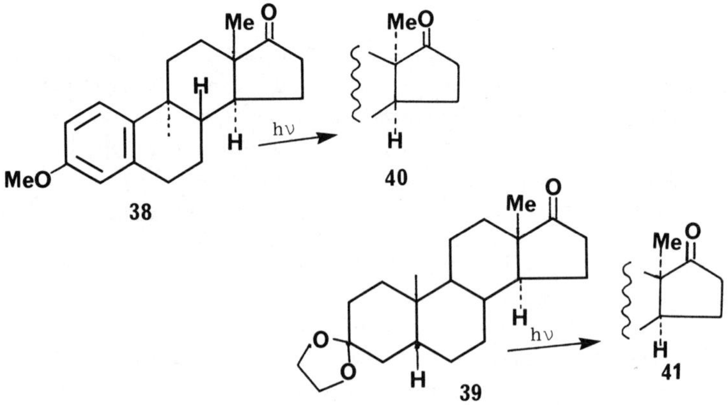

2.2 Norrish Type II Processes

The photoprocess of intramolecular H-abstraction by a carbonyl functionality and subsequent reaction to yield cleavage products or cyclobutanol derivatives has been studied and exploited with many steroidal systems.

The cleavage reaction has been used for side chain degradation in steroids and to introduce unsaturation into the C and D rings. Thus the side chain in the lanosterol (42) is converted into a propene unit,[13] and the androstanol (43) which has the benzophenone moiety essential for the H-abstraction process attached to an oxygen atom at C-17, yields a mixture which on treatment with $Pb(OAc)_4$ followed by hydrolysis affords the ethylenes (44) and (45).[14] The product ratio is dependent on the photolysis reaction solvent and varies from 2:1 respectively in benzene to 4:1 in trifluorotrichloroethane. Type II fragmentation of thioesters of homoallylic alcohols has been elegantly employed as a mild procedure for the dehydration of the alcohol to yield a 1,3-diene. This procedure adapted to the steroid series allows the smooth conversion of the cholesteryl derivative (46) into (47) in over 90% yield.[15]

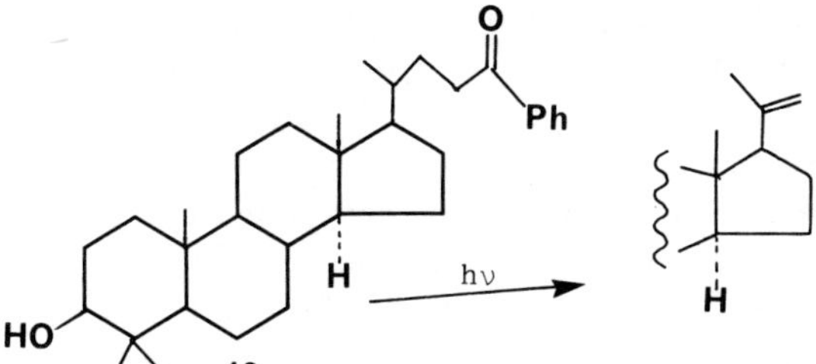

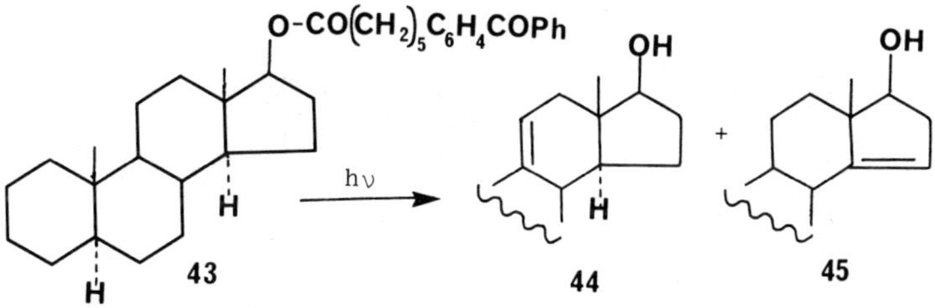

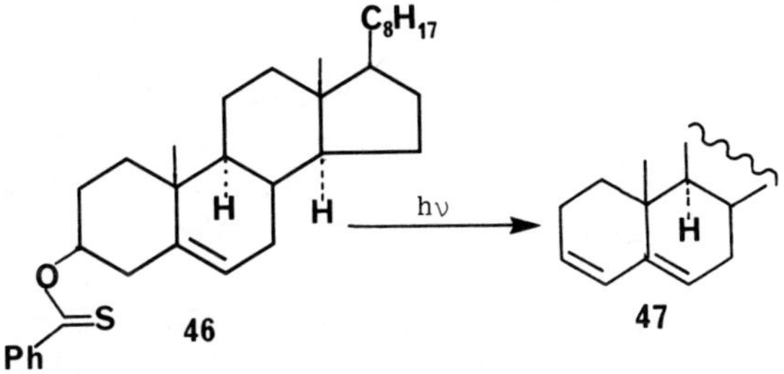

Irradiation of α,β-unsaturated carbonyl compounds may, dependent on the molecular structure, yield the β,γ-isomer by initial intramolecular γ-hydrogen abstraction. With some steroid systems the geometry appears to preclude such abstraction and photoenol formation subsequent to the deconjugation process is most unlikely.

Nevertheless there are examples of reactions which do involve the photo-migration of an ethylenic bond from the A to the B ring in the steroid and vice versa without skeletal rearrangement. Thus 0-acetyl-10α-testosterone (48) is isomerised to 3-oxo-17β-acetoxy-Δ^5-10α-androstene (49).[16] In such cases, however, it does appear by analogy with other enone rearrangements that this reaction in the steroid series may be a consequence of inter-molecular H-abstraction. For the D ring enone (50) irradiation not only results in deconjugation but also in ring fission and the diketone (51) results.[17]

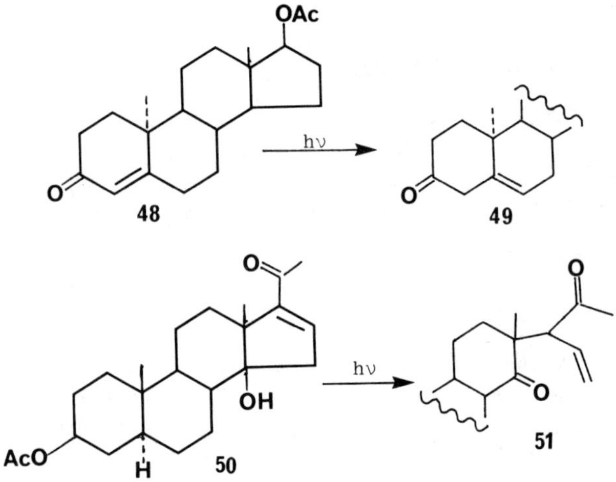

OAc
O
48
hν
O
49
O
OH
AcO
H
50
hν
O
O
51

A common repercussion of γ-hydrogen abstraction by photoexcited ketones is the formation of cyclobutanols by intramolecular combination of the 1,4-diradical. This reaction is common among steroid systems and has been observed for a variety of molecular structures. Processes involving H-abstraction by carbonyl compounds from angular methyl groups are not uncommon and for example irradiation of (52) gives, as well as the de-acetylated product (53), 35 and 22% yields of (54) and (55) respectively, and similarly (56) gives (57) but in this case the yield is 97%.[19] Minor structural differences, however, may result in H-abstraction from other positions even though the molecular

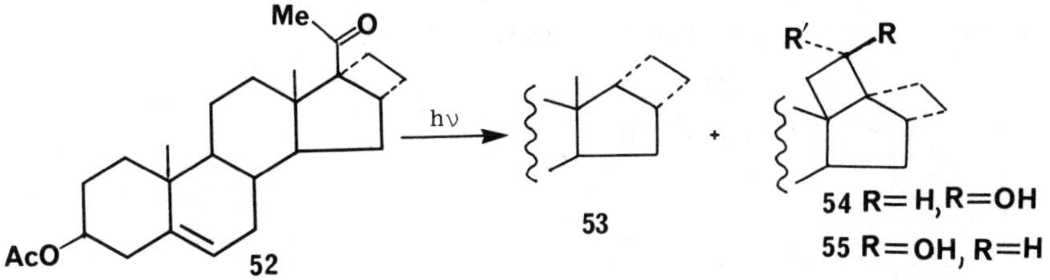

52 53 54 R=H,R=OH 55 R=OH, R=H

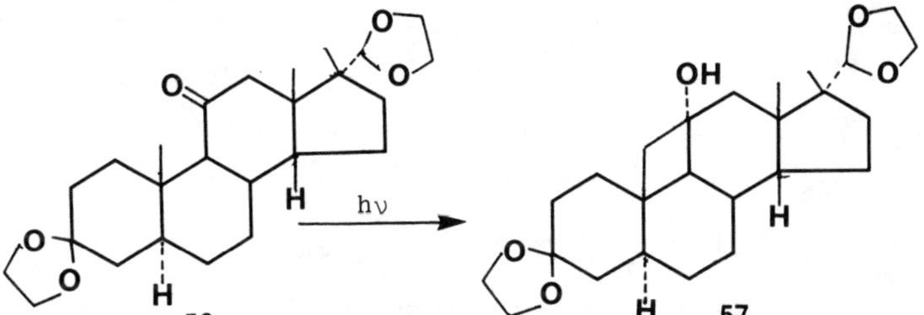

56 57

geometry may appear to favour the involvement of an angular methyl group. Thus irradiation of the α-diketone (58) leads to γ-hydrogen abstraction from the ring methylene by the carbonyl α to the methyl group to give the cyclobutanol (59), rather than the formation of (60) by the alternative reaction involving the other carbonyl moiety and the angular methyl group.[20] Such preference for H-abstraction from a ring methylene is reported for (61) which has a closely similar structure to (56). In this case, despite the availability of γ-hydrogens on the angular methyl group, the H-abstraction occurs from the β-position and the cyclopropanol (62) is subsequently obtained in

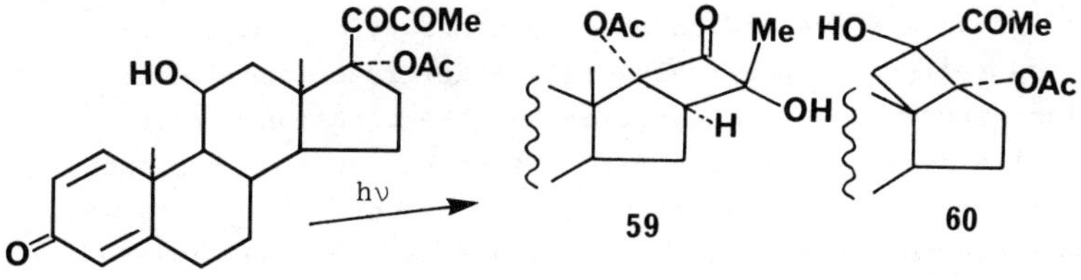

58 59 60

yields over 70% based on recovered (61).[21] The Norrish Type II cyclisation can also be used to form azetidinols and the conversion of (63) to (64) (38% yield) is cited here to illustrate the process with a steroidal system.

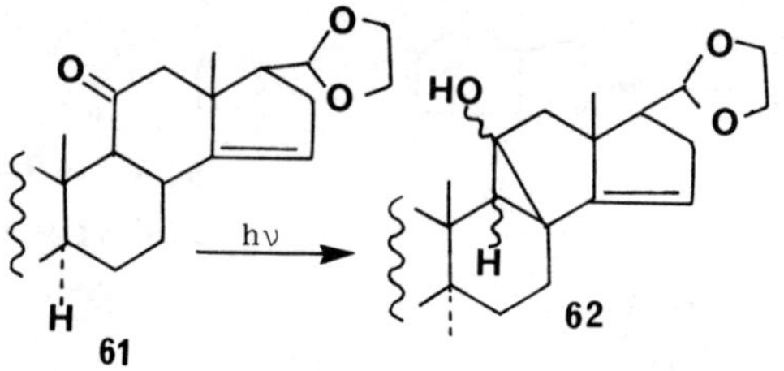

2.3 Addition Reactions

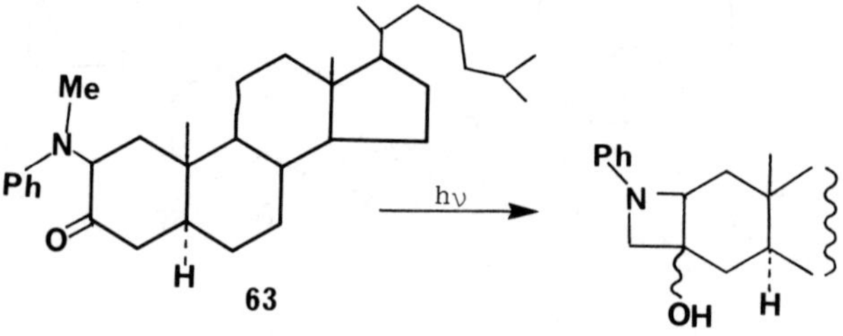

There are many examples in the steroid series of cycloaddition reactions of ethylenes to enone systems and of oxetan formation: those cited below are to illustrate the breadth of interest in these processes. Enones and dienones undergo ($2\pi+2\pi$) regiospecific reaction with various ethylenes and for example the two trans adducts (65) and (66) and the cis isomer (67) are formed from the addition of methyl acrylate to (68).[23] For enediones such as (69) the major product (70) reflects a ($4\pi+2\pi$) cycloaddition.[24] In the presence of 1,3-dienes, other reaction opportunities are available to dienones and the photoaddition products (71) and (72), (73), and (74) which have been observed from steroidal systems reflect respectively Diels-Alder reactions, two ($2\pi+2\pi$) cycloadditions and cross addition.[25] The reaction mixtures are generally dominated by the first of these types of cycloaddition product (e.g. > 60%) but with the cyclopropyl enone (75) and 2,3-dimethylbutadiene, the cross addition compound (76) is the sole adduct isomer and is obtained in 69% yield.

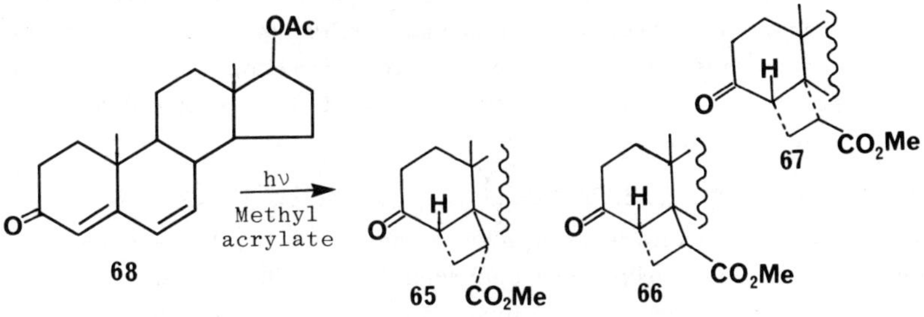

OAc

68 —hν, Methyl acrylate→ 65, 66, 67

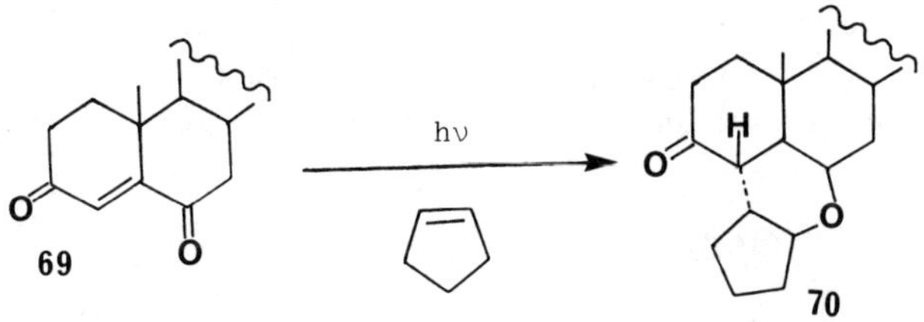

H, CO_2Me (65); H, CO_2Me (66); H, CO_2Me (67)

69 —hν, cyclopentene→ 70

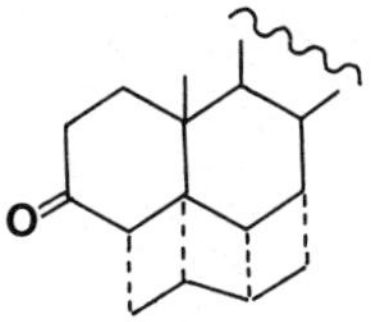

73 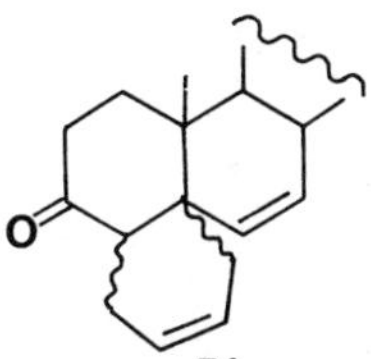 71 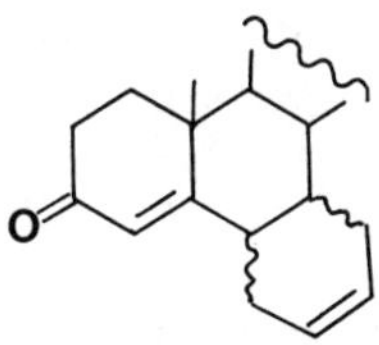 72 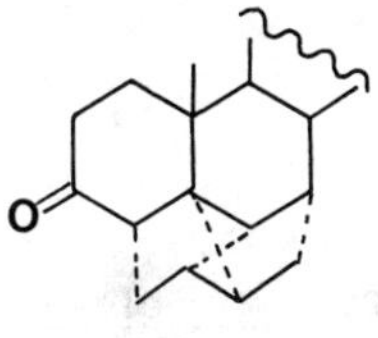 74

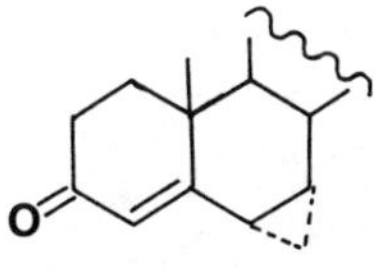

75 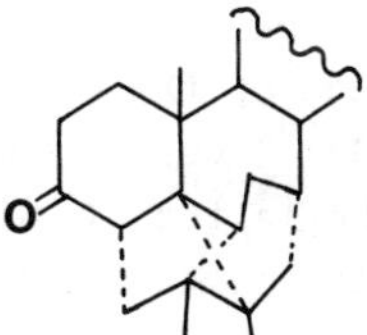 76 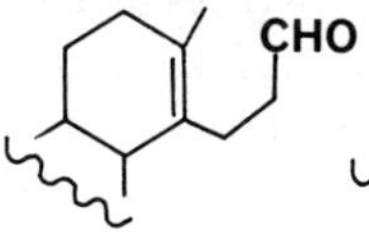 78 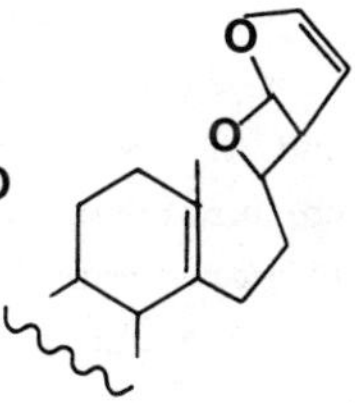 77

CHO (78)

Oxetan formation is a very facile and generally high yield photoprocess. With steroidal compounds, this reaction is exemplified by the formation of (77) in 20% from irradiation of the photolysis product (78) (see Section 2.1) of 3β-acetoxyandrost-5-en-17-one and furan.[26]

2.4 Di-π-Methane Rearrangements

Both oxadi-π-methane rearrangements and those involving cross-conjugated dienones have been observed in steroid systems. An example of the former type of process is given in Section 2.1 with the conversion of (27) into (29) by triplet sensitisation. Rearrangements involving the latter class of compound are illustrated here by the reaction of prednisone acetate (79).[27] In neutral media (79) gives a 65% yield of (80) which on further irradiation in protic solvents produces the novel compound (81) in a 47% yield.

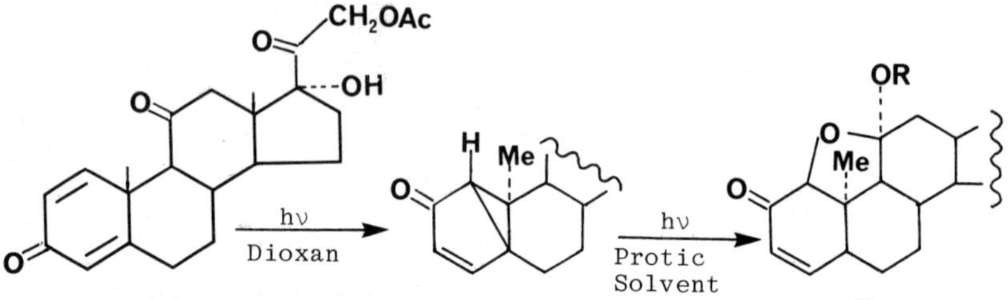

2.5 Miscellaneous Reactions

There are several reported reactions of steroidal carbonyl compounds which cannot readily be classified under the above headings. The two examples of such reactions given here of the formation of the tetrahydrofuran derivative (82) from (83),[28] and of the photo-induced conversion of the enone (84) in concentrated sulphuric acid to (85)[29] were chosen since the former occurs in 65% yield and the latter offers a route to reduced azulene analogues of biologically active steroids.

MeO OMe

hν

83 82

hν
H_2SO_4

84 85

References

1. H.Nozaki, M.Kurita, T.Mori, and R.Noyori, Tetrahedron, 1968, 24, 1821.

2. M.L.Mihailovic, L.Lorenc, N.Popov, and J.Kalvoda, Helv.Chim.Acta, 1971, 54, 2281.

3. G.Quinkert and H.Stark, Angew.Chem., Internat.Ed.Engl., 1983, 22, 637.

4. C.Ouannes and R.Beugelmans, Bull.Soc.Chim.France., 1972, 4275.

5. A.Canovas and J-J.Bonet, Helv.Chim.Acta, 1980, 63, 2390.

6. A.Canovas, J.Fonrodona, J-J.Bonet, M.C.Brianso, and J.L.Brianso, Helv.Chim.Acta, 1980, 63, 2380.

7. K.Kojima, K.Sakai, and K.Tanabe, Tetrahedron Letters, 1969, 3399.

8. J.R.Williams and H.Ziffer, Tetrahedron, 1968, 24, 6725.

9. R.J.Chambers and B.A.Marples, J.C.S.Chem.Comm., 1972, 1122.

10. J.Brussee and H.J.C.Jacobs, Org.Photochem.Synth., 1976, 2, 67.

11. R.Jacquesy and H.L.Ung, Tetrahedron, 1976, 32, 1375.

12. D.L.Smith and T.A.Spencer, J.Heterocycl.Chem., 1979, 16, 807.

13. J-M.Bernassau and M.Fetizon, Synthesis, 1975, 795.

14. R.Breslow and P.Kalicky, J.Amer.Chem.Soc., 1971, 93, 3540.

15. S.Achmatowicz, D.H.R.Barton, P.D.Magnus, G.A.Poulton, and P.J.West, J.C.S.Perkin I, 1973, 1567.

16. S.Kuwata and K.Schaffner, Helv.Chim.Acta, 1969, 52, 173.

17. F.Marti, H.Wehrli and O.Jeger, Helv.Chim.Acta, 1973, 56,1078.

18. P.Sunder-Plassman, P.H.Nelson, P.H.Boyle, A.Cruz, J.Iriarte, P.Crabbe, J.A.Zderic, J.A.Edwards, and J.H.Fried, J.Org.Chem., 1969, 34, 3779.

19. P.Gull, Y.Saito, H.Wehrli and O.Jeger, Helv.Chim.Acta, 1974, 57, 863.

20. M.Obayashi, E.Mizuta, and S.Noguchi, Chem.Pharm.Bull., 1979, 27, 1679.

21. P.Gull, H.Wehrli, and O.Jeger, Helv.Chim.Acta, 1971, 54,2158.

22. A.A.El-Hammany, J.Hill, and J.Townend, J.Chem.Soc.Perkin I, 1983, 573.

23. G.R.Lenz and L.Swenton, J.Chem.Soc.Chem.Comm., 1979, 444.

24. G.R.Lenz, J.Org.Chem., 1979, 44, 1597.

25. See for example G.R.Lenz, J.Org.Chem., 1979, 44, 1382.

26. S.Jarosz and A.Zamojski, Tetrahedron, 1982, 38, 1453.

27. J.R.Williams, R.H.Moore, R.Li, and J.F.Blout, J.Amer.Chem.Soc., 1979, 101, 5019.

28. H.Karvas, F.Mark, H.Wehrli, K.Schaffner, and O.Jeger, Helv.Chim.Acta, 1974, 57, 1851.

29. P.Lupon, J.C.Ferrer, J.Piniella, and J-J.Bonet, J.Chem.Soc.Chem.Comm., 1984, 25, 1255.

6
Carbonyl Compounds: Cycloaddition

By H.A.J. Carless
DEPARTMENT OF CHEMISTRY, BIRKBECK COLLEGE, LONDON WCIE 7HX, UK

Introduction and Scope

The photoaddition of carbonyl compounds to alkenes, to generate the four-membered oxetane ring, was first discovered by Paterno and Chieffi in 1909 (1).

PhCHO + Me$_2$C=CHMe $\xrightarrow{h\nu}$ oxetane (1)

64%, Φ = 0·45

The reaction was not reinvestigated until the work of Büchi and co-workers in 1954,[1] but the literature of the last thirty years has seen over a thousand examples of these [2+2] photocycloadditions, nowadays often referred to as Paterno-Büchi reactions. The versatility of the cycloaddition in synthesis lies not only in its generality, but also in the variety of methods developed for the transformation of the strained oxetane ring to produce other functional groups. There is still considerable scope for research here, in the same way that enone/alkene [2+2] photocycloadditions to give cyclobutanes, and their subsequent transformations, have been exploited in natural product syntheses.[2]

Three pertinent reviews of this topic have appeared,[3-5] the most recent covering the literature to 1981. Details, especially of mechanistic work, can be found in these. Since then, there have been some notable applications of the reaction to the field of organic synthesis, and it is with these developments that the present survey culminates.

For experimental work, it is relevant that the carbonyl component is almost always both the light-absorbing and the reactive species in these cycloadditions. Competitive light absorption by the oxetane products is minimal, and the reactions can therefore be taken to completion without problems of secondary photolysis. In parallel with photochemical hydrogen abstractions, it is only those carbonyl compounds which have lowest n,π^* excited states that tend to undergo efficient [2+2] photocycloaddition.

Aliphatic carbonyl compounds often react both by singlet and triplet excited states, whereas aromatic carbonyl compounds undergo intersystem crossing to the triplet state so rapidly that reactivity for them is observed exclusively from the triplet species. Several examples have shown how the stereochemical outcome of the reaction is strongly dependent on the excited state involved, with the singlet species showing much greater stereoselectivity.

The regiochemistry of the photocycloaddition to simple alkenes has been extensively examined, and the earliest generalisation (due to Büchi) is still a useful and relevant one. The dominant oxetane isomer in the addition to an unsymmetrical alkene is that derived from attack by the n,π* carbonyl oxygen atom to generate the <u>more stable biradical</u> intermediate [<u>e.g.</u> (2)].

(2)

93%

Only recently has direct spectroscopic evidence for 1,4-biradicals such as (3) been found.[6] Before that time, evidence for biradical involvement came from

(i) formation of stereoisomeric oxetanes from photocycloaddition to single geometric isomers of alkenes,

(ii) interception of biradicals by SO_2, or O_2 (4), and

(iii) formation of alternative products, as in (5), by competition with ring closure of the 1,4-biradical.

(3)

(4)

(5)

1,5-H transfers

In some examples, particularly those arising from carbonyl singlet excited states, and for attack on electron-deficient alkenes, there is evidence for extremely rapid (certainly non-biradical) reaction, proceeding without any loss of stereochemistry (6).

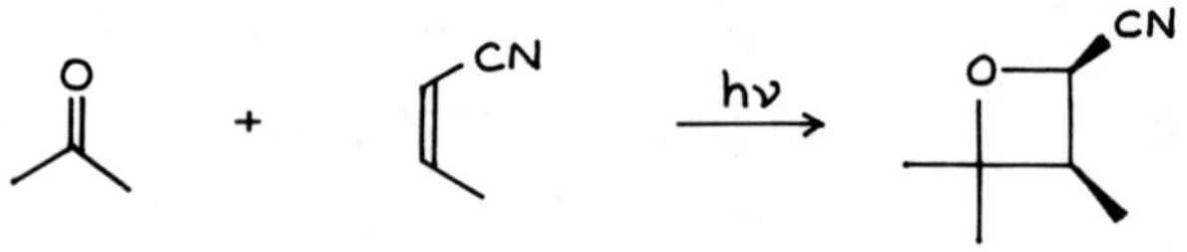

(6)

The Paterno-Büchi addition is applicable to a wide range of substituted alkenes: the reaction rate (and efficiency) increases as the ionisation energy of the alkene decreases, so that only ethene and occasionally 1-alkenes make unsuccessful simple addends. Conjugated dienes can be usefully added, (7), although in these cases the reaction yield may be reduced by competitive triplet-sensitised dimerisation of the diene.[4]

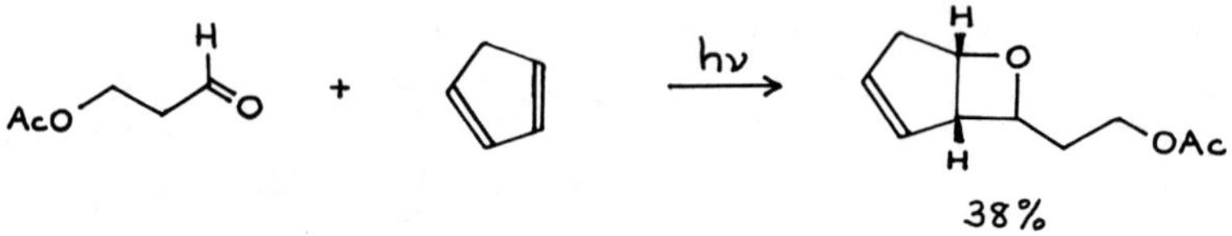

(7)

38%

Photoexcited carbonyl compounds will add to allenes in a highly regioselective reaction (8), and the 1:1 adducts may react further by addition to the remaining double bond.[7]

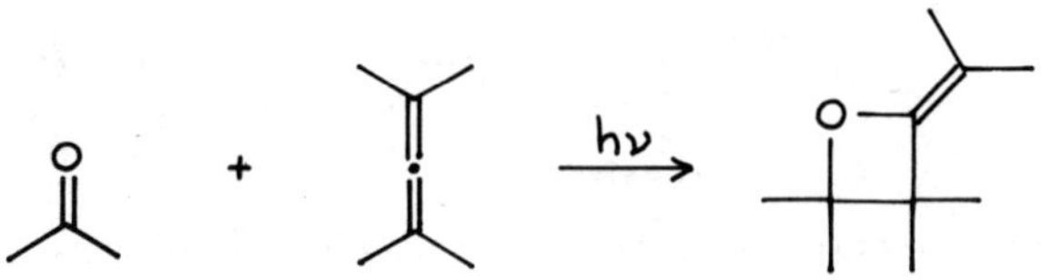

(8)

Ketenimines will similarly react at the alkenyl bond, with the regioselectivity of addition depending upon the substitution pattern of the reactants. There is useful reactivity in the strained oxetanes so produced; thus, the α-imino-oxetane (9) formed from fluorenone and the ketenimine $Me_2C{=}C{=}NPh$ will rearrange in high yield to the β-lactam on treatment with lithium bromide, presumably via ring opening and alternative reclosure.[8]

NPh LiBr MeCN 70°C Ph N O (9) 91%

The keto group of α-dicarbonyl compounds and of α-ketoesters (see later examples) reacts well in these [2+2] photocycloadditions. Ester groups in themselves are generally inert [e.g. the acetate in (7)], although the carbonyl group of benzoates can make a respectable reactant. The visible-wavelength absorption of α-dicarbonyls makes selective excitations possible in which the monoketone products are photostable. Thus, biacetyl reacts regio- and stereo-selectively with the unsaturated ether (10), to give a cis functionalised 2-alkoxyoxetane ring that is a central structural feature of the important thromboxane A_2 molecule.[9]

hν (10) 21%

Irradiation of quinones in the presence of alkenes can lead to oxetanes in excellent yields, as shown for the example (11) involving p-benzoquinone.[10] Furthermore, the product can be subjected to an acid-catalysed dienone-phenol rearrangement to yield the fused phenol (12) in quantitative yield.

(11) 88% (12)

Oxetane synthesis from enones is fortunately rarer. Whilst acyclic enones generally dissipate excitation energy by *cis-trans* isomerisation, cyclic enones react reliably with alkenes to lead to cyclobutane formation. Only in circumstances where the alkene is electron-rich, and the enone flanked by electron-withdrawing groups, does oxetane formation become competitive.

An interesting example of synthesis involving the formation and subsequent opening of an oxetane ring is provided by the irradiation of m-acetoxybenzaldehyde and the optically active cyclopentene (13), as an ingenious approach to the aromatic prostaglandin analogue (14).[11] Reductive (benzylic) cleavage of the oxetane ring, using palladium and hydrogen, and further synthetic transformations, afford the optically active prostanoid (14).

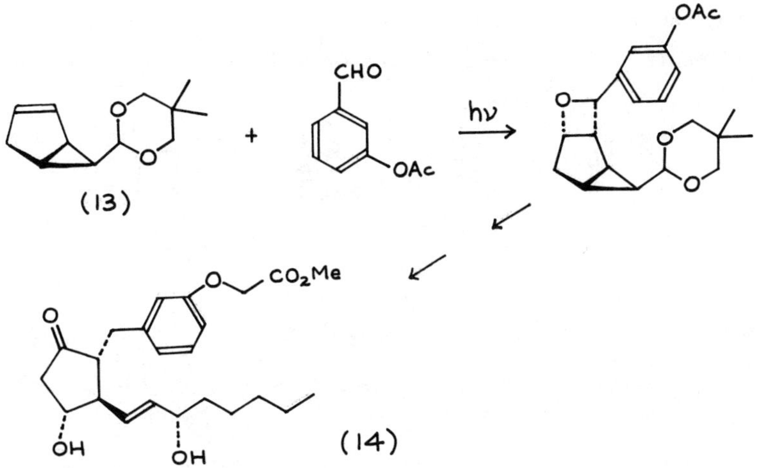

(13)

(14)

Metathesis Reactions

It is feasible to form an oxetane by [2+2] photocycloaddition, and then to cleave the ring in the alternative mode, resulting in a metathesis reaction. As an illustration of this procedure, irradiation of benzophenone in the presence of norbornene gives the tricyclic oxetane (15) in 80% yield, thermolysis of which cleanly provides the unsaturated aldehyde (16).[12] The oxetane ring can also be cleaved by acid catalysis or, under milder conditions, by treatment with rhodium(I) or silver(I) complexes.

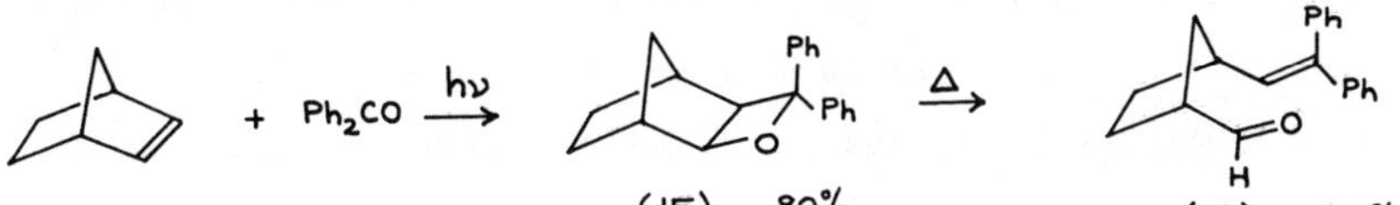

$+ Ph_2CO \xrightarrow{h\nu}$ (15) 80% $\xrightarrow{\Delta}$ (16) 100%

Jones[13] has applied such metathesis to the synthesis of E-non-6-enol, a sex attractant of the Mediterranean fruit fly. Photocycloaddition of propanal to cyclohexa-1,3-diene gives the exo-isomer of the bicyclic oxetane (17) in good yield.

Hydrogenation and stereoselective cleavage of the oxetane ring by a rhodium(I) catalyst produces <u>E</u>-non-6-enal, which can be reduced to the desired unsaturated alcohol.

hν, MeCN; (17) 62%

Rh(I), C_6H_6, Δ; $LiAlH_4$

Occasionally, and often for reasons of developing conjugation, the oxetanes may undergo fission so easily that metathesis products are directly isolated from a photochemical reaction, as in the case of irradiation of the phthalimide derivative (18).[14]

(18) hν → → 64% + CH_2O

Intramolecular Reactions

One of the most powerful uses of the Paterno-Büchi reaction for synthesis involves intramolecular attack: these additions are exceptionally rapid, because the reactants are juxtaposed, and they can give a high degree of stereocontrol in the bi- or polycyclic oxetane that is produced. γδ-Unsaturated carbonyls, for example, can lead to bicyclic products arising from the two possible orientations of attack: in general, there is a preference for cyclopentane over cyclobutane ring formation (19).[15]

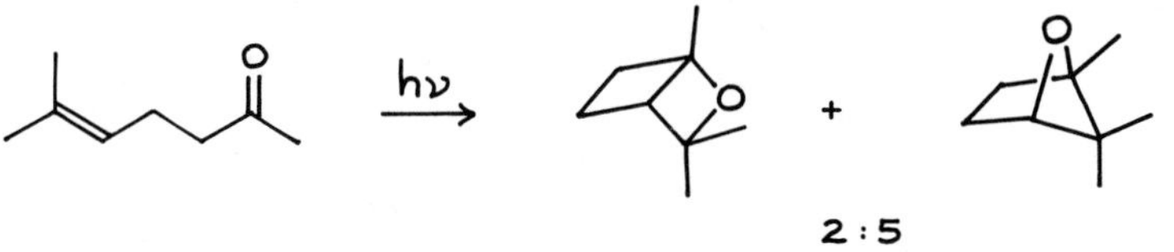

hν (19)

2 : 5

The unsaturated carbonyls may themselves be derived from photolysis of bicyclic or tricyclic ketones, via the Norrish type I reaction. Prolonged irradiation of the bicyclohexanone (20) leads to the cyclobutenyl aldehyde, and thence to an oxetane in 60% overall yield.[16]

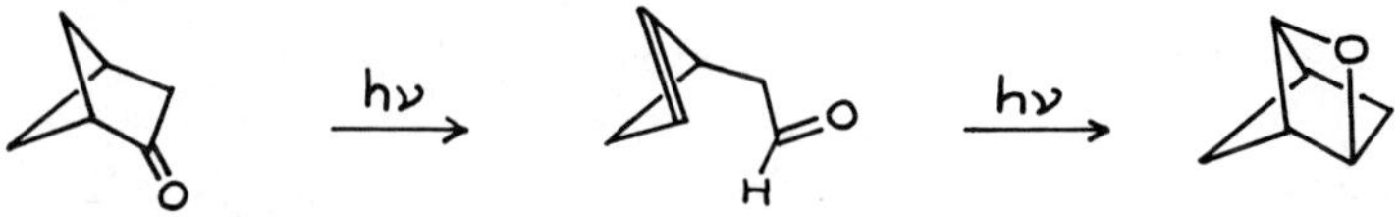

hν hν

(20)

The chemical reactivity of the oxetane ring makes a variety of targets attainable. Thermolysis above 100°C of an oxetane produced from 2-allylcycloheptanone converts it (in ca 90% yield) to a mixture of unsaturated alcohols and the corresponding dienes having the azulene skeleton, with the possibility of dehydrogenation over palladium to yield azulene (21) itself (25% yield).[17] Reduction of oxetanes by lithium aluminium hydride or lithium/liquid ammonia is a convenient route to alcohols, as shown by the photocycloaddition of the cyclohexenyl aldehyde (22) to yield the tricyclic oxetane, and its subsequent reduction to a specific isomer of bicyclo[3.2.1]octan-8-ol.[18]

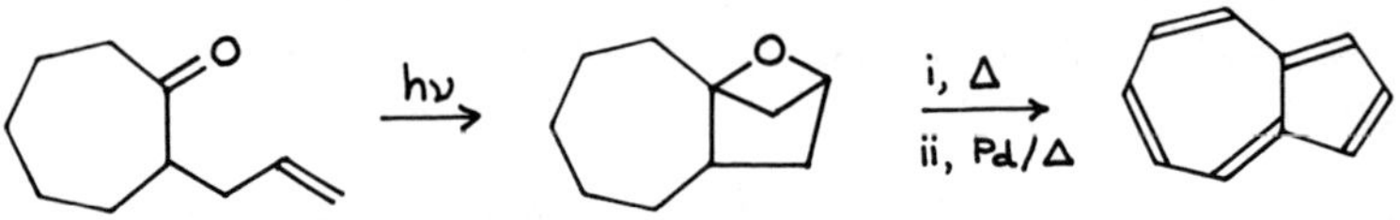

hν
i, Δ
ii, Pd/Δ
(21)

(22)
hν
LiAlH$_4$

Similar cycloaddition-cleavage reactions are vital steps in the synthesis of optically active tricyclo-octanone (24), starting from the chiral endo-norbornenyl ketone (23).[19]

hν
56%
(23)
(24)

These intramolecular Paterno-Büchi reactions are not restricted to carbonyl compounds having unsaturation at the $\gamma\delta$-position. Thus, irradiation of the $\epsilon\zeta$-unsaturated ketone (25) gives a bicyclic intramolecular adduct (26) as one of two oxetane products.[20] This route represents a photochemical approach to ring analogues having the strained dioxabicyclo[n.1.1]alkane system found in the thromboxane A_2 molecule. In a fluorine-substituted example, the orientation of intramolecular addition is such as to give oxetanes having the [3.2.0] ring system (27), rather than the desired [3.1.1] naturally occurring structure.[21]

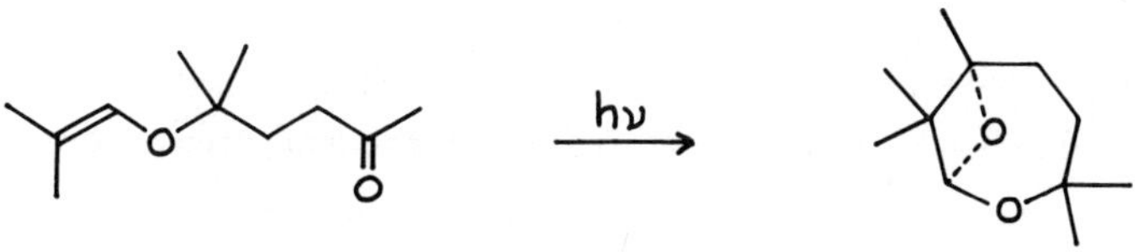

(25) $\xrightarrow{h\nu}$ (26)

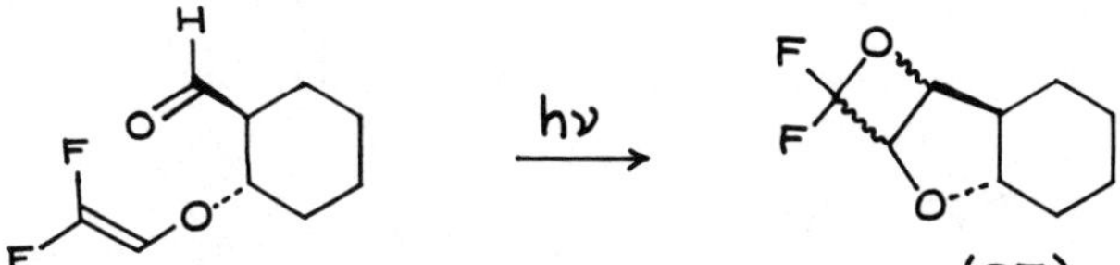

$\xrightarrow{h\nu}$ (27)

1,4-Biradical Interception

There are occasions when the involvement of a 1,4-biradical intermediate in the Paterno-Büchi reaction can allow a fate other than ring closure [*e.g.* (5)]. These alternative reactions, through serendipity or intention, may lead to useful syntheses from carbonyl/alkene photochemistry. Selective (singlet) excitation of the dihydropyranyl ketone (28) is a case in point: the irradiation product is the bicyclic acetal (29), which can be hydrogenated to give a very short route to the beetle attractant *exo*-brevicomin (31).[22] The reaction presumably involves a 1,4-biradical (30), which undergoes a 1,5-hydrogen shift rather than closure to the very strained tricyclic oxetane.

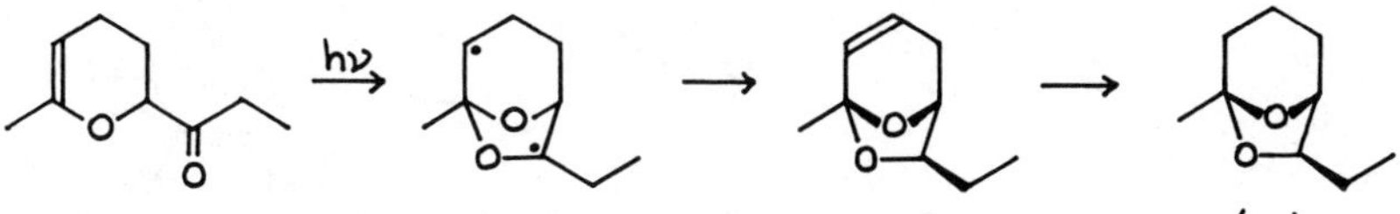

(28) $\xrightarrow{h\nu}$ (29) $\rightarrow$ (30) $\rightarrow$ (31)

Irradiation of substituted ynones in the presence of alkenes can lead to an interesting [3+2] photocycloaddition in which dihydrofurans such as (33), rather than oxetanes, can be the main products.[23] There is evidence to show that they are formed as a result of closure of one radical centre of the biradical on to the adjacent triple bond, to give a carbene (32), followed by hydrogen shifts. It may be possible to control the product distribution, because the oxetane is formed by a singlet pathway, but the dihydrofuran arises from an alkynone triplet.

hν

(32)

(33)

8%

77%

Related Photocycloadditions

A reaction closely related to oxetane ring formation, and first observed by Büchi, involves photocycloaddition of carbonyl compounds to alkynes. The strained oxete ring product, which can be stable at low temperatures,[24] undergoes [2+2] cycloreversion at normal temperatures to generate $\alpha\beta$-unsaturated carbonyl compounds; for example, benzaldehyde and but-2-yne generate an oxete (34), which ring-opens above $-35^{\circ}C$ to the E-enone (35).

hν

Ph H

Ph

(34)

Ph

43% (35)

Thiocarbonyl compounds show a well-defined photochemistry analogous to that of carbonyl compounds;[25] a range of thietanes, or even stable thietes [*e.g.* (36)],[26] are available by photocycloaddition to alkenes and alkynes.

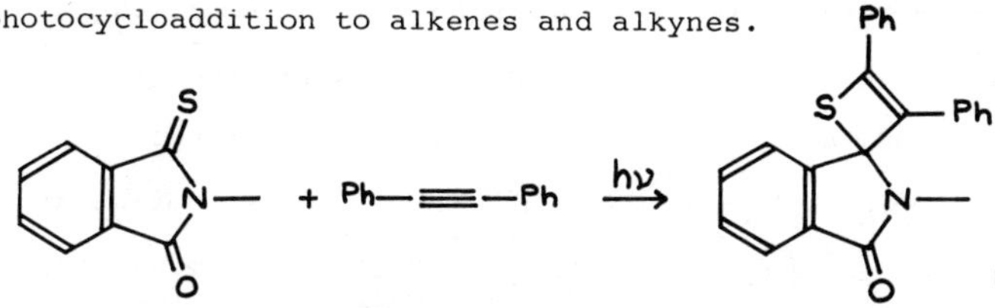

Ph
S
Ph
S
N—
+ Ph—≡—Ph
hν
N—
52%
O
O
(36)

Intramolecular attack can also occur in the photochemistry of unsaturated thiocarbonyls, as in the remarkable reaction of the thioimide (37) to afford a fused β-lactam.[27] The additional ability of thiocarbonyl compounds to form long-lived second singlet (S_2, $\pi\pi^*$) excited states upon u.v. irradiation makes wavelength-dependent photochemical reactions a common possibility for them.

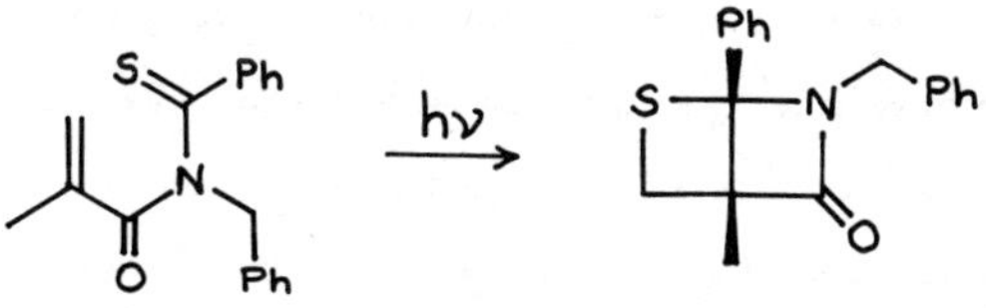

S Ph
N
O Ph
(37)
hν
Ph
S N Ph
O
95%

Vinyl Ether Additions

Simple aliphatic and aromatic carbonyl compounds give efficient photocycloaddition reactions to vinyl ethers. It is the unusually rapid formation and reaction from an excited state complex,[4,5] with vinyl ether as electron donor and excited carbonyl compound as electron acceptor, which would appear to be responsible for these efficient photochemical reactions. This high reactivity makes it possible to perform selective photoaddition of carbonyl compounds to vinyl ethers, even in the presence of an alkenyl function within the same molecule, as in

(10).[9] Although the yields of these reactions are often excellent, the regioselectivity of addition is not always so notable: thus, benzophenone reacts with ethyl vinyl ether to give 2-alkoxyoxetane (38) and 3-alkoxyoxetane (39) in the ratio 25:75, and in a combined yield of 99%.[28] Methyl substitution at the β-position of the vinyl ether (i.e. $Me_2C{=}CHOR$) can reverse this effect, making the 2-alkoxyoxetane the dominant product (70-85%).[29]

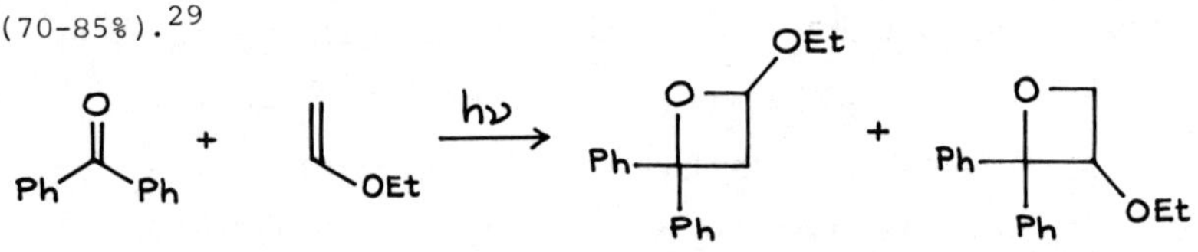

(38) (39)

There is an essential difference in reactivity between these two oxetane isomers: the 2-alkoxyoxetane, as a cyclic acetal, reacts quantitatively with water and alcohols to give β-hydroxyaldehydes and related species, or in high yield with Grignard reagents to give hydroxyethers (40).[28] The 3-alkoxyoxetanes can survive such conditions, and are thus easily separated from these mixtures.

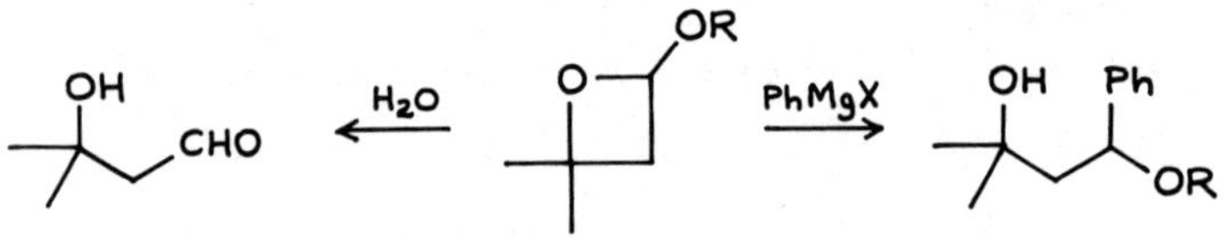

(40)

Intramolecular examples of carbonyl/vinyl ether additions proceed well [see also (26) and (27)], as shown by the synthesis of the simple dioxabicycloheptane (41).[30]

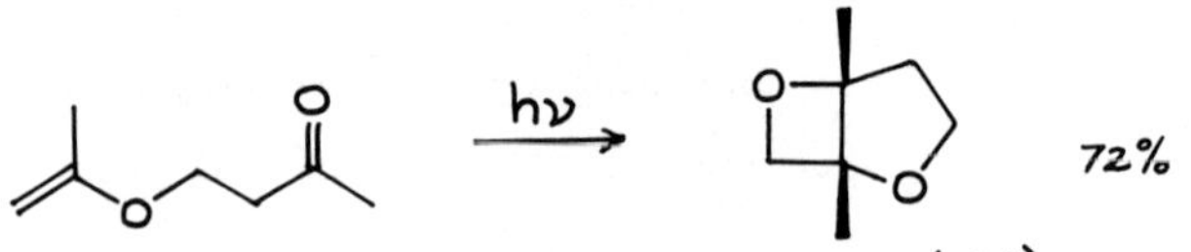

72%

(41)

By conversion of carbonyl compounds to their enol derivatives (either silyl ethers[31] or acetates[32]), it is possible to add the latter as alkenes in [2+2] photocycloadditions (42).[31] Subsequent removal of the protecting group then generates an oxetanol, in a process which is formally the addition of an 'enol equivalent' to the photoexcited carbonyl group.

AcO–CH$_2$–CO–CH$_2$–OAc + CH$_2$=CH–OSiMe$_3$ $\xrightarrow{h\nu}$ oxetane (AcOCH$_2$)$_2$, OSiMe$_3$ 57% (42)

Vinylene carbonate makes a robust addend in photocycloadditions to carbonyl compounds, as illustrated by an approach to the branched-chain sugar, DL-apiose (45).[33] Irradiation of the diacetoxyketone (43) in the presence of vinylene carbonate gives an oxetane which can be treated with base, to induce deacetylation and ring-opening, via the hemiacetal (44).

(43) + vinylene carbonate $\xrightarrow{h\nu}$ oxetane 81% $\xrightarrow{OH^-/H_2O}$ (44) $\longrightarrow$ (45) CHO, H–C–OH, HOCH$_2$–C–OH, CH$_2$OH

In a related reaction, photocycloaddition of acetone to dioxole (46), followed by acid treatment, gives the glyceraldehyde derivative (47) in 65% overall yield.[34]

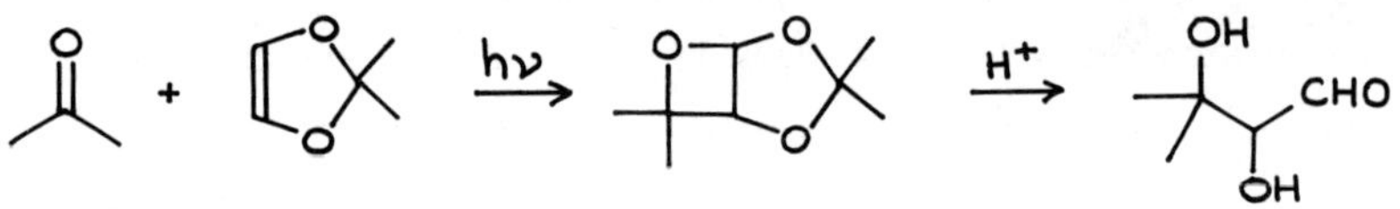

(46) (47)

One promising application of the Paterno-Büchi reaction has developed from earlier work by Gotthardt[35] in which chiral menthyl (48a) esters of phenylglyoxylic acid were reacted photochemically with alkenes such as 2,3-dimethylbut-2-ene to induce chirality (up to 53% diastereomeric excess) in the resulting oxetane ring. Scharf and co-workers[36] have used crowded (-)-8-phenylmenthyl (48b) phenylglyoxylate esters to give excellent yields (99%, using a falling-film photoreactor; with a 96% diastereomeric excess) of oxetanes by addition to the dioxole (46). There is control of chirality at two centres, because the ester is introduced specifically exo to the isopropylidene group of (49). The chiral alcohol can be recovered, and the oxetane was converted to the L-apiose derivative (50).

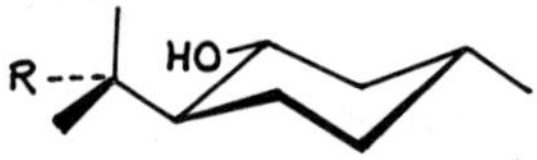

(48) a, R = H
b, R = Ph

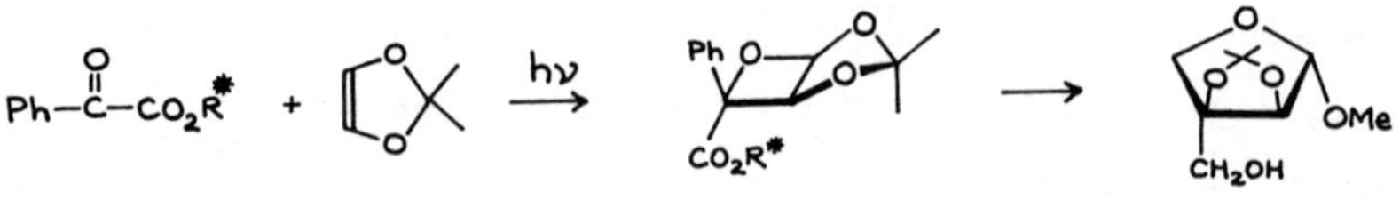

(46) (49) (50)

Heterocyclic Aromatic Additions

The photocycloaddition of benzophenone to an excess of furan occurs in an excellent yield and in an entirely regiospecific manner (51).[37] On prolonged irradiation, benzophenone may react further with the 1:1 adduct to give 2:1 adducts in which the second photoreaction is much less regioselective.

$Ph_2C=O$ + [furan] $\xrightarrow{h\nu}$ [oxetane adduct, Ph, Ph] (51)

94%

Similar additions of aliphatic aldehydes to furans (usually occurring in greater than 70% yield) have proved exceptionally useful in synthesis. Access to 3-substituted furans is not normally easy, but an interesting route to them involves treatment of the above oxetane photoproducts with an acid catalyst in aprotic solvents (52).[38] Oxidation of the resulting alcohol (52), where $R=CH_2CH_2CHMe_2$, gives a short synthesis of the naturally occurring 3-furyl ketone, perillaketone.

$\xrightarrow{H^+}$ $\longrightarrow$

(52)

There must be scope for synthetic applications in the fact that derivatives of many other heterocyclics such as thiophen, pyrrole, imidazole and isoxazole are capable of participating in these cycloadditions. The oxetane adducts arising from the corresponding pyrroles are less stable than those from furans, and they proceed directly to 3-(1-hydroxyalkyl)pyrroles. Jones[39] has further shown the versatility of this reaction by preparing a 1:1 adduct from aldehyde and pyrrole, and then irradiating in the presence of a second aldehyde, to provide 3,4-unsymmetrically disubstituted pyrroles (53).

OH OH
i, hν, MeCHO
ii, hν, PrCHO
N N
(53) 50%

Photocycloaddition of acetaldehyde to furan, followed by hydroxylation with potassium permanganate, has enabled a recent synthesis of 3-deoxy-DL-streptose (54) derivatives.[40] In this case, ring opening of the oxidised bicyclic adduct leads to recyclisation of the furanose ring, with spontaneous epimerisation at C-3 of the sugar.

+ MeCHO hν → KMnO$_4$ → OH OH

OH OH OHC CHO → CHO OH 3 OH

(54)

The [2+2] photoaddition of an aldehyde to a furan can lead to the equivalent of a stereoselective aldol reaction. Thus, irradiation of aliphatic or aromatic aldehydes in the presence of 2,5-dimethylfuran gives oxetanes in good yield (82% for R=Et).[41] The exo stereochemistry of the bicyclic adducts is responsible for the useful control of relative configuration in the product, and the strained cyclic acetal grouping of (55) can be deprotected to give threo-aldol products (56) in 88-92% yield, by treatment with very mild aqueous acid (0.01 N HCl, room temperature, 0.5h.).

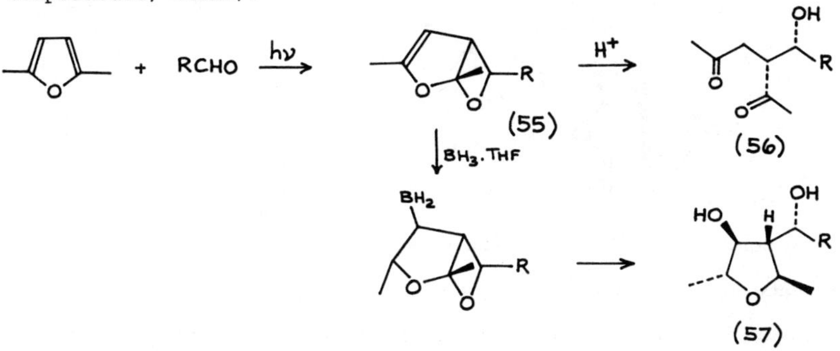

The remaining functionality in the bicyclic adduct (55) can be used to introduce further chiral centres: for example, a hydroboration-oxidation reaction of the double bond in (55) occurs exclusively from the exposed exo face of the molecule. An accompanying hydrogenolysis of the oxetane acetal ring gives the tetrahydrofuranol (57), with excellent control of relative stereochemistry at five chiral centres.

These photocycloadditions of aldehydes to furans, followed by synthetic transformations on the oxetane ring, can be directly applied to natural product targets, as shown by Schreiber's route to avenaciolide (59),[42] an antifungal metabolite. Moreover, these particular photochemical reactions are not restricted in scale, as evidenced by the production of the initial aduct (58) in nearly quantitative yield whilst working on a 50 g scale!

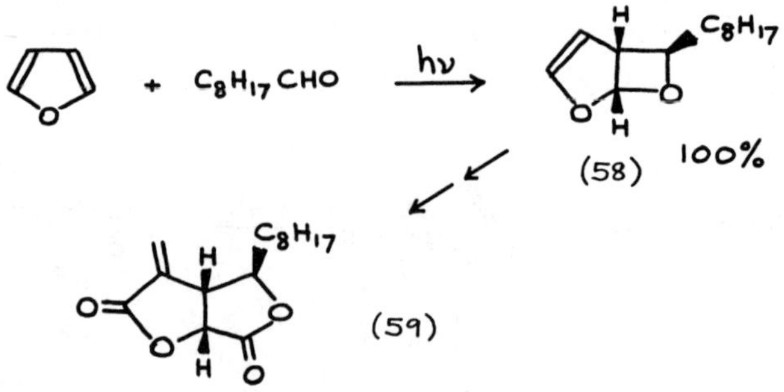

The potential of this approach is shown by the key opening sequence in the synthesis of asteltoxin (61), in which an aldehyde-furan photocycloaddition reaction assembles three contiguous chiral centres of the target molecule.[43,44] Asteltoxin is a mycotoxin isolated from mouldy maize, and it owes its toxicity to its action as a potent inhibitor of oxidative phosphorylation. Peracid treatment of the double bond in adduct (60), and stereocontrolled introduction of the ethyl group, set up the tetrahydrofuranol ring ready for the subsequent manipulations (including attachment of the pyronyltriene side chain!) that complete this elegant synthesis.

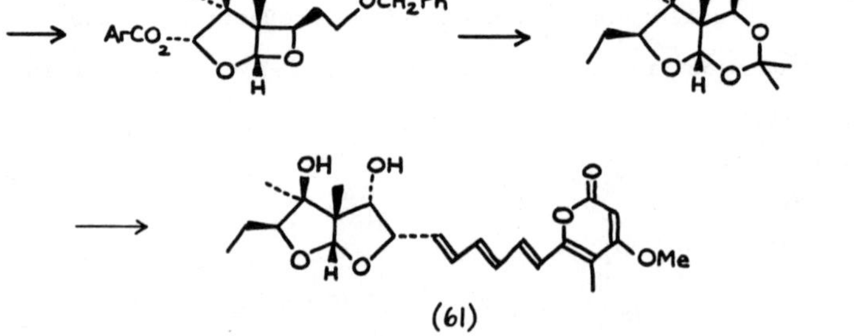

References

1. G. Büchi, C.G. Inman and E.S. Lipinsky. J. Am. Chem. Soc., 1954, 76, 4327.

2. S.W. Baldwin, Org. Photochem., 1981, 5, 123; W. Oppolzer, Acc. Chem. Res., 1982, 15, 135.

3. D.R. Arnold, Adv. Photochem., 1968, 6, 301.

4. G. Jones, II, Org. Photochem., 1981, 5, 1.

5. H.A.J. Carless in "Synthetic Organic Photochemistry", ed. W.M. Horspool, Plenum Press, New York, 1984, p 425.

6. S.C. Freilich and K.S. Peters, J. Am. Chem. Soc., 1985, 107, 3819.

7. H. Gotthardt, R. Steinmetz and G.S. Hammond, J. Org. Chem., 1968, 33, 2774.

8. J.A. Green, II, and L.A. Singer, Tetrahedron Lett., 1974, 4153.

9. H.A.J. Carless and G.K. Fekarurhobo, Tetrahedron Lett., 1985, 26, 4407.

10. D. Bryce-Smith, A. Gilbert and M.G. Johnson, J. Chem. Soc. (C), 1967, 383.

11. D.R. Morton and R.A. Morge, J. Org. Chem., 1978, 43, 2093.

12. G. Jones, II, S.B. Schwartz and M.T. Marton, J. Chem. Soc., Chem. Commun., 1973, 374.

13. G. Jones, II, M.A. Acquadro and M.A. Carmody, J. Chem. Soc., Chem. Commun., 1975, 206.

14. P.H. Mazzocchi, L. Klingler, M. Edwards, P. Wilson and D. Shook, Tetrahedron Lett., 1983, 24, 143.

15. N.C. Yang, M. Nussim and D.R. Coulson, Tetrahedron Lett., 1965, 1525.

16. J. Meinwald and R.A. Chapman, J. Am. Chem. Soc., 1968, 90, 3218.

17. P. Jost, P. Chaquin and J. Kossanyi, Tetrahedron Lett., 1980, 21, 465.

18. J. Meinwald and A.T. Hamner, J. Chem. Soc., Chem. Commun., 1969, 1302.

19. M. Nakazaki, K. Naemura and Y. Kondo, J. Org. Chem., 1976, 41, 1229.

20. H.A.J. Carless and G.K. Fekarurhobo, J. Chem. Soc., Chem. Commun., 1984, 667.

21. J. Fried, S. Kittisopukil and E.A. Hallinan, Tetrahedron Lett., 1984, 25, 4329.

22. P. Chaquin, J-P. Morizur and J. Kossanyi, J. Am. Chem. Soc., 1977, 99, 903.

23. S. Saba, S. Wolff, C. Schröder, P. Margaretha and W.C. Agosta, J. Am. Chem. Soc., 1983, 105, 6902; see also V.B. Rao, S. Wolff and W.C. Agosta, ibid, 1985, 107, 521.

24. L.E. Friedrich and J.D. Bower, J. Am. Chem. Soc., 1973, 95, 6869.

25. J.D. Coyle, Tetrahedron, 1985, 41, 5393.

26. J.D. Coyle, P.A. Rapley, J. Kamphuis and H.J.T. Bos, Tetrahedron Lett., 1985, 26, 2249.

27. M. Sakamoto, Y. Omote and H. Aoyama, J. Org. Chem., 1984, 49, 396.

28. S.H. Schroeter and C.M. Orlando, J. Org. Chem., 1969, 34, 1181, and following paper.

29. S.H. Schroeter, J. Chem. Soc., Chem. Commun., 1969, 12.

30. H.A.J. Carless and D.J. Haywood, J. Chem. Soc., Chem. Commun., 1980, 1067.

31. Y. Araki, J. Nagasawa and Y. Ishido, Carbohydrate Res., 1981, 91, 77.

32. H. Ruotsalainen and T. Kärki, Acta Chem. Scand., 1983, B37, 151.

33. Y. Araki, J. Nagasawa and Y. Ishido, J. Chem. Soc., Perkin Trans. I, 1981, 12.

34. H.-D. Scharf and J. Mattay, Tetrahedron Lett., 1976, 3509.

35. H. Gotthardt and W. Lenz, Angew. Chem. Internat. Ed. Engl., 1979, 18, 868.

36. A. Nehrings, H.-D. Scharf and J. Runsink, Angew. Chem. Internat. Ed. Engl., 1985, 24, 877.

37. G.O. Schenck, W. Hartmann and R. Steinmetz, Chem. Ber., 1963, 96, 498.

38. A. Zamojski and T. Kozluk, J. Org. Chem., 1977, 42, 1089.

39. G. Jones, II, H.M. Gilow and J. Low, J. Org. Chem., 1979, 44, 2949.

40. T. Kozluk and A. Zamojski, Tetahedron, 1983, 39, 805.

41. S.L. Schreiber, A.H. Hoveyda and H.-J. Wu, J. Am. Chem. Soc., 1983, 105, 660.

42. S.L Schreiber and A.H. Hoveyda, J. Am. Chem. Soc., 1984, 106, 7200.

43. S.L. Schreiber and K. Satake, J. Am. Chem. Soc., 1983, 105, 6723; 1984, 106, 4186.

44. S.L. Schreiber, Science, 1985, 227, 857.

7
Enone and Dienone Rearrangements

By H.A.J. Carless
DEPARTMENT OF CHEMISTRY, BIRKBECK COLLEGE, LONDON WCIE 7HX, UK

$\alpha\beta$-Unsaturated Carbonyl Compounds

The strong u.v. absorption spectra of conjugated enones have made them popular candidates for photochemical reactions: there is intense (ϵ >10,000) $\pi\pi^*$ absorption at 210-250 nm, overlapping with weaker (100 > ϵ > 10) $n\pi^*$ absorption at 310-330 nm. A wide variety of photochemical processes is possible, reflecting the presence of both alkenyl and carbonyl groupings. The reaction of cis-trans isomerisation tends to dominate for acyclic enones and enals, whereas for the more restricted cyclic enones (especially cyclopentenones and cyclohexenones), [2+2] photocycloaddition of the C=C double bond to alkenes occurs, and has been much exploited in synthesis.[1] These points are not taken further in the present article.

Enones can also reflect aspects of carbonyl photochemistry - such as photoreduction, oxetane formation, and the Norrish type I and type II photoreactions - and all these processes have been observed in competition with the reactions which are more characteristic of the enone excited states. A substantial review has summarised the photochemical rearrangements of enones, with emphasis on an understanding of the mechanisms of these reactions.[2] Chapman and Weiss have covered earlier examples involving synthetic applications of $\alpha\beta$-unsaturated ketone photochemistry.[3]

The rearrangement which occurs on irradiation of 4,4-dimethylcyclohex-2-enone (1) is very typical: a bicyclo[3.1.0]hexanone or so-called 'lumiketone' (2) is produced, alongside a cyclopentenone derivative.[4]

hν
Bu^{t}OH
(1)
(2) 60%

The reaction, referred to as a 'type A' rearrangement, is characterised as involving a $\pi\pi^*$ triplet state. It proceeds in low quantum yield [e.g. Φ = 0.0065 for (2)][3] - in stark contrast to the related rearrangements of cyclohexa-2,5-dienones to be described below. Enone rearrangement to lumiketone is restricted to those cyclohexenones having two alkyl substituents at C-4. Because the rate constant for rearrangement is relatively small [$k = 5 \times 10^5$ s^{-1} for (2)], other photochemical reactions such as dimerisation and reduction can compete. In many subsequent examples, t-butanol has been chosen as the reaction solvent which best gives rearrangement at the expense of these other competitive processes.

The photoreaction of the chiral phenanthrenone (3) shows the stereochemical aspect of this conversion.[5] Inversion occurs at the chiral centre (C-10), to give a product having at least 95% retention of optical purity.

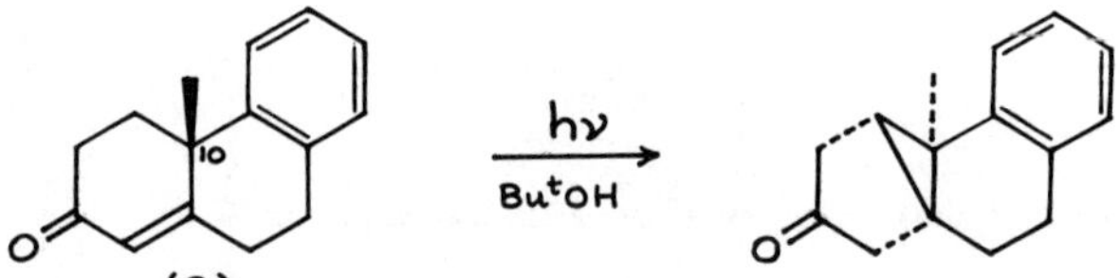

A concerted pathway, or dipolar intermediates, or even diradical species, have all been invoked in different examples of the rearrangement. The dipolar scheme outlined by (4) is most useful in predicting products from such photochemical reactions: irradiation of the enone (1) in acetic acid solution, for example, gives the ketoacetate (5) as the main product (30-40%), probably <u>via</u> trapping of the polar intermediate.[4]

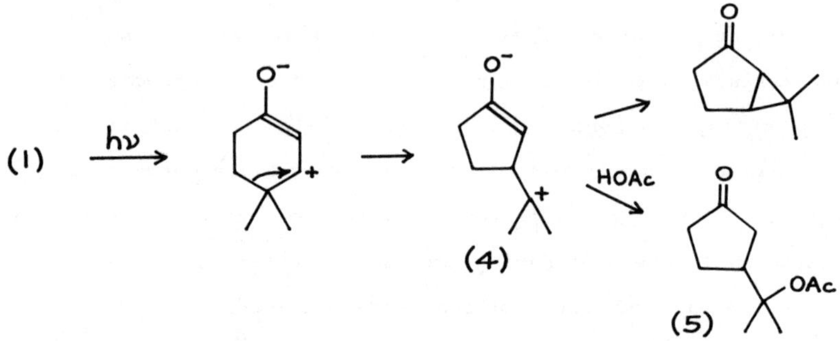

The preparative use of strong acids as solvents in enone photochemical reactions should not be overlooked,[6] and there are rearrangements related to the type A process which can proceed by photo-excited 1,2-shifts from the protonated states of enones. Irradiation at 300 nm of the 10-methyloctalone (6) in concentrated sulphuric acid gives a cyclopentenone (7) after neutralisation.[7] In the steroid field, this photoreactivity of protonated enones can rearrange the [6,6] A/B rings found in androst-1-en-3-one derivatives to the corresponding [5,7] ring-fused abeo-steroids.[8]

(6) (7)

When at least one of the groups at C-4 of the cyclohexenone is aromatic, an alternative, and more efficient, rearrangement involving phenyl migration takes place: a bicyclo[3.1.0]hexanone is formed by a reaction which is analogous to the di-π-methane rearrangement. The enone photoreaction, occurring from the n,π* triplet state, is remarkably stereoselective; there is a 143:1 preference for formation of the endo-phenyl isomer (8) over (9), although there are not large differences in steric effects between the two isomers.[9]

(8) Φ = 0·043 (9) Φ = 0·0003 Φ = 0·0002

Larger ring enones, such as cyclohept-2-enone and cyclooct-2-enone, isomerise on irradiation to yield strained trans-isomers which are reactive towards dimerisation and nucleophilic addition. The overall process may be highly selective, and not attainable by ground-state chemistry: thus, 2-methylcycloheptenone (10) undergoes isomerisation to the E-isomer on irradiation in methanol, followed by syn-addition of solvent. The result is a stereospecific and regiospecific addition to yield methoxyketone (11), whereas the normal base- or acid-catalysed Michael addition of methanol to the enone (10) is non-stereospecific.[10]

hν
MeOH
OMe
(10) (11)

Photochemically induced radical-type additions to the C=C bond of enones can also be regiospecific and stereoselective. The synthetic utility of this reaction is best illustrated by the work of Fraser-Reid and associates, in the carbohydrate field, for the addition of polyfunctional alcohols to hexenopyranosuloses such as (12).[11]

$OCPh_3$ hν, 350 nm Ph_2CO HO OH OH OH OEt OEt
(12) 75-79%

Sometimes, the tendency of excited enones to abstract hydrogen in an intramolecular sense can be put to use. As might be expected, the bicyclic enone (13), on irradiation at 300 nm, gives formation of lumiketone and cis-trans isomerisation of the propenyl group. However, the cis-enone (14), on irradiation at 254 nm (S_2 state) gives a respectable yield of propellane (15), formed by intramolecular hydrogen abstraction and radical combination reactions.[12]

(13) hν (300 nm) ⇄ (14) hν (254 nm) → (15) 47%

Intramolecular hydrogen abstraction is also involved in a photochemical approach to angularly fused tricyclics, by irradiation of the bicyclic enone (16).[13] In this case, hydrogen abstraction by the β-carbon atom of the enone is followed by transannular C-C bond formation in the diradical intermediate. Such a method of connection of five-membered rings is in interesting contrast to that which starts from βγ-enone photochemistry, as outlined in the next section.

(16) hν → 55% + 19%

There have been several recent developments in the photochemical rearrangement of αβ-unsaturated enones to their βγ-unsaturated isomers. This deconjugation reaction, which is fairly general for acyclic enones with a γ-alkyl group, is believed to involve intramolecular γ-hydrogen abstraction by the excited carbonyl group. The resultant dienol (17) can be trapped by silylation, in a useful synthesis of (Z)-siloxydienes [e.g. (18)].[14] Otherwise, the dienol may protonate at α- or γ-positions to give βγ-enone or starting material respectively.

(17) X = H
(18) X = $SiMe_3$

On a preparative basis, it has often been difficult to form high (or sometimes, any!) yields of $\beta\gamma$-isomer, but the technique of irradiation of the $\alpha\beta$-enone in the presence of a weak base (pyridine or 1,2-dimethylimidazole) overcomes this difficulty.[15] The base probably deprotonates the dienol, and so prevents a straightforward suprafacial shift of hydrogen back to the γ-site. A similar mechanism holds for the photoisomerism of $\alpha\beta$-unsaturated esters. $\alpha\beta$-Isomers that were reported to be inert towards photochemical deconjugation can, in fact, be converted to their $\beta\gamma$-isomeric esters in good yield (>80%) by irradiation in the presence of a base such as 1,2-dimethylimidazole.[16] The isomerisation of α-alkylidene-lactones to α-vinyl-lactones has been reported, in almost quantitative yields, on irradiation in methanol (19). It is worth noting that, by utilisation of a chiral base, (-)-ephedrine, this route can be persuaded to give enantioselective deconjugation.[17]

(19)

85%

$\beta\gamma$-Unsaturated Carbonyl Compounds

It has been ten years since the appearance of Houk's impressive review, covering the photochemistry and spectroscopy of $\beta\gamma$-unsaturated carbonyl compounds.[18] The present section is an attempt to highlight some of the more recent synthetic applications in this field. It was the early research of Givens and co-workers[19] that clearly laid the basis for applications of $\beta\gamma$-unsaturated ketone photochemistry. These workers found that direct irradiation of the bicyclo[2.2.2]octenone (20) gave a 1,3-acyl shift to form the cyclobutanone (21), whereas triplet-sensitised irradiation yielded the oxa-di-π-methane rearrangement product (22) which was the formal result of a 1,2-acyl shift.

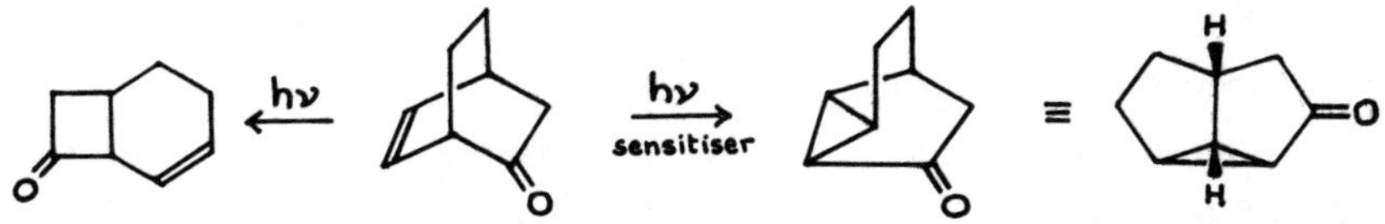

Subsequent mechanistic work[2] supports the idea of the 1,3-acyl shift occurring from the $n\pi^*$ excited states of the enone (S_1 and T_2), whilst the lowest triplet state of the enone (T_1, $\pi\pi^*$) leads to the oxa-di-π-methane rearrangement. Hence, for selective tricyclic ketone formation, a triplet sensitiser should be chosen which lies between the T_1 and T_2 energy levels: acetone (as solvent) or acetophenone have been commonly used.

Demuth, Schaffner and their colleagues have done much to develop this $\beta\gamma$-enone photochemistry in synthesis.[20] It certainly provides a convenient route to an assembly of fused five-membered rings of the type found in several classes of natural product, such as prostaglandins, pentalenolactones and coriolins. The bicyclic ketone (20) is readily prepared by hydrolysis of the Diels-Alder adduct from cyclohexa-1,3-diene and 2-chloroacrylonitrile, and an optimised yield of 80-85% has been obtained for the photolysis (20)→(22).[21] One of the most fascinating applications of this chemistry depends on the fact that either of the bicyclic (20) or tricyclic (22) ketones can be resolved into enantiomers, after formation of acetals with chiral diols. Because the photochemical 1,2-acyl shift is stereospecific, chirality can be totally controlled by resolution either before or after photolysis. Useful conversions of the tricyclooctanone photoproduct include almost quantitative isomerisation to the unsaturated ketone (23) by the use of Nafion-TMS, a trimethylsilylated version of a perfluorinated sulphonic acid resin, and the high-yielding, stereospecific alkylation of the ketone to provide the methylated analogue (24). In fact, the latter compound, in chiral form, (-)-(24), has been the branching point in two recent enantiospecific syntheses of iridodial (25),[22] a biosynthetic intermediate to iridoid glucosides in plants, and of loganin aglucon 6-acetate (26) in 17 steps and 7% overall yield from cyclohexa-1,3-diene.[23]

H H O NaH MeI (22) Nafion-TMS H H O

98%

(24) (23)

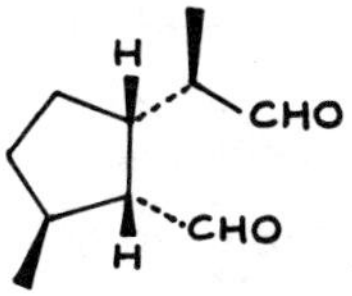

(25) 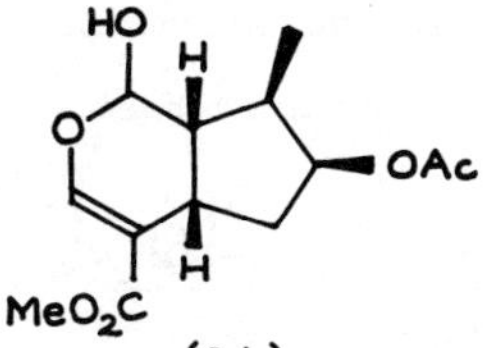 (26)

Independently, Yates and Stevens[24] have applied the oxa-di-π-methane rearrangement of the bicyclic ketone (27) in an approach to the synthesis of the sequiterpene cedrol (29). Photolysis of the βγ-unsaturated ketone (27) in the presence of acetophenone as sensitiser (and solvent!) led to the desired product (28) in 76% yield. A subsequent homoconjugate addition of lithium dimethylcuprate introduces the required methyl group into ring A and gives the correct A/B ring skeleton for final conversion to cedrol.

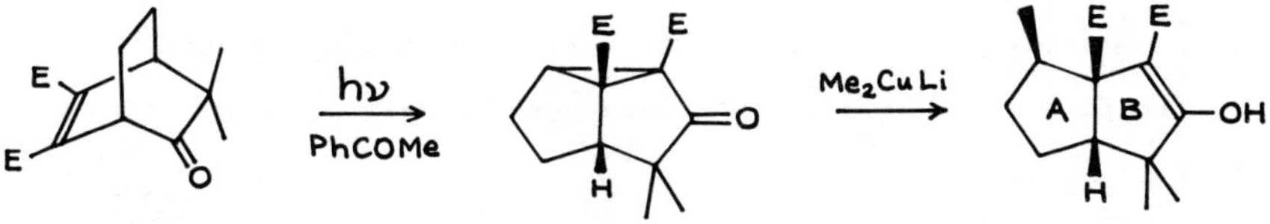

(27) (28)

E = CO_2Me

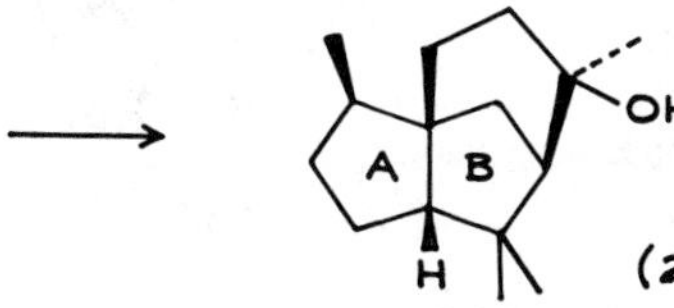

(29)

A classic example of the oxa-di-π-methane rearrangement of bicyclodecenone (30) shows potential for application to synthesis, in that irradiation gives the expected tricyclic compound, and ensuing cyclopropyl ring opening with acids can lead to the bridgehead-substituted ketone (31).[25] The [5,6] ring system is also approachable using the photochemical 1,3-acyl shift of $\beta\gamma$-enones, as discussed below.

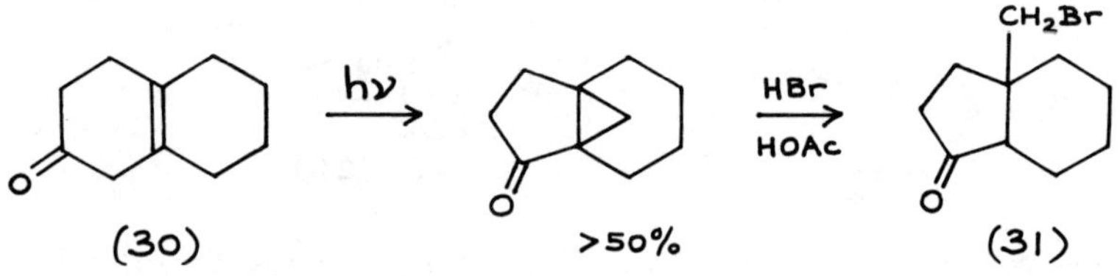

hν HBr HOAc CH_2Br

(30) >50% (31)

A highly regioselective rearrangement of the unsaturated diketone (32) provides a short and convenient route towards the A/B rings of the anti-tumour sesquiterpene coriolin (34), in racemic form.[26] Here, it is noticeable that predominantly one of the two $\beta\gamma$-enone units is rearranged, either on direct or sensitised irradiation. The Norrish type I α-cleavage of (32) causes rapid interconversion of the two epimers at C-3, so that the tricyclic diketone (33) is also formed as an epimeric mixture of C-4 isomers; however, this point does not affect the subsequent synthetic steps, which include cleavage of the cyclopropyl ring by Birch reduction.

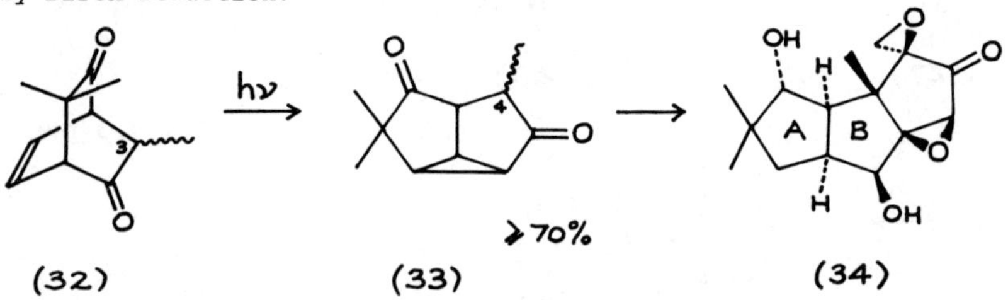

hν >70%

(32) (33) (34)

The recent approach of Schultz et al.,[27] which looks most promising, has involved the generation of $\beta\gamma$-enones such as (35) by intramolecular Diels-Alder reactions of 6-alkenylcyclohexa-2,4-dienones, followed by their irradiation to give rearranged polyquinane rings (36).

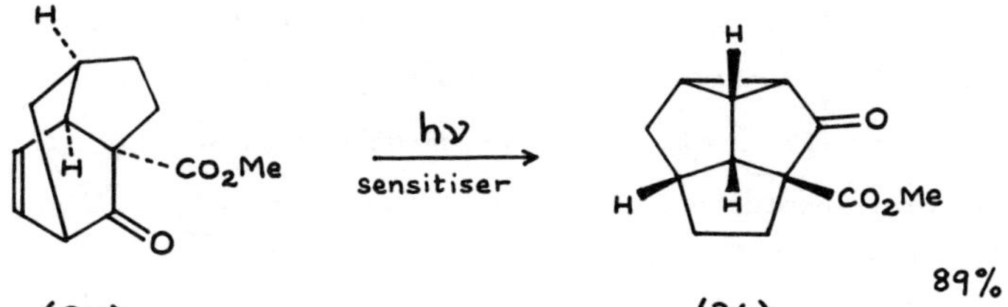

(35) → hν sensitiser → (36) 89%

Upon direct irradiation, acyclic βγ-unsaturated ketones often dissipate their energy exclusively by cis-trans isomerisation.[28] In some examples, other photoreactions associated with the carbonyl group can dominate: e.g. α-cleavage, especially when stabilised allylic radicals are produced. This α-cleavage reaction can also lead to 1,3-acyl shifts via recombination at the opposite end of the allylic radical.

In restrained cyclic βγ-unsaturated ketones, the 1,3-acyl shift occurring on photolysis has been used more reliably as a means of changing ring size or connectivity. For a chiral cyclopent-2-enyl methyl ketone, the 1,3-acyl shift is largely enantioselective, and can occur only in part by a radical pair which loses stereochemistry.[29] The photoisomerisation is reversible in some examples, as shown by the equilibrium (1:1) established in attaining the A/B ring spiro-steroid system of (37).[30]

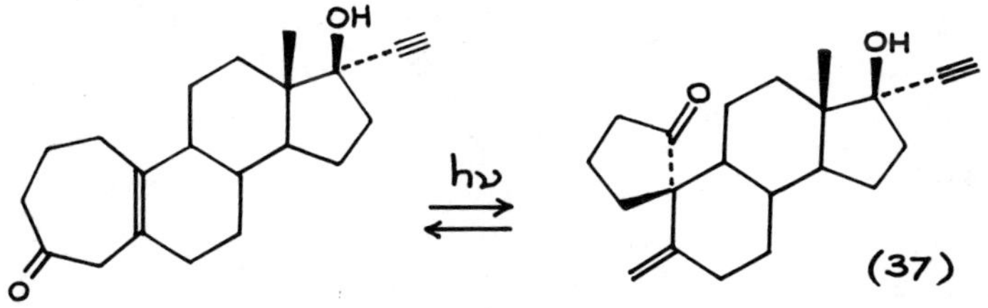

hν ⇄ (37)

Nevertheless, such 1,3-shifts can be stereospecific, as evidenced by the conversion of the chiral cyclopentenyl ketoacid (38) into the hydroazulene (39) skeleton, in which epimerisations are found as a much slower photochemical process, important only at high conversion.[31]

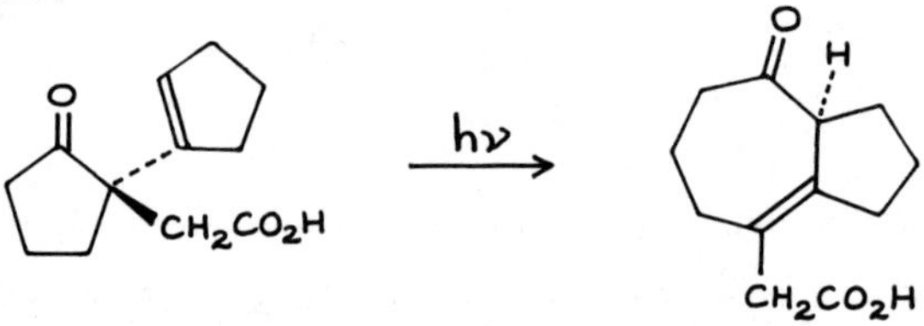

$\xrightarrow{h\nu}$

CH_2CO_2H CH_2CO_2H

(38) (39)

A key step in the total synthesis of the sesquiterpene, pinguisone (41), in racemic form, involves a light-induced 1,3-acyl shift in the bicyclo[3.2.2]nonenone (40).[32] Even more recently, a similar photochemical reaction has been used to create the [5,6] cis-fused ring system also found in ptilocaulin (42), the antimicrobial cyclic guanidine isolated from a Caribbean sponge.[33] In this latter case, the starting material for photolysis is reasonably available, in four steps, from tropolone.

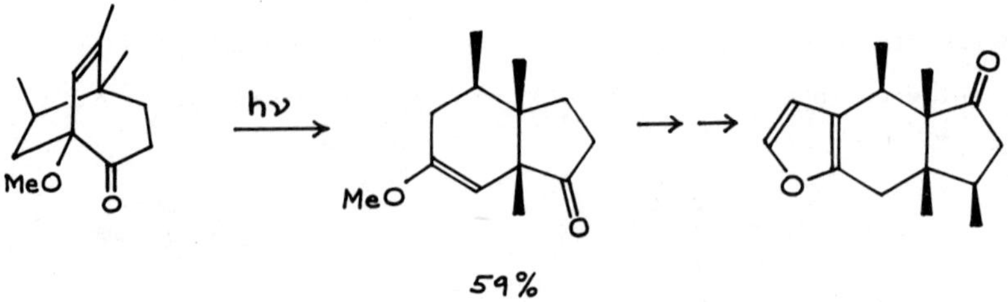

MeO $\xrightarrow{h\nu}$ MeO $\rightarrow\rightarrow$

59%

(40) (41)

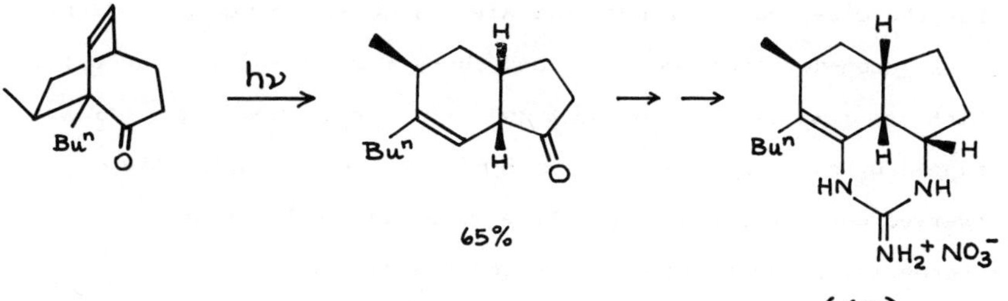

65%

(42)

Cross-conjugated Dienones

Cyclohexa-2,5-dienones having two substituents at C-4 undergo a 'type A' rearrangement on photolysis, similar to that already discussed for enones [(1)⟶(2)]. A typical example is that of rearrangement of the diphenyldienone (43) to a bicyclo[3.1.0]hexenone (45). However, significant differences arise in comparison with the enone reaction, in that the dienone $n\pi^*$ triplet state is responsible for rearrangement, and the reaction is very efficient, both in terms of rate constant and quantum yield [$k > 2 \times 10^{10}$ s^{-1}, $\Phi = 0.85$ for (43)]. The dienone rearrangement has been the subject of intensive research,[34] and the mechanistic pathway outlined in (43)⟶(45) gives a useful idea of the consequences of reaction.

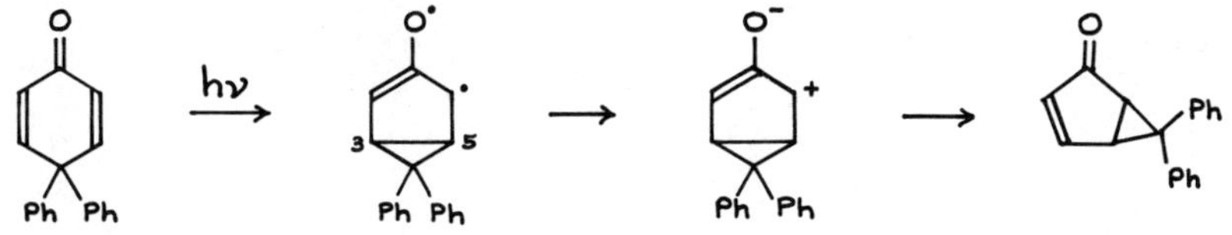

(43) (44) (45)

Thus, the triplet excited state of (43) leads to C-3—C-5 bond formation, and subsequently to the zwitterion (44).[35] The intermediate involvement of a zwitterion accounts for the fact that the observed rearrangements are characteristic of migrations to electron-deficient centres, and there are several instances in which the positive site of such zwitterionic intermediates can be trapped by nucleophiles.[36] In other examples, including the observed gas-phase reaction, there is evidence that free radical intermediates can be involved in the rearrangement.

The earliest recorded example (1830) of this cyclohexadienone photochemistry concerned the photolability of α-santonin (46), an anthelmintic isolated from the Wormwood bush. By 1960, at least four major structures and photochemical processes involved in this system had been established.[34,37] The type A product, lumisantonin (47), can be formed in reasonable yields by irradiation of (46) in neutral solvents such as dioxane; however, the photoproduct itself readily undergoes secondary photolysis to give a linearly conjugated dienone, mazdasantonin (48), as well as phenolic products. Irradiation of (48) in water leads to a ring-opened acid, photosantonic acid (49), by a mechanism to be described in the next section. In contrast, in aqueous acidic solutions, α-santonin gives a characteristic ring rearrangement to the [5,7] hydroazulene skeleton of isophotosantonic lactone (50).

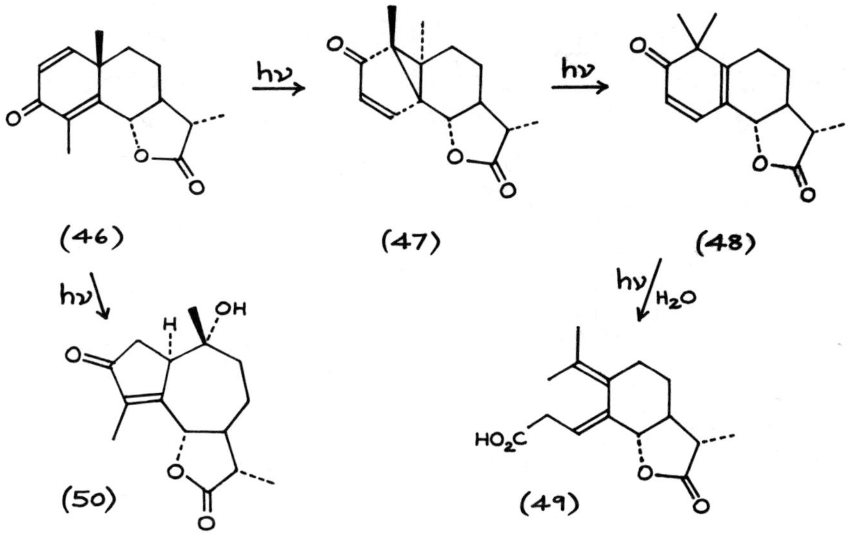

When other cyclohexadienones which are unsubstituted at C-4 are irradiated under these conditions, they tend to give both hydroazulene products and spiro-ketones. The formation of these products can be understood on the basis of interception of the zwitterion by protonation, followed by alternative ring openings of a cyclopropylcarbinyl cation centre in (51). Any cation-stabilising groups, such as 4-methyl, encourage bond breaking (a), with attack by nucleophile to produce hydroazulene (52). However, cation-stabilising groups at C-2 can promote the alternative cyclopropyl bond fission (b), resulting in exclusive spiro-ketone formation (53). The easy availability of dienones has led to the exploitation of both of these reaction pathways in natural product synthesis.

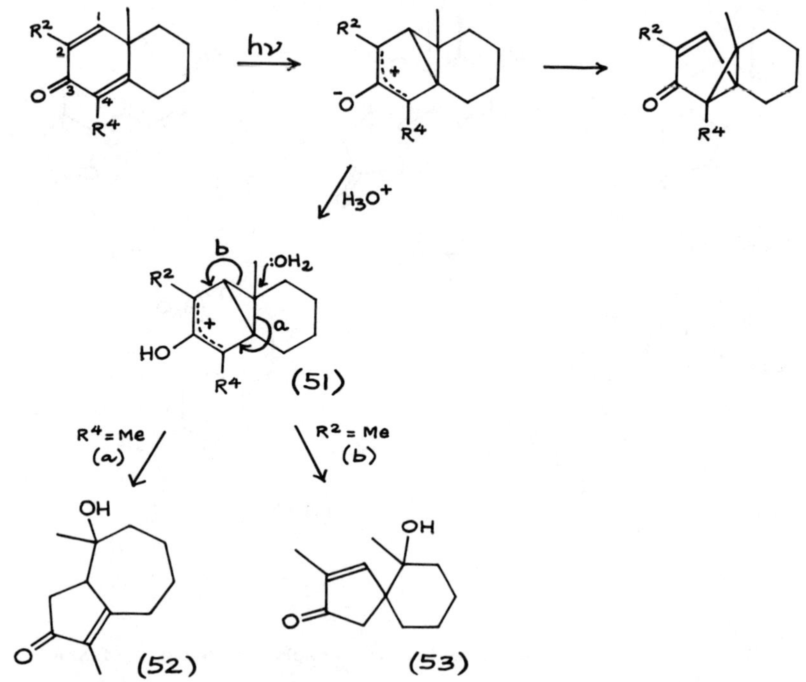

The conversion of 6,6-fused to 5,7-fused rings is illustrated by the photorearrangement of the 4-methyl substituted dienone (54) in aqueous acetic acid, leading to hydroxyketone and then in two steps to the tricyclic sesquiterpene, 4-epi-globulol.[38]

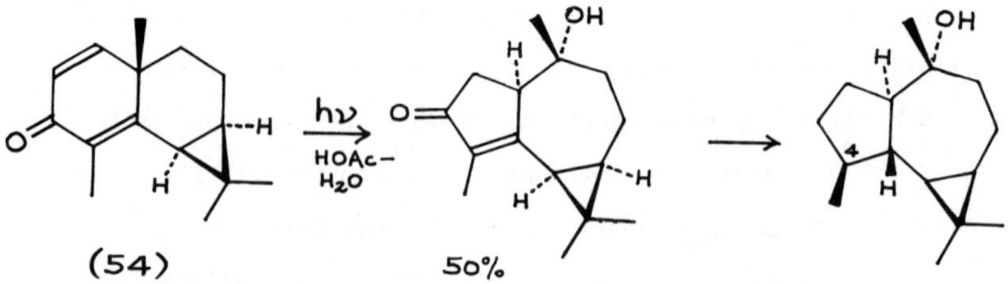

This approach of simultaneous ring-contraction and ring-expansion is also applicable to different ring-size precursors, as shown by the conversion of the [6,5]-ring dienone (55) into a [5,6] ring-fused photoproduct, and thence to the sesquiterpene oplopanone (56).[39]

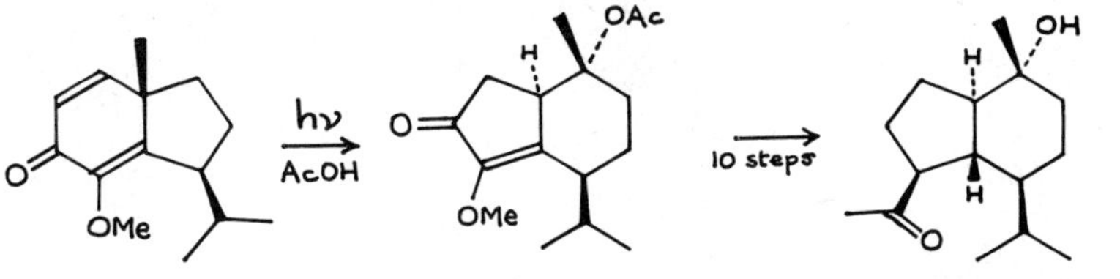

(55) hν AcOH 91% 10 steps (56)

As an example of the spiro-ketone pathway, the 2-methoxy-dienone (57) has been irradiated in glacial acetic acid to give the acetoxyketone (58) in excellent yield, and then by seven subsequent steps, α-vetispirene (59).[40]

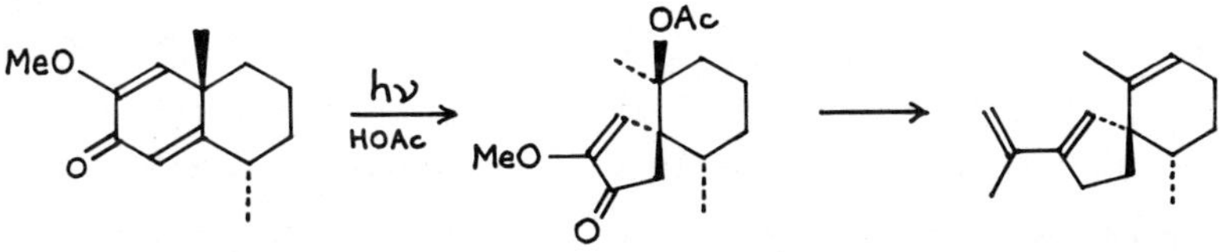

(57) hν HOAc (58) 89% (59)

One of the advantages of photochemical reactions is that they can often be carried out under essentially neutral conditions, and are thus capable of being applied to the production of sensitive organic target molecules. A clear example of this comes from the work of Barton and Hulshof on the synthesis of the mould metabolite terrein (62),[41] which is sensitive both to acid and to base. Ring contraction of the 4-pyrone (60) occurs on irradiation in water, to give the cyclopentenedione (61). The *in situ* reduction of the dione with sodium cyanoborohydride gives terrein (62) directly.

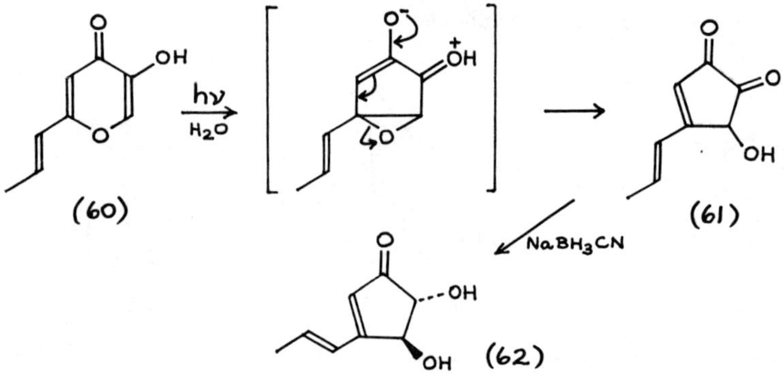

(60) (61) (62)

Linearly Conjugated Dienones

One of the most reliable photochemical reactions of cyclohexa-2,4-dienones is that of ring opening to yield a dienylketene [*e.g.* (63)].[42] The reaction, which can be considered to be a consequence of α-cleavage, occurs from the nπ* singlet excited state. The ketene may revert thermally to starting material, or may be trapped by the presence of nucleophiles such as amines to afford *Z*-dienyl derivatives (64).

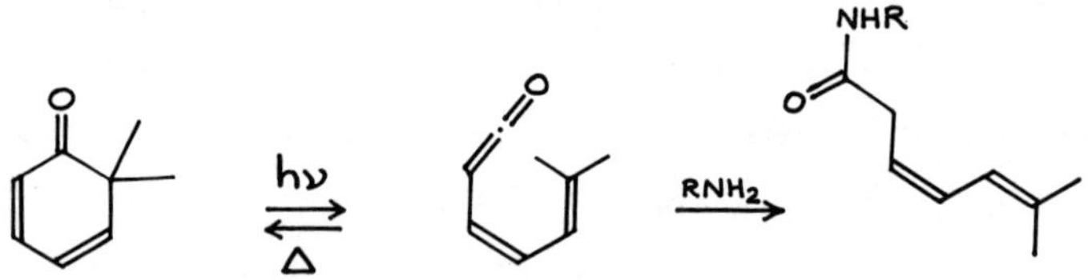

(63) (64)

Since the starting cyclohexadienones are easily prepared from aromatic compounds (*e.g.* allyl phenols, by Claisen rearrangement, *etc.*), this route gives a worthwhile straight-chain synthesis from benzenoids. The most spectacular example here has been the synthesis of the saffron pigment, dimethyl crocetin (66), in three steps, by Barton, Quinkert *et al.*[43] The key stage is the double photochemical ring opening of the bis-cyclohexadienone (65) to a bis-ketene, which is trapped *in situ* by methanol.

Dehydrogenation (with dichlorodicyanoquinone) yielded the fully conjugated heptaene diacid structure found in crocetin.

hν

(65)

DDQ

MeO_2C CO_2Me

(66)

Quinkert and co-workers have continued to find applications for this cyclohexadienone to dienylketene rearrangement. The presence of a non-proton donating nucleophile, such as 1,5-diazabicyclo[2.2.2]octane (DABCO), activates the ketene towards reaction with weaker nucleophiles than those it would normally react with. In these circumstances, exemplified by the photochemistry of (67), intramolecular attack by a hydroxyl group can occur readily to generate a macrocyclic lactone in encouraging yield.[44]

For dienones having shorter hydroxymethylene chains at C-6, intramolecular attack is less favourable in comparison with bimolecular reaction: thus irradiation of the dienone (68) under these conditions produces a 22-membered dilactone (69) in 30% yield, along with a trilactone (6%).

(67) $\xrightarrow[\text{DABCO}]{h\nu}$ 73%

(68) $\xrightarrow[\text{DABCO}]{h\nu}$ (69) 30%

References

1 A.C. Weedon, in "Synthetic Organic Photochemistry", ed. W.M. Horspool, Plenum Press, New York, 1984, p. 61

2 D.I. Schuster, in "Rearrangements in Ground and Excited States", Vol. 3, ed. P. de Mayo, Academic Press, New York, 1980, p. 167

3 O.L. Chapman and D.S. Weiss, Org. Photochem., 1973, 3, 197; see also D.I. Schuster, M.M. Greenberg, I.M. Nunez and P.C. Tucker, J. Org. Chem., 1983, 48, 2615

4 O.L. Chapman, T.A. Rettig, A.A. Griswold, A.I. Dutton and P. Fitton, Tetrahedron Lett., 1963, 2049

5 O.L. Chapman, J.B. Sieja and W.J. Welstead, Jr., J. Am. Chem. Soc., 1966, 88, 161; see also D.I. Schuster, R.H. Brown and B.M. Resnick, ibid., 1978, 100, 4504

6 R.F. Childs, Rev. Chem. Intermed., 1980, 3, 285

7 D.G. Cornell and N. Filipescu, J. Org. Chem., 1977, 42, 3331

8 P. Lupon, J.C. Ferrer, J.F. Piniella and J.-J. Bonet, J. Chem. Soc., Chem. Commun., 1983, 718

9 H.E. Zimmerman and K.G. Hancock, J. Am. Chem. Soc., 1968, 90, 3749; for the most recent developments, cf. H.E. Zimmerman, C.E. Caufield and R.K. King, ibid., 1985, 107, 7732

10 H. Hart, B. Chen and M. Jeffares, J. Org. Chem., 1979, 44, 2722

11 N.L. Holder, Chem. Rev., 1982, 82, 287

12 F. Nobs, U. Burger and K. Schaffner, Helv. Chim. Acta, 1977, 60, 1607

13 Y. Tobe, T. Iseki, K. Kakiuchi and Y. Odaira, Tetrahedron Lett., 1984, 25, 3895

14 C.S.K. Wan and A.C. Weedon, J. Chem. Soc., Chem. Commun., 1981, 1235

15 R. Ricard, P. Sauvage, C.S.K. Wan, A.C. Weedon and D.F. Wong, J. Org. Chem., 1986, 51, 62

16 R.M. Duhaime, D.A.Lombardo, I.A. Skinner and A.C. Weedon, J. Org. Chem., 1985, 50, 873

17 F. Henin, R. Mortezaei, J. Muzart and J.-P Pete, Tetrahedron Lett., 1985, 26, 4945

18 K.N. Houk, Chem. Rev., 1976, 76, 1

19 R.S. Givens, W.F. Oettle, R.L. Coffin and R.G. Carlson, J. Am. Chem. Soc., 1971, 93, 3957

20 For a review, see M. Demuth and K. Schaffner, Angew. Chem. Internat. Ed. Engl., 1982, 21, 820

21 M. Demuth, P.R. Raghavan, C. Carter, K. Nakano and K. Schaffner, Helv. Chim. Acta, 1980, 63, 2434

22 P. Ritterskamp, M. Demuth and K. Schaffner, J. Org. Chem., 1984, 49, 1155

23 M. Demuth, S. Chandrasekhar and K. Schaffner, J. Am. Chem. Soc., 1984, 106, 1092

24 P. Yates and K.E. Stevens, Tetrahedron, 1981, 37, 4401

25 J.R. Williams and H. Ziffer, Tetrahedron, 1968, 24, 6725

26 M. Demuth, P. Ritterskamp and K. Schaffner, Helv. Chim. Acta, 1984, 67, 2023

27 A.G. Schultz, F.P. Lavieri and T.E. Snead, J. Org. Chem., 1985, 50, 3086

28 A.J.A. van der Weerdt and H. Cerfontain, J. Chem. Soc., Perkin Trans. II, 1980, 592

29 D.E. Sadler, J. Wendler, G. Olbrich and K. Schaffner, J. Am. Chem. Soc., 1984, 106, 2064

30 J.R. Williams and G.M. Sarkisian, J. Org. Chem., 1980, 45, 5088

31 R.L. Coffin, R.S. Givens and R.G. Carlson, J. Am. Chem. Soc., 1974, 96, 7554; for examples of 1,3-shift without side-chain isomerisation, see R H van der Veen and H Cerfontain, Tetrahedron, 1985, 41, 585

32 T. Uyehara, Y. Kabasawa, T. Kato and T. Furuta, Tetrahedron Lett., 1985, 26, 2343

33 T. Uyehara, T. Furuta, Y. Kabasawa, J. Yamada and T. Kato, J. Chem. Soc., Chem. Commun., 1986, 539

34 K. Schaffner and M. Demuth, in "Rearrangements in Ground and Excited States", Vol. 3, ed. P. de Mayo, Academic Press, New York, 1980, P. 281

35 H.E. Zimmerman and D.I. Schuster, J. Am. Chem. Soc., 1962, 84, 4527; for a recent example, see H.E. Zimmerman and D.C. Lynch, ibid., 1985, 107, 7745

36 D.I. Schuster and K.-C. Liu, Tetrahedron, 1981, 37, 3329; intramolecular trapping of a zwitterion by azide ion has been reported by A.G. Schultz, S.O. Myong and S. Puig, Tetrahedron Lett., 1984, 25, 1011

37 D.H.R. Barton, P. de Mayo and M. Shafiq, J. Chem. Soc., 1958, 140

38 D. Caine and J.T. Gupton, III, J. Org. Chem., 1975, 40, 809

39 D. Caine and F.N. Tuller, J. Org. Chem., 1973, 38, 3663

40 D. Caine, A.A. Boucugnani, S.T. Chao, J.B. Dawson and P.F. Ingwalson, J. Org. Chem., 1976, 41, 1539

41 D.H.R. Barton and L.A. Hulshof, J. Chem. Soc., Perkin Trans. I, 1977, 1103

42 G. Quinkert, Pure Appl. Chem., 1973, 33, 285

43 G. Quinkert, K.R. Schmieder, G. Dürner, K. Hache, A. Stegk and D.H.R. Barton, Chem. Ber., 1977, 110, 3582

44 G. Quinkert, G. Fischer, U.-M. Billhardt, J. Glenneberg, U. Hertz, G. Dürner, E.F. Paulus and J.W. Bats, Angew. Chem. Internat. Ed. Engl., 1984, 23, 440

8
Alkenes: Isomerisation and Rearrangement

By A. Gilbert
CHEMISTRY DEPARTMENT, UNIVERSITY OF READING, WHITEKNIGHTS, PO BOX 224, READING, BERKS. RG6 2AD, UK

The lowest energy electronic transition of simple alkenes involves excitation of an electron from the highest occupied π orbital to the lowest unoccupied π^* orbital. The transition is allowed and occurs by absorption in the 170-190 nm region of the electromagnetic spectrum. Many systems of interest to photochemists and the types of alkene derivatives which may be considered for synthetic exploitation often have conjugative substituents and these cause a bathochromic shift of the $\pi\pi^*$ absorption to more accessible wavelengths (e.g. 250 nm). Intersystem crossing between the singlet and triplet states of alkenes is inefficient but since the energy difference between these states is large, the triplet state of alkenes is readily accessible by sensitisation. This method of excitation is widely used for the study of alkene triplet processes, and for the initiation of the photoreactions of non-conjugated alkenes.

The photochemical reactions of alkenes which are considered in this Chapter are *cis-trans* isomerisations, sigmatropic shifts, di-π-methane rearrangements, and electrocyclic reactions.

1. cis-trans Isomerisations

Since there is effectively no π bond in the $\pi\pi^*$ excited state of alkenes, the central bond may rotate and a 90° rotation to an orthogonal arrangement (1) produces a species with the minimum electronic interaction and hence the lowest energy conformation. Such a species is formed in both the singlet and triplet manifolds and is accessible by excitation of either the *cis* or *trans* isomer of the alkene. On decay to the ground state this non-planar form of the excited alkene can yield either geometrical isomer (see Figure 1). Thus photo *cis-trans* inconversion of alkenes is a common process and is important in several biological

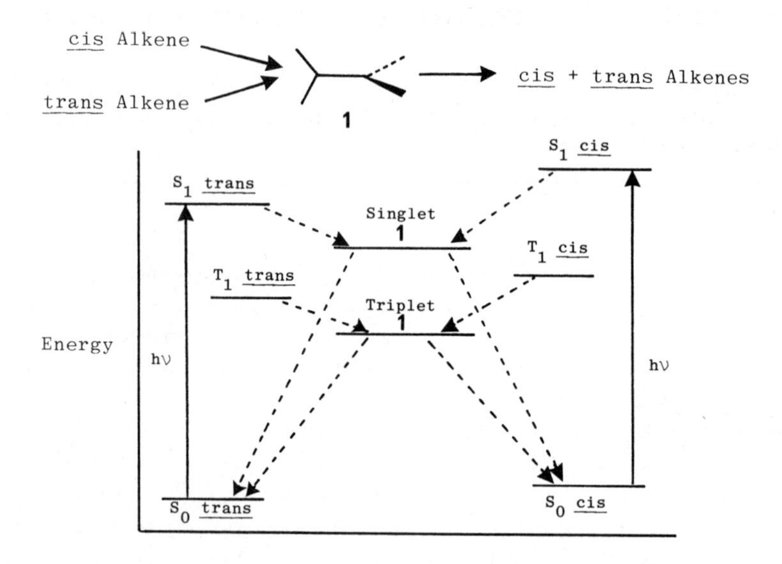

Figure 1 Decay of singlet and triplet non-planar intermediates to *cis* and *trans* alkene isomers

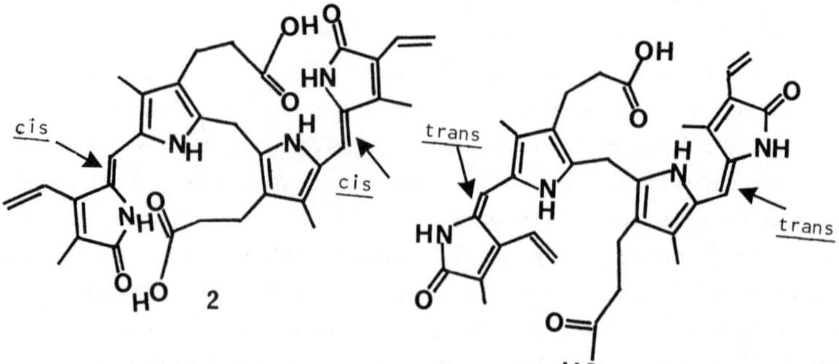

systems including the photoreactions of visual pigments and in the photo-treatment of neonatal jaundice. In the latter case the photoreaction involves the conversion of the water insoluble cis,cis isomer (2) of the neurotoxin bilirubin into the soluble trans,trans isomer (3) which can then be excreted.[1]

Before considering the synthetic utility of this isomerisation, it is necessary to appreciate that direct and sensitised irradiation of alkenes may lead to different ratios of the two geometrical isomers in the photostationary state. For direct irradiation, the photostationary state is dependent on the relative absorptivities (extinction coefficients) of the cis and trans isomers at the particular exciting wavelength and on the partitioning ratios for decay to the ground state of the involved excited state(s). In the case of stilbenes, for example, this ratio is approximately unity and irradiation of the cis or trans isomer or mixtures of the two in the 313 nm region yields a photostationary state having over 90% of the cis isomer: this reflects the far higher absorptivity of the trans (ε_{313} 16,300 l mol^{-1}) than cis (ε_{313} 2,300 l mol^{-1}) stilbene at these wavelengths.

In the sensitised cis-trans isomerisation of alkenes, the composition of the photostationary state is dependent on the triplet energy of the sensitiser and upon the particular alkene system. For stilbenes, the ratio of cis to trans isomer increases from 5 to 14 for sensitisers of energy 150 to approximately 210 kJ mol^{-1} then decreases with increase in sensitiser energy until by 260 kJ mol^{-1} the value is 2 which remains approximately constant up to triplet energies of 300 kJ mol^{-1} (see Figure 2a). These features result from the ability of the sensitisers to populate the triplet states of the cis and trans alkene isomers and the branching ratio for decay of the non-planar intermediate (Figure 1) to the ground state isomers. Thus for sensitisers which have triplet energies greater than those of both the cis and trans isomers, the photostationary state reflects the branching ratio but lower energy sensitisers which can only populate the triplet of the trans isomer efficiently (lower energy than the cis) reduce the proportion of the trans isomer in the ground state and so the cis:trans ratio increases. When the triplet energy of the sensitiser is below those of both geometrical isomers but above that of the non-planar triplet alkene state, the cis:trans ratio falls as the branching ratio from this intermediate state becomes a dominant factor again.[2]

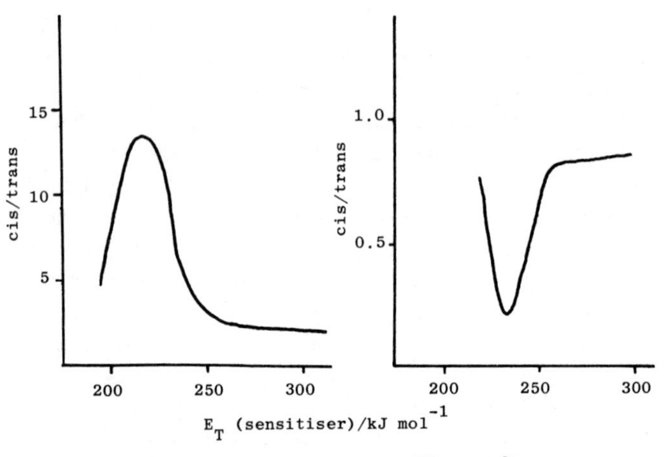

Figure 2a
Stilbene isomerisation

Figure 2b
Penta-1,3-diene isomerisation

Sensitised irradiation of conjugated dienes produces a pattern for the *cis*:*trans* ratio which is essentially the reverse of that of stilbenes, and for penta-1,3-diene, for example, as shown in Figure 2b the ratio is less than unity regardless of the triplet energy of the sensitiser. It should also be noted that with some systems intramolecular photosensitisation can occur. Thus irradiation of 4-phenylbut-2-ene with 266 nm radiation yields the *cis* isomer with a quantum yield of 0.66.[3]

The synthetic potential of *cis*-*trans* photoisomerisation lies in the possibility of transforming the more stable geometrical isomer into the less stable isomer. Thus, for example, irradiation of the *trans* cinnamic acid (4) produces the *cis* isomer (5) in 74% yield[4] and simple cycloalkenes such as *cis* cýclo-octene yield the *trans* product on sensitisation with benzenoid compounds.[5] In some cases the *trans* cycloalkene so produced is so reactive that in alcoholic solvents or water, protonation occurs followed by reaction of the carbonium ion to yield ethers or alcohols respectively. By this reaction 1-methylcycloheptene produces (6) and limonene selectively adds water to the cyclic ethylene moiety to give a 90% yield of the alcohol (7).[6]

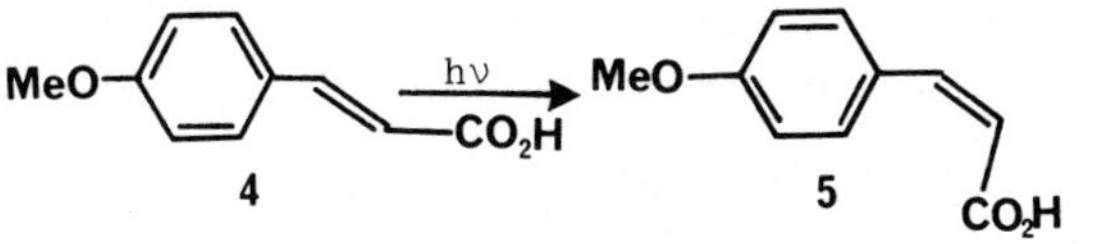

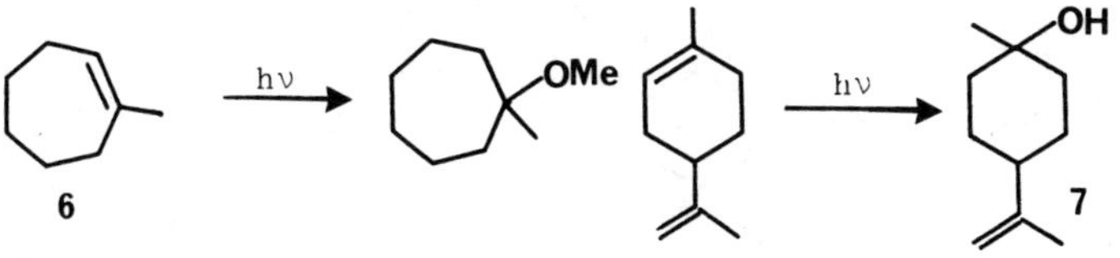

This type of isomerisation in diene and polyene systems has been extensively used particularly where the cis isomer is required for subsequent cyclisation. The most widely studied examples of this process involve the formation of polynuclear aromatic compounds from trans 1,2-disubstituted alkenes. The overall process is illustrated in Scheme 1 by the formation of phenanthrenes from trans stilbene. A second photoreaction is involved in which the cis isomer cyclises to a readily oxidisable dihydro intermediate (see Section 4 for examples of this process).

Scheme 1

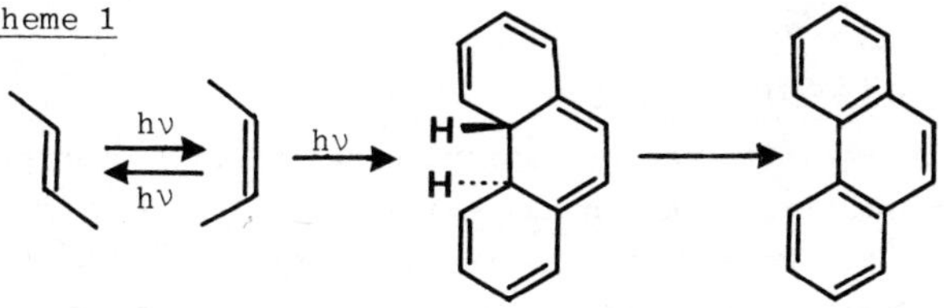

Further examples of the uses of cis-trans photoisomerisations include the conversion of the trans dienone (8) into cis-2-(cyclohexylideneethylidene)-cyclohexanone (9) which was required for the synthesis of simple calciferol analogues,[7] and in the formation of the all trans form (10) of Vitamin A from the mixture of 11-cis (11) and all trans isomers produced from the Wittig reaction of (12) and (13). Isomer (10) is required for its nutritional properties and for use in the pharmaceutical industry.

8 hν 9

12 $\dot{P}Ph_3$ + CHO 13 OAc

11 OAc + 10 OAc

hν

2. Sigmatropic Shifts

These rerarrangements can be induced both thermally and photochemically and involve a σ bond moving across a π system to a new site. There are several types of sigmatropic shifts which may occur dependent on the nature and extent of the π system:[8] those of particular interest in photochemistry involve [1,3], [1,5], and [1,7] rearrangements as illustrated in Schemes 2, 3 and 4 respectively. The shift may occur in either a suprafacial or an antarafacial manner: that is to say that the σ bond is broken and made on the same or opposite side of the π system respectively. Photochemically induced [1,3] and [1,7] shifts proceed suprafacially and [1,5] shifts antarafacially. The reverse situation is observed in thermal processes although antarafacial thermal

Scheme 2

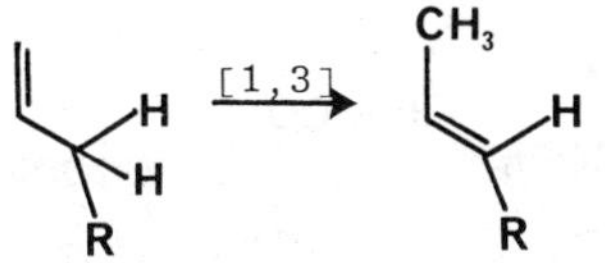

Scheme 3

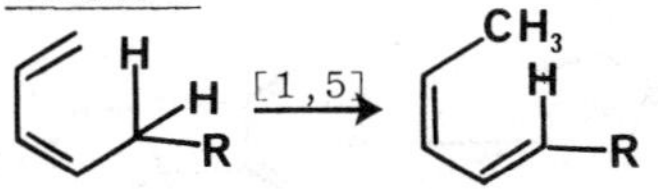

Scheme 4

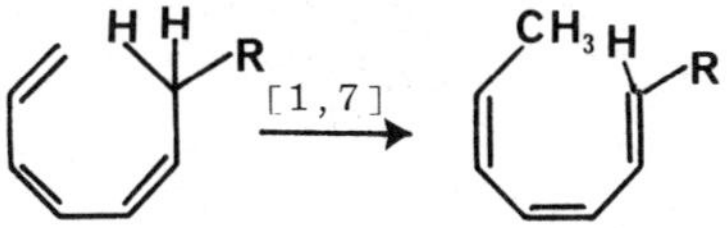

[1,3] sigmatropic shifts are virtually unknown which may be accounted for by the low probability of maintaining and developing the necessary orbital overlaps during the process.[9] Photochemically induced [1,3] shifts are facile and although occurring with simple alkenes have more commonly been reported for 4-arylalkenes and hexa-1,5-dienes which have cyano substituents to increase the wavelength of absorption. The photoreactions of (14)[10] and (15)[11] exemplify the [1,3] sigmatropic rearrangement of these two classes of compound and the former case also illustrates that where the migrating substituent is not a single atom, in the photo-induced process the configuration of the group is retained: this is inverted in a [1,3] thermal reaction. The photorearrangements of the perfluoro compound (16)[12] and of mortonin C (17) are cited here in order to indicate the breadth and generality of the [1,3] process.

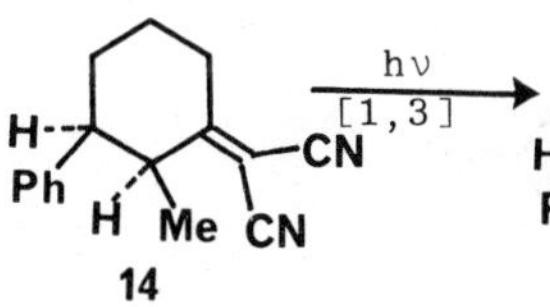

14

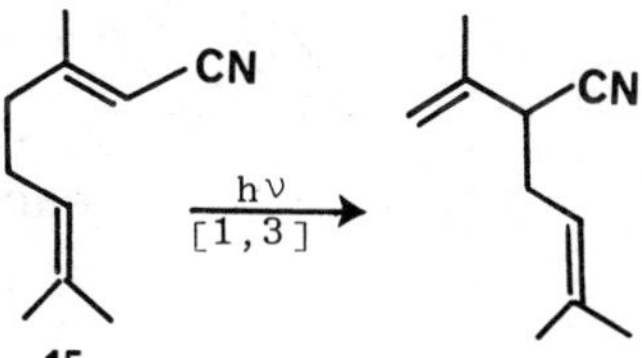

15

Perfluoro Perfluoro

hν [1,3]

16 55%

PhCOO OCOPh hν [1,3] PhCOO OCOPh R R^1

HO CO_2H 17 HO CO_2H

R=CH=CMe_2 R'=H

R'=CH=CMe_2 R=H

Photochemically induced sigmatropic shifts are singlet state processes and occur in a concerted manner. Triplet photosensitisation of [1,3] sigmatropic shift systems leads to quite different reaction pathways and products. For example direct irradiation of bicyclo[2.2.2]octadienone derivatives (18) yields bicyclo[4.2.0]octa-2,4-dien-6-ones (19) via a [1,3] shift whereas in the presence of acetophenone an oxadi-π-methane reaction (see Section 3) to give tricyclo[$3.3.0.0^{2,8}$]octan-3-ones (20) occurs,[14] and similarly the hexa-1,5-diene (21) yields either (22) or (23) dependent on the absence or presence of a sensitiser respectively.[15] It should also be noted that the nature of the reaction solvent may have a profound effect upon the type of photoproduct which is preferred. Thus while the [1,3] benzyl shift compound (24) results from irradiation of (25) in cyclohexane, in the polar solvent acetonitrile the cyclised product (26) is also formed probably via a dipolar intermediate.[16]

hν [1,3] 19

18 hν PhCOMe 20

Photochemical [1,5] and [1,7] sigmatropic shifts are not as widely reported as the [1,3] rearrangement. Nevertheless the former process is a photochemical feature of several 1,3-diene systems and its occurrence on irradiation of 2-methylstyrenes provides a useful route towards apolignans as shown in Scheme 5. The o-xylylene produced by a [1,5] shift on irradiation of the 2-methylstyrene is formed stereoselectively and can be trapped stereospecifically with reactive dienophiles such as maleic anhydride.[17]

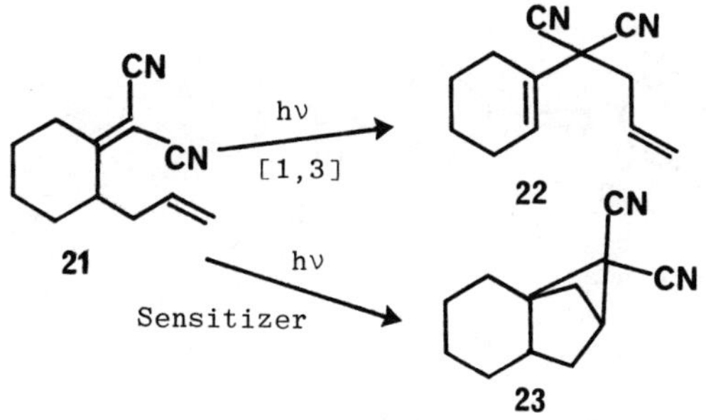

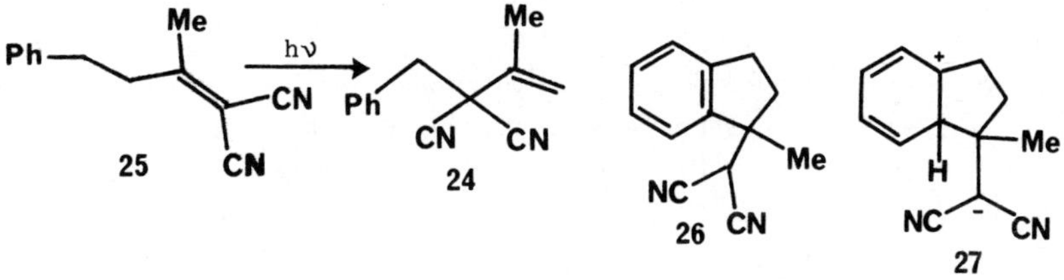

Scheme 5

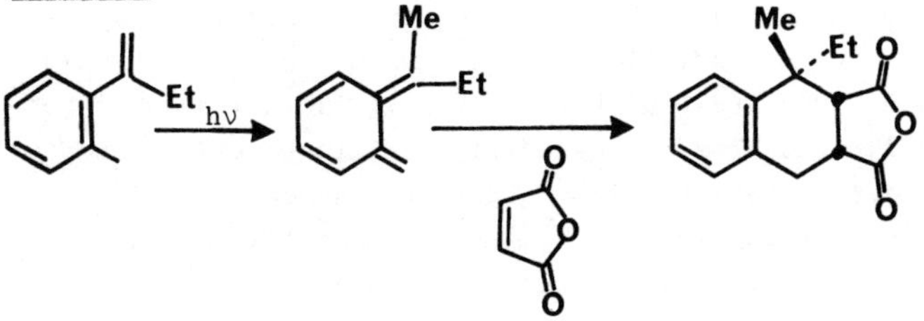

A major feature in the photochemistry of cycloheptatrienes is a sequence of [1,7] sigmatropic hydrogen shifts (see Scheme 6)[18] and while this must be borne in mind for such compounds, as yet this rearrangement has little synthetic application.

Scheme 6

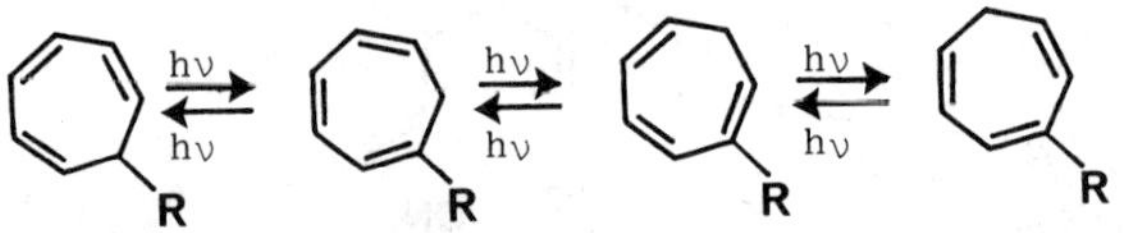

3. Di-π-Methane Rearrangement[19]

Scheme 7

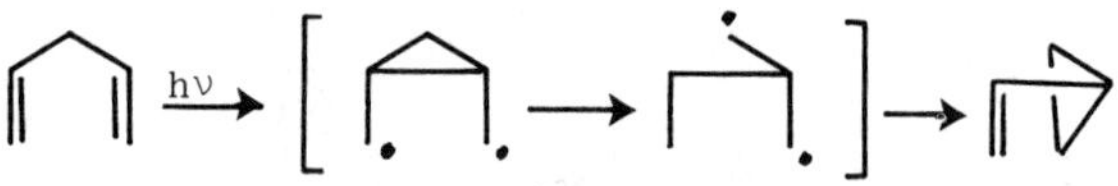

The essential feature of this photoprocess is the rearrangement of penta-1,4-dienes to ethenyl cyclopropane systems. The reaction formally involves a 1,2-shift with cyclisation as outlined in Scheme 7 but although this analysis is useful for the prediction of products from a particular system, the diradicals drawn do not represent intermediates (i.e. energy minima on the potential energy surface) in the process. The mechanism and efficiency of the process is very dependent on the nature of the system and for acyclic and monocyclic compounds the rearrangement occurs via the singlet state but in all cases, substitution at the C-3 position is necessary for an efficient reaction. The rearrangement of such compounds appears to be concerted and is stereospecific at all centres with, in particular, retention of configuration at C-1 and C-5. Thus direct irradiation of the styrene derivatives (27)[20] and (28)[21] yields (29) and (30) respectively. The reaction is not only stereospecific but also regiospecific as illustrated by the exclusive conversion of (31) to (32). There are numerous reports of the di-π-methane rearrangement of 1,4-dienyl moieties and these may be present in a variety of molecular environments. The generality of the process is illustrated here by the photoreactions of compounds (33)-(37) and the synthetic applicability of this rearrangement is indicated by the formation of the sesquiterpene taylorione (38) from the 1,4-diene (39).[28] It must, however, be appreciated that a particular molecular feature may favour an alternative pathway and, for example, the di-π-methane reaction of the allene (40) to give (41) is the minor process and the 2π+2π cycloaddition to form (42) is preferred.[29]

hν 27 29 hν 28 30

hν 31 32 hν 33 38%

hν 34 Ref. 24 hν 35 Ref. 25

hν 36 Ref. 26

hν 37 Ref. 27

hν 39 38

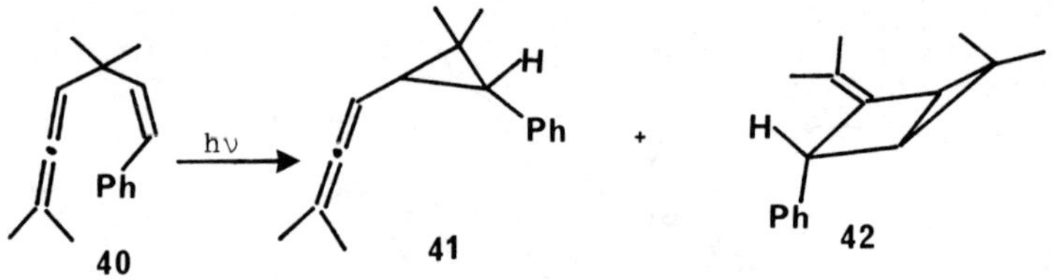

Triplet sensitisation of acyclic 1,4-dienes simply yields cis and trans isomers since the relaxation of this excited state to the non-planar triplet is rapid. For bicyclic compounds, however, because of the extra rigidity, an orthogonal triplet state is not formed and these systems undergo a non-concerted sensitised di-π-methane rearrangement. Thus acetone sensitisation of barrelene (43) produces the fluxional compound semibullvalene (44)[30] and similar irradiation of the tricyclic diene (45) yields the pentacyclic compound (46).[31]

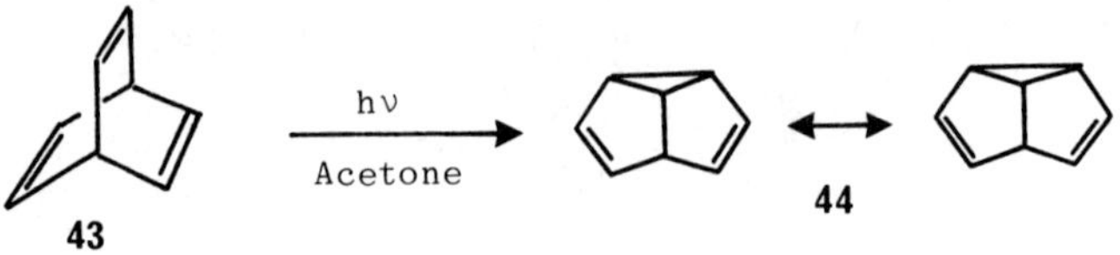

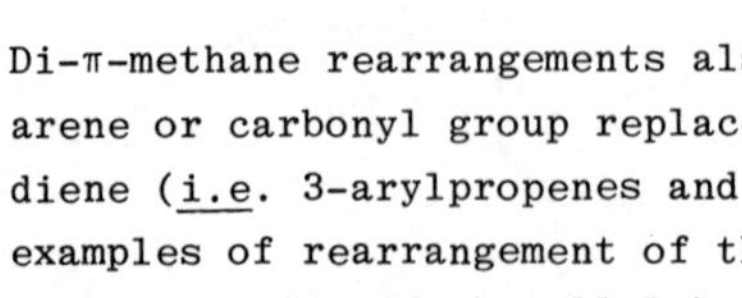

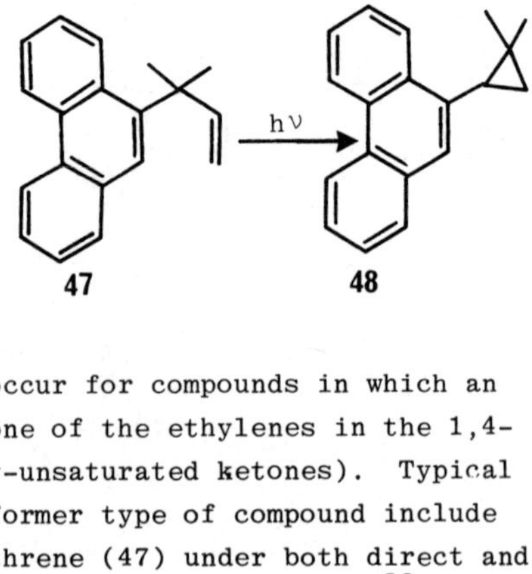

Di-π-methane rearrangements also occur for compounds in which an arene or carbonyl group replaces one of the ethylenes in the 1,4-diene (i.e. 3-arylpropenes and β,γ-unsaturated ketones). Typical examples of rearrangement of the former type of compound include the conversion of the allylphenanthrene (47) under both direct and sensitised irradiation to the cyclopropyl derivative (48)[32] and acetophenone sensitisation of (49) which produces (50) in 95% yield.[33] Under the latter conditions, bicyclo[2.2.2]octenones (18) undergo an oxadi-π-methane rearrangement to give high yields of tricyclo[3.3.0.0^{2,8}]oct-3-ones (20) (cf. direct irradiation of (18) to (20) in Section 2).[14] There are numerous published reports of the rearrangement of βγ-unsaturated ketones but the conversion of (18) to (20) is of considerable interest since it has been elegantly used as the key step for the enantiospecific total synthesis of cyclopentanoid natural products. Thus the syntheses of iridodial (51)[34] and (±) coriolin (52)[35] have been achieved using the photoconversions of (53) and (54) respectively as the route to produce the essential tricyclo[3.3.0.0^{2,8}]octan-3-one intermediate.

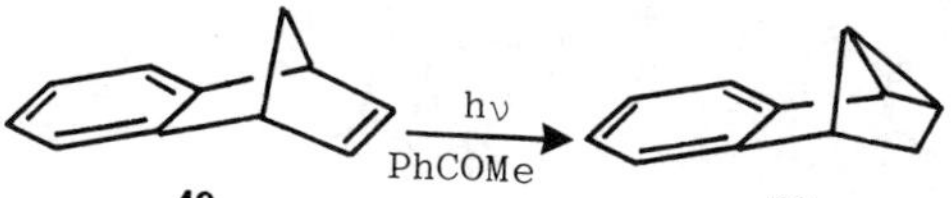

hν
PhCOMe
49
50 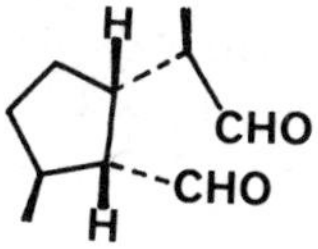
H
CHO
CHO
H
51 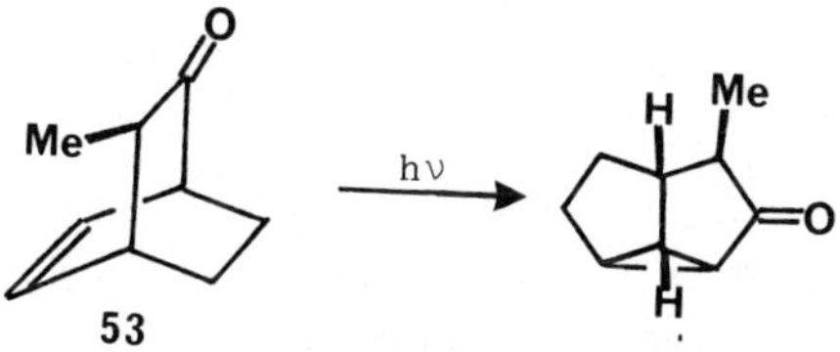
Me
O
53
hν
H Me
O
H
Me
Me
O
Me
O
54
hν
O H Me
Me
Me
4
O
H
C-4 epimers
HO
H Me
O
Me
O
Me
O
H
52

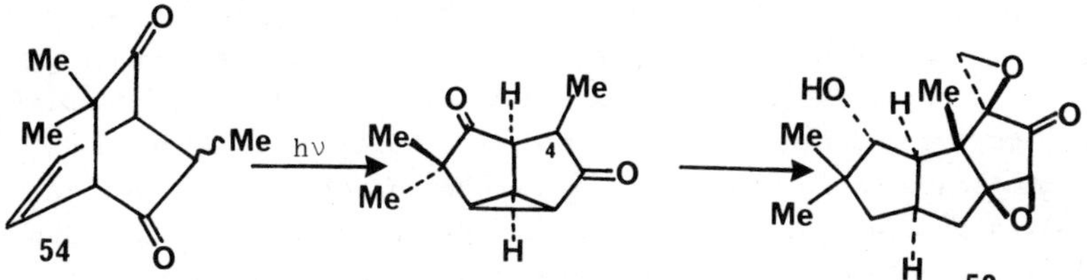

Formation of ethenylcyclopropane systems also arises on irradiation of dienone systems. Details of this process are given in a later Chapter but it is relevant to note here that these rearrangements are of considerable synthetic applicability and have been used as a key step for spiro annulation and (6,6) to (5,7) and (6,5) to (5,6) ring transformations. Examples of the use of the reaction for these purposes are given below for the synthesis of β-vetivone (55),[36] grayanotoxin II (56),[37] and α-cardinol (57).[38]

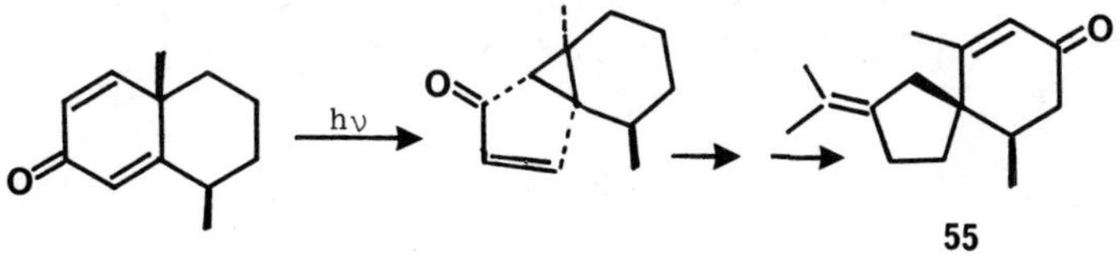

O
hν
O
O
55

hν
AcOH

56

hν
AcOH

57

4. Electrocyclic Reactions

These reactions are those pericyclic processes (i.e. concerted and cyclic) in which cyclisation or ring-opening occurs. It is thus evident that the relationship of the stereochemistries of the product and reactant is determined by whether the ring opening or closure reaction involves a conrotatory or disrotatory mode. Which of the two reaction modes will operate in particular cases may be deduced on the basis of the Woodward-Hoffmann rules.[8] For photochemical processes if the number of electrons which is involved in the ring-closure or ring opening reaction is 4n then a disrotatory mode will be observed and if the number of such electrons is (4n+2) then the process will occur by a conrotatory mode. The reverse situation is encountered for the thermally induced processes. Thus photochemically the 1,3-diene (58) yields the cis disubstituted cyclobutene (59). The ring opening reaction of cyclic alkenes photochemically has been little studied largely because of their short wavelength absorption but the corresponding

hν
Disrotatory

Thermal
Conrotatory

58 59 60

thermal process to yield the diene isomer (60) by a conrotatory mode is well known. The diene photoclosure reaction is a very useful method for the synthesis of cyclobutenes and is also a recommended procedure for the parent compound. Typical examples of this process are the conversions of (61)[39] and (62)[40] into (63) and (64) respectively and the corresponding reactions of (65)[41] and (66)[42] illustrate the versatility of the reaction and its applicability to heterocyclic and polycyclic systems.

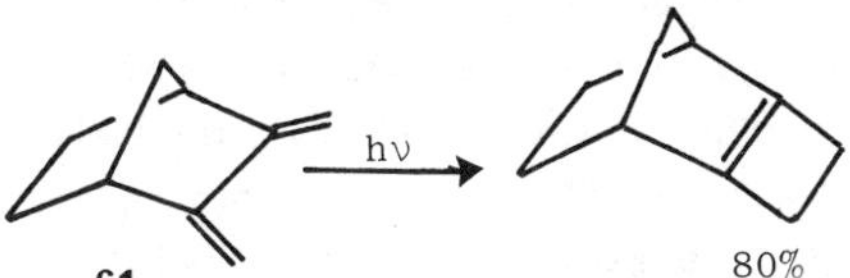

hν 61 → 63 80%; 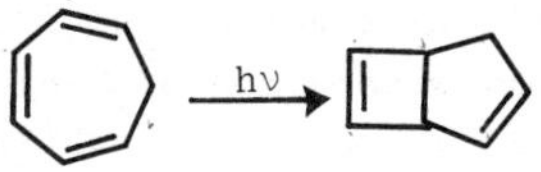 62 hν → 64

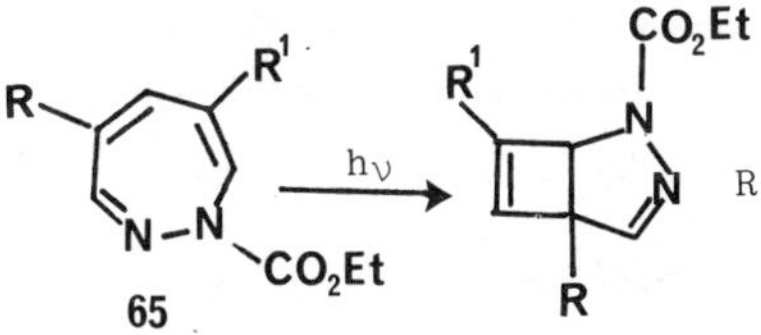

65 hν → R' = H,Me,OMe, NHAc,NEt_2

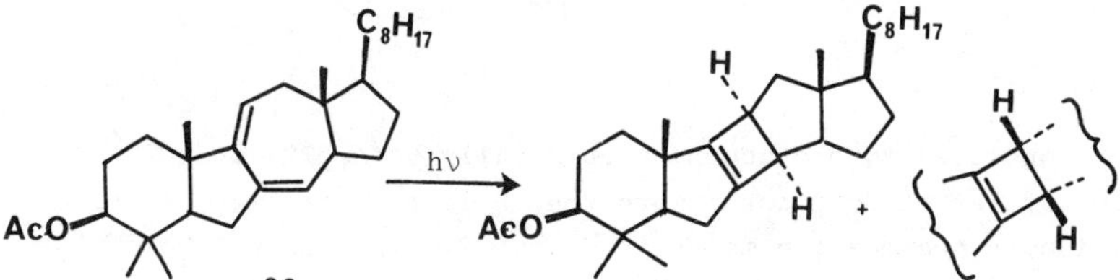

66 hν →

The electrocyclic interconversion between cyclohexa-1,3-dienes and hexa-1,3,5-trienes is readily achieved photochemically and consistent with the above rules occurs in a conrotatory manner. Thus the 254 nm radiation of the optically active dienes (67) and (68) respectively produces the trienes (69) and (70) with retention of chirality.[43] The reverse process occurs with 300 nm radiation and again involves no loss of optical activity. The use of different wavelengths to induce the ring closure and ring opening processes of isomers is a common feature in the photochemistry of these compounds and further, can be employed to direct the reaction to a required product. For example, from 254 nm irradiation of bicyclo[4.3.0]nona-2,4-diene (71), initially the major product is the triene (72) and the minor product the cyclobutene (73).[44] This situation is explained by the higher absorptivity of the diene than the triene at this wavelength and so the ring opening reaction is more favoured than the reverse process. With 300 nm radiation, however, the triene (72) now has the higher absorptivity and is readily converted to the diene

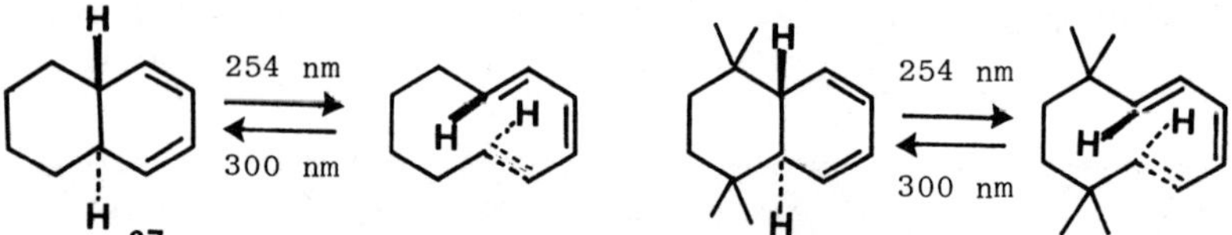

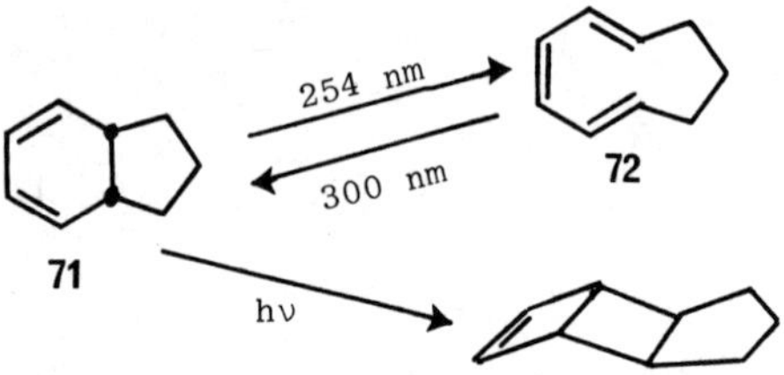

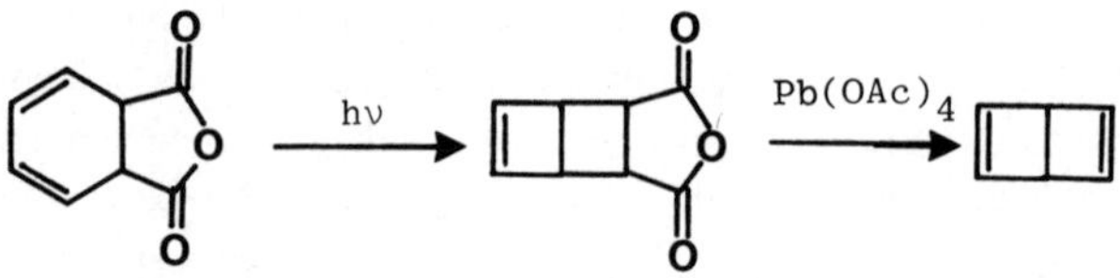

which continues to produce photochemically both (72) and (73). Since the latter does not absorb the incident radiation, in time this isomer becomes the major product. Cyclisation to the cyclobutene may also be preferred for some cyclohexa-1,3-dienes which on ring opening would yield a severely strained product. This feature has been used in a synthesis of Dewar benzene from the diene (74). On irradiation (74) yields (75) rather than the triene and treatment of (75) with lead tetraacetate produces the Dewar isomer.[45]

Despite the above features which favour production of cyclobutenes from some cyclohexa-1,3-dienes, the ring opening reaction to the triene is observed in many systems and forms the basis of the best known examples of the application of photochemistry to fine chemical manufacture: this involves the synthesis of vitamin D_3 (76). The key reaction involves the ring opening of cholesta-5,7-dien-3-ol (77) photochemically to give pre-vitamin D_3 (78). The photo process is followed by a 1,7-hydrogen shift at 50-80°C to give (76). As a result of the absorption features of dienes and trienes mentioned above, it is necessary to stop the reaction at 50% conversion and isolate the pre-vitamin D_3 otherwise cyclisation of (78) to (79) becomes increasingly significant.[46]

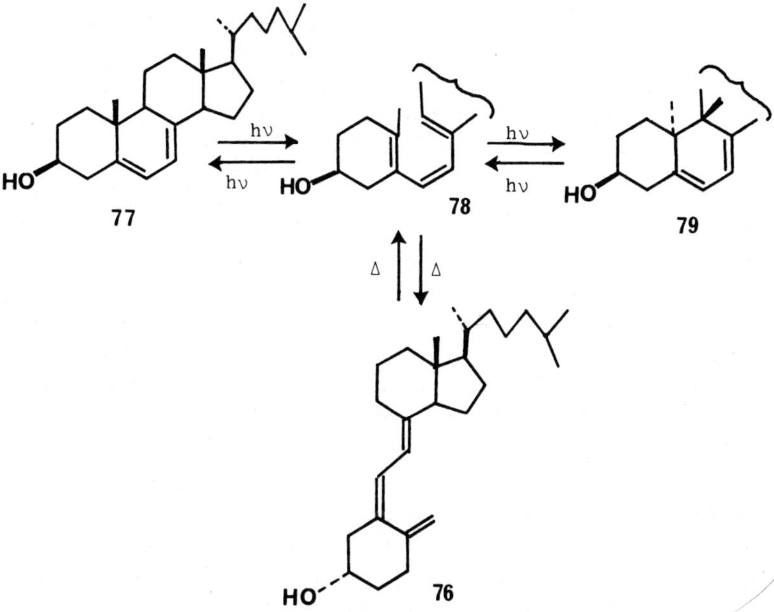

As noted above, the photocyclisation of hexa-1,3,5-trienes can be a facile process. The reaction also occurs with 1,2-diaryl ethylenes and this forms the basis of the photochemical synthesis of polynuclear aromatic compounds. The essential features of the reaction are the photoisomerisation of the trans 1,2-diaryl ethylene to the cis isomer (see Section 1) followed by conrotatory photocyclisation to give the trans 4a,4b-dihydrophenanthrene intermediate (see Scheme 1). In the presence of oxygen or iodine the intermediate is rapidly converted to the polynuclear arene.[47] The reaction is widely applicable to a variety of systems. The examples of the conversions of the α- and β-naphthyl compounds (80) and (81) into chrysenes and benzo[c]phenanthrenes with yields of

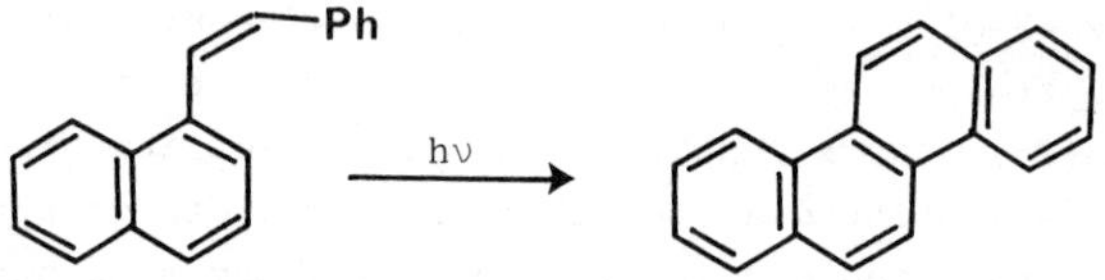

Ph

hν

80

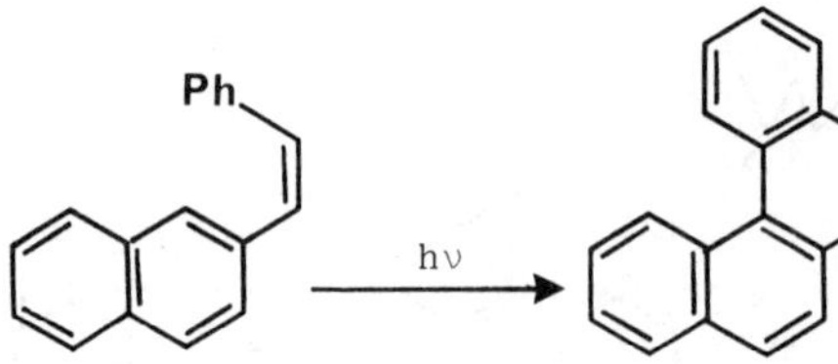

Ph

hν

81

66-89%[48] illustrate the synthetic potential of the process. This cyclisation has been used by several workers as a route to helicenes,[47] and the dimethylhexahelicene produced from the dimethylstyryl arene (82) has been resolved and on further elaboration provides a route to an optically active [2.2]paracyclophanohexahelicene.[49] For systems in which the 1,2-diaryl ethylene unit is held in a cis geometry by the molecular structure, the photocyclisation to polynuclear arenes is common and facile. For example pyrrolinones (83) give 55-70% yields of the phenanthropyrrolinones (84).[50] The reaction occurs with a wide variety of different types of arenes and for those which have methoxy substituents at the site(s) of cyclisation, the elements of methanol are eliminated in the photoprocess and the reaction still occurs with reasonable yields of products. These aspects of the process are respectively exemplified by the conversion of 1,2 bis

82

83

$R = CO_2Et$, OCH_2Ph, OEt, $OCHMe_2$

84

85

(pyrazyl)ethylene to 1,4,5,8-tetra-azaphenanthrene (85) in 90% yield,[51] and the formation of 43% of the phenanthrene (86) from irradiation of the stilbene (87).[52] The purpose of the latter process was to produce a phenanthrene suitable for elaboration to 4-hydroxypiloquinone (88).

87

86 $R = CO_2Me$

88 $R = COCH_2CH_2CHMe_2$

References

1. For a fuller account of photoreactions important in biological systems see "The Science of Photobiology", Ed. K.C. Smith, Plenum Press, New York, 1977.

2. G.Beddard in "Light, Chemical Change and Life", The Open University Press, Milton Keynes, 1982, Chapter 5.3.

3. M.Comtet, J.Amer.Chem.Soc., 1969, 91, 7761.

4. J.Bergman, K.Osaki, G.M.J.Schmidt, and F.I.Sonntag, J.Chem.Soc., 1964, 2021.

5. J.S.Swenton, J.Org.Chem., 1969, 34, 3217.

6. For a review of such processes see P.J.Kropp, Organic Photochemistry, 1979, 4, 7.

7. I.T.Harrison and B.Lythgoe, J.Chem.Soc., 1958, 837.

8. For a full discussion of this topic see I.Fleming "Frontier Orbitals and Organic Chemical Reactions", Wiley Interscience, London, 1976.

9. Reference 8, p. 99.

10. R.C.Cookson, J.Hudec, and M.Sharma, J.Chem.Soc.Chem.Comm., 1971, 107 and 108.

11. R.C.Cookson, Quart.Rev., 1968, 22, 423.

12. J.V.Drayton, W.T.Flowers, R.N.Haszeldine, and T.A.Parry, J.Chem.Soc.Chem.Comm., 1976, 490.

13. A.Sanchez, M.Martinez, and Rodriguez-Hahn, Tetrahedron, 1984, 40, 1005.

14. M.Demuth, P.R.Raghavan, C.Carter, K.Nakano, and K.Schaffner, Helv.Chim.Acta, 1980, 63, 2434.

15. R.F.C.Brown, R.C.Cookson, and J.Hudec, Tetrahedron, 1968, 24, 3955.

16. R.C.Cookson, D.E.Sadler and K.Salisbury, J.Chem.Soc.Perkin I, 1981, 774.

17. J.M.Hornback and R.D.Barrow, J.Org.Chem., 1982, 47, 4285.

18. A.R.Bremberg, V.C.Freestone, A.A.Gorman and J.B.Sheridan, Tetrahedron, 1979, 35, 2311.

19. For a detailed discussion of the mechanistic aspects of this rearrangement see H.E.Zimmerman, Acc.Chem.Res., 1982, 15, 312.

20. H.E.Zimmerman and A.C.Pratt, J.Amer.Chem.Soc., 1970, 92, 6267.

21. H.E.Zimmerman, P.Baeckstrom, T.Johnson and D.W.Kurtz, J.Amer.Chem.Soc., 1972, 94, 5504.

22. H.E.Zimmerman and A.A.Baum, J.Amer.Chem.Soc., 1971, 93, 3646.

23. P.S.Mariano and J.Ko, J.Amer.Chem.Soc., 1972, 94, 1766.

24. T.Sasaki, S.Eguchi, M.Ohno and T.Umemura, Tetrahedron Letters, 1970, 3895.

25. H.E.Zimmerman, P.Hackett, D.F.Juers, J.M.McCall and B.Schroder, J.Amer.Chem.Soc., 1971, 93, 3653.

26. P.S.Mariano and R.B.Steitle, J.Amer.Chem.Soc., 1973,95,6114.

27. J.Perreten, D.M.Chihal, G.W.Griffin and N.S.Bhacca, Tetrahedron Letters, 1973, 95, 3427.

28. D.I.Schuster and B.R.Sckolnick, Org.Photo Syn., 1971, 1, 39.

29. D.C.Lankin, D.M.Chihal, G.W.Griffin and N.S.Bhacca, Tetrahedron Letters, 1973, 4009.

30. H.E.Zimmerman and G.L.Grunewald, J.Amer.Chem.Soc., 1966, 88, 183.

31. L.A.Paquette, J.D.Kramer, P.B.Lavrik and M.J.Wyvratt, J.Org.Chem., 1977, 42, 503.

32. H.E.Zimmerman and G-S.Wu, Canad.J.Chem., 1983, 61, 866.

33. J.R.Edman, J.Amer.Chem.Soc., 1966, 88, 3454.

34. P.Ritterskamp, M.Demuth, and K.Schaffner, J.Org.Chem., 1984, 49, 1155.

35. M.Demuth, P.Ritterskamp and K.Schaffner, Helv.Chim.Acta, 1984, 67, 2023.

36. D.Caine, S.L.Graham and T.T.Vora, J.Org.Chem., 1980, 45, 3798.

37. S.Gasa, N.Hamanaka, S.Matsunaga, T.Okuno, N.Takeda and T.Matsumoto, Tetrahedron Letters, 1976, 553.

38. D.Caine and A.S.Frobese, Tetrahedron Letters, 1977, 3107.

39. D.H.Aue and R.N.Reynolds, J.Amer.Chem.Soc., 1973, 95, 2027.

40. R.Srinivasan, J.Amer.Chem.Soc., 1962, 84, 3432.

41. T.Tsuchiya, J.Kurita and A.Kojima, J.Chem.Soc.Chem.Comm., 1980, 444.

42. J.J.Jagodzinski and R.R.Sicinski, Tetrahedron Letters, 1981, 22, 3901.

43. B.Matuszewski, A.W.Burgstahler and R.S.Givens, J.Amer.Chem.Soc., 1982, 104, 6874.

44. W.G.Dauben and M.S.Kellogg, J.Amer.Chem.Soc., 1971, 93, 3805.

45. E.E. van Tamelen, S.P.Pappas and K.Kirk, J.Amer.Chem.Soc., 1971, 93, 6092.

46. W.G.Dauben and R.B.Phillips, J.Amer.Chem.Soc., 1982, 104, 5780.

47. For authoritative reviews on this cylisation see W.H.Laarhoven, Recl.J.R.Neth.Chem.Soc., 1983, 102, 185,241; F.B.Mallory and C.W.Mallory, Org.React. (N.Y.), 1984, 30, 1.

48. D.L.Nagel, R.Kupper, K.Antonson and L.Wallcave, J.Org.Chem., 1977, 42, 3626.

49. M.Nakazaki, K.Yamamoto and N.Maeda, J.Org.Chem., 1980, 45, 1985.

50. K.Yakushijin, M.Kozuka and H.Furukawa, Chem.Pharm.Bull., 1980, 28, 2178.

51. S.C.Shim and S.K.Lee, Synthesis, 1980, 116, and Bull.Korean Chem.Soc., 1980, 1, 68.

52. R.G.F.Giles, P.R.K.Mitchell and M.V.Sargent, J.Chem.Soc.Perkin I, 1983, 2147.

9
Alkenes: Cycloaddition

By P.A. Wender
DEPARTMENT OF CHEMISTRY, STANFORD UNIVERSITY, STANFORD, CALIFORNIA 94305, USA

INTRODUCTION

The photoinduced [2+2] cycloaddition of alkenes represents a potent strategy level reaction for organic synthesis. Its value derives in part from the element of convergency and stereochemical control which it imparts to a synthesis plan, allowing for the conjunction of two alkenes to produce a cyclobutane product with up to four stereogenic centers (eq 1). Since its discovery

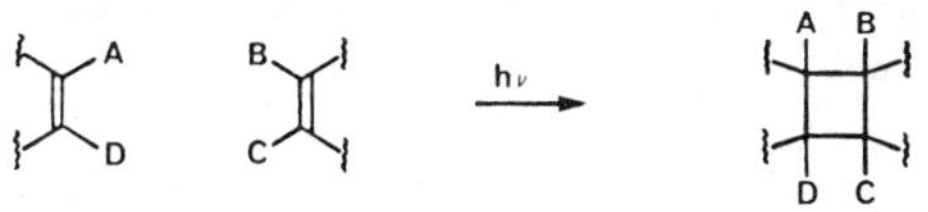

(1)

in the early part of this century,[1] this reaction has found increasingly frequent and effective use in complex molecule synthesis as chronicled with proportionate frequency in numerous publications and reviews on the subject.[2] It is the objective of this chapter to provide an instructive overview of this reaction class with an emphasis on its use in synthesis.

DISCOVERY AND EVOLUTION

The discovery of the photo[2+2]cycloaddition occurred in the early 1900's and is attributable to a group of Italian chemists led by Ciamician.[1] Taking advantage of their Mediterranean climate, these investigators used sunlight as their light source to induce the transformation of various organic compounds, most notably carvone, but at the time they were not able to establish unequivocally the photoproduct structures. A half century later Büchi and coworkers, coincidentally from another geographical region known for its abundant solar radiation, reported that exposure of carvone to California sunlight provided a product which they were able to characterize conclusively as the [2+2] cycloadduct, photocarvone (eq 2).[3] Between these events,

(2)

photoinduced cycloaddition reactions of other pi-systems were discovered including the photodimerization of stilbene and pyrone derivatives, heralding the generality of this reaction. In the late fifties and early sixties, contributions appeared in rapid succession from several laboratories including those of Zavarin, Cookson, Yates, Dauben, Hammond, Schenck, de Mayo, Eaton, Corey, Chapman, and Zimmerman.[2] By the mid-sixties, this reaction had captured the attention of a range of scientists interested in its theoretical, mechanistic, biological and synthetic ramifications. In the past two decades, the reaction has evolved into a versatile and practical tool for organic synthesis and a subject for basic and applied research.

GENERAL CONSIDERATIONS

In the context of this analysis, the study and implementation of this reaction in synthesis, three considerations inevitably arise. How might the reaction be used? What is its mechanism(s)? What other reactions might compete with the cycloaddition? These queries will be addressed briefly below and their consequences explored in more detail in the following section.

The applications of [2+2] photocycloaddition extend significantly beyond its obvious value in the synthesis of cyclobutane derivatives. As catalogued in Scheme I, the photoadduct can undergo a) ring expansion, b) ring contraction, c) one bond cleavage, or d) two bond cleavage reactions. Through these processes, rings of 3 members and larger are accessible along with acyclic systems (vide infra). Moreover, while these possibilities are illustrated only for carbon networks, it is important to recognize that photo-induced [2+2] cycloadditions can be accomplished with heteroatomic pi-systems, most notably involving oxygen,[4] nitrogen[5] and sulfur[6] atoms (eqs 3-5). Consequently, access to a range of heterocyclic and carbocyclic as well as acyclic systems is possible.

Scheme I

expansion
a
b
contraction
2-bond cleavage
d
c
1-bond cleavage

(3)

(4)

(5)

The mechanism of the photo[2+2]cycloaddition is still under investigation. However, from its study one can glean several important working hypotheses. As illustrated in Scheme II, direct excitation of an alkene (1) generally produces a singlet excited state which can return to the ground state singlet (-1), intersystem cross to a triplet state (2), or react with a ground state alkene to produce an exciplex (3) or cycloadduct _via_ a concerted path (4). The exciplex can proceed to product, ground state alkenes or a biradical intermediate which has a similar fate. When the triplet state is populated through direct excitation and intersystem crossing (1, 2) or through triplet sensitization (15), cycloaddition is possible _via_ an exciplex (9), _via_ a diradical derived from this exciplex (11) or directly from interaction between triplet and ground state alkenes (not shown). The mechanism for a given cycloaddition will then be a function of the specific substrates and reaction conditions

Scheme II

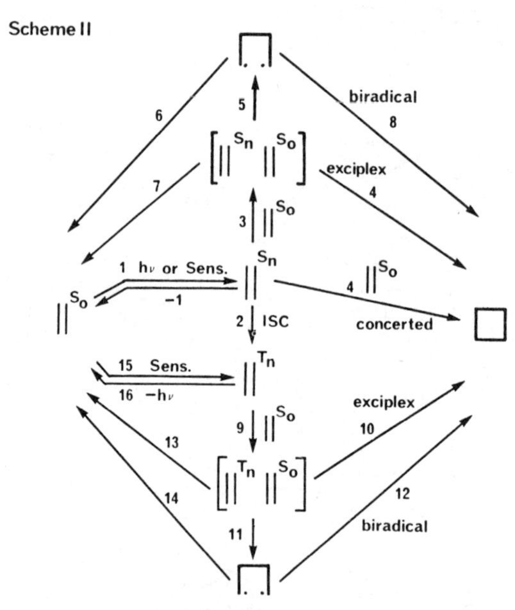

employed. It follows from this analysis that the majority of intermolecular cycloaddition reactions proceed via a triplet excited state. Three factors contribute to this situation. The singlet excited state is short-lived relative to the triplet state. For many systems, intersystem crossing to the triplet state is fast and efficient. Finally, sensitization can be used to populate the triplet state directly.

The third general consideration, the identification of alternative modes of reaction, is an issue which must be weighed along with use (Scheme I) and mechanism (Scheme II). Because of the richness of alkene photochemistry, this analysis is the one which frustrates the newcomer and worries even the experienced photochemist. For this reason, it is appropriate to examine those reactions which most frequently complicate or prevent photo[2+2]cycloaddition. As illustrated with specific examples[7] in Scheme III, these reactions are numerous; however, through a consideration of their mechanisms and that of the cycloaddition (Scheme II), most of these potentially competing reactions may be suppressed or avoided. In general, control over these processes can be achieved through changes in the reaction conditions, selective excitation, and/or modest modification of the substrate.

Scheme III

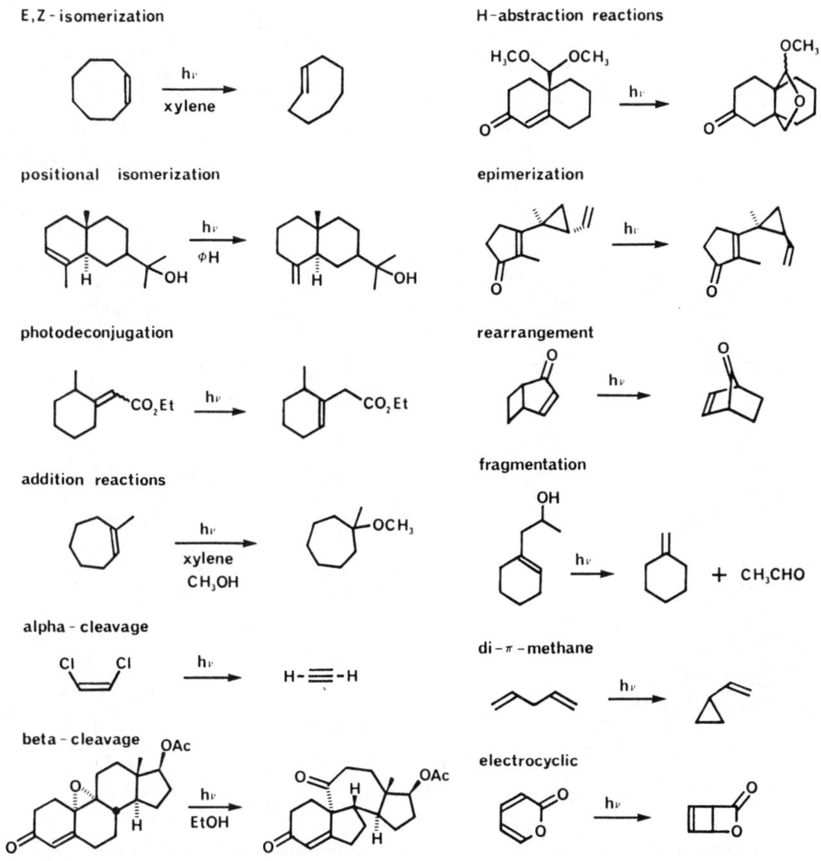
E,Z-isomerization
hν
xylene
H-abstraction reactions
H₃CO
OCH₃
hν
OCH₃
O
positional isomerization
hν
ϕH
H
OH
epimerization
photodeconjugation
CO₂Et
rearrangement
addition reactions
xylene
CH₃OH
OCH₃
fragmentation
OH
+ CH₃CHO
alpha-cleavage
Cl
H-≡-H
di-π-methane
beta-cleavage
OAc
EtOH
electrocyclic

SYNTHETIC APPLICATIONS

Retrospective analysis of the past 25 years reveals that the photo[2+2]cycloaddition has been a spectacularly successful tool for organic synthesis. In Scheme IV are illustrated only a few of the many compounds which have been synthesized through the use of this strategy level reaction during this time.[8-33] In this section, we will analyze the factors contributing to the synthetic success of this reaction, including its mode, regio- and stereoselectivity.

Scheme IV

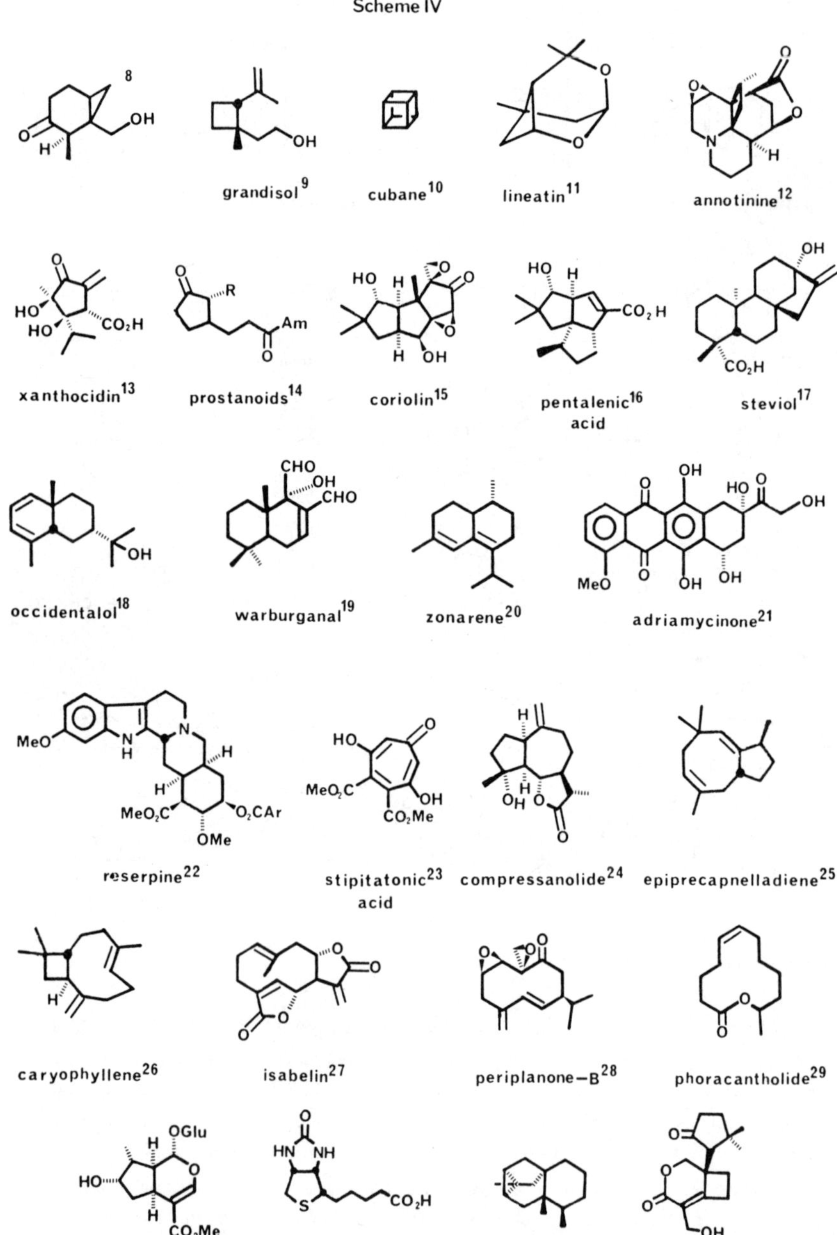
8
OH
O
H
grandisol[9]
OH
cubane[10]
lineatin[11]
O
annotinine[12]
N
H
xanthocidin[13]
HO
CO₂H
prostanoids[14]
R
Am
coriolin[15]
HO H
H OH
pentalenic[16] acid
CO₂H
steviol[17]
OH
CO₂H
occidentalol[18]
warburganal[19]
CHO
zonarene[20]
adriamycinone[21]
MeO
reserpine[22]
MeO₂C
O₂CAr
OMe
stipitatonic[23] acid
MeO₂C
CO₂Me
compressanolide[24]
epiprecapnelladiene[25]
caryophyllene[26]
isabelin[27]
periplanone–B[28]
phoracantholide[29]
loganin[30]
OGlu
CO₂Me
biotin[31]
HN NH
S
CO₂H
ishwarane[32]
fomannosin[33]

Simple Alkenes. Isolated pi-systems generally absorb in the high energy end of the UV region. This property has been exploited in their use as ground state components in photo[2+2]-cycloadditions, where selective excitation of a second alkene, say, an enone, is desired. Numerous mechanistic studies on these systems have been reported which serve as a useful introduction to the subject and a starting point for our analysis. Direct excitation ($\lambda \sim 230$ nm) of Z-2-butene gives at low conversion only cis-anti-cis and cis-syn-cis products (eq 6).[34] Similar reaction of E-2-butene provides only cis-anti-cis and all trans products. These results are in accord with the mechanism outlined in Scheme II involving a concerted cycloaddition (path 4) from an initially formed singlet state alkene or reaction *via* an exciplex (path 8) derived from the singlet. These cases illustrate that the stereochemistry of the alkene can be conserved for singlet state cycloadditions when the concentration of trapping agent is sufficiently high to intercept the short-lived excited state species. Ordinarily, however, unimolecular E-Z-isomerization favorably competes with the bimolecular cycloaddition reaction.

(6)

Sensitized excitation can be used to produce an alkene triplet state which, like the excited singlet state, can dissipate energy by E-Z isomerization. However, when the alkene is part of a 3- to 7-membered ring, relatively efficient triplet sensitized cycloaddition can be realized (eqs 7, 8).[35,31] With larger rings, configurational isomerization is again a competing reaction; for example, Z-cyclooctene undergoes photoequilibration with its E-isomer.[36] Nevertheless, the cycloadducts derived from 3- to 7-membered ring precursors service a range of the most commonly encountered problems in synthesis. Problems

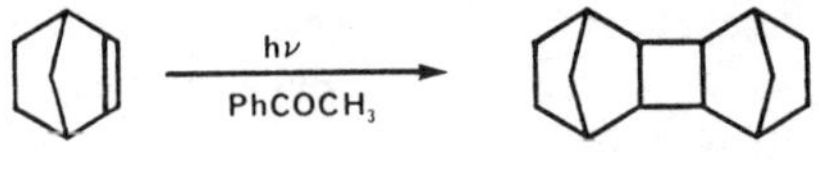

(7)

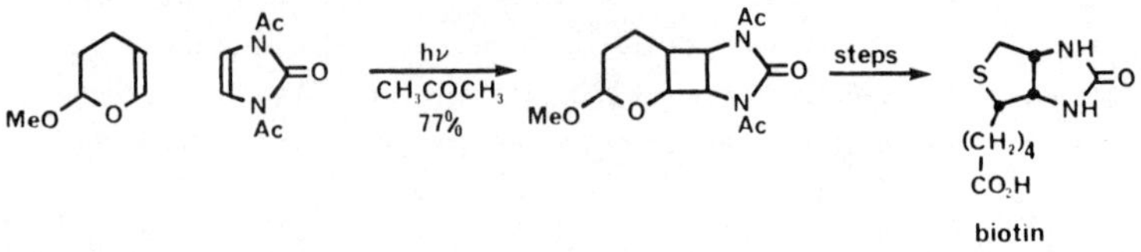

(8)

biotin

do arise in using this process for crossed-cycloadditions since dimerization is generally a competing process. Changes in reactant concentration can be employed to ameliorate this situation.[31]

The intramolecular version of this reaction provides a better means for controlling stereo- and regiochemistry in the cycloaddition and "crossed" cycloaddition.[2m,27] Particularly noteworthy is the recent demonstration that copper (I) catalysis can be used to improve the efficiency and enhance or reverse the stereoselectivity of the cycloaddition.[38] For example, bis allylic ethers are converted in generally high yield to the 3-oxabicyclo[3.2.0]heptanes _via_ a bidentate copper complex (eq 9). Moreover, in certain cases allylic alcohols have been converted to the _endo_ cycloadduct, presumably arising from a tridentate complex. This process has recently been utilized in a novel synthesis of grandisol (eq 10).[9] Of further note is its utility in crossed cycloadditions where selective coordination of different alkenes to a common copper template is used to preorganize reactant alkenes (eq 11).[39]

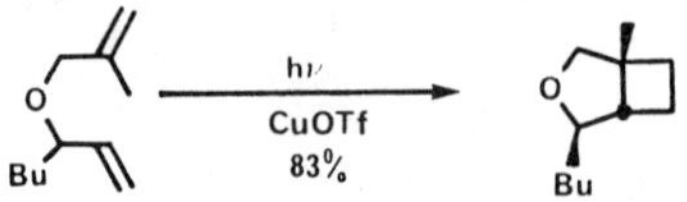

(9)

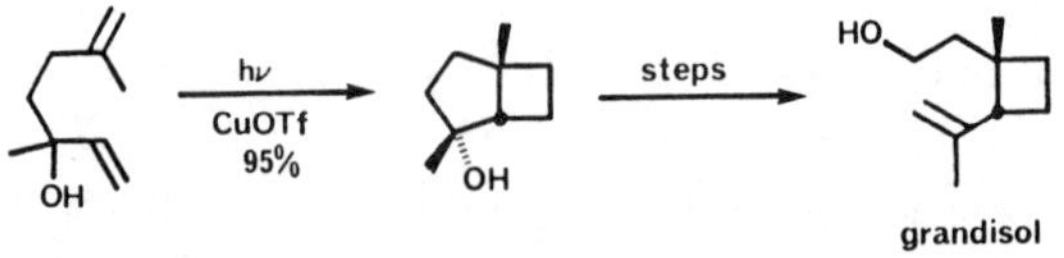

(10)

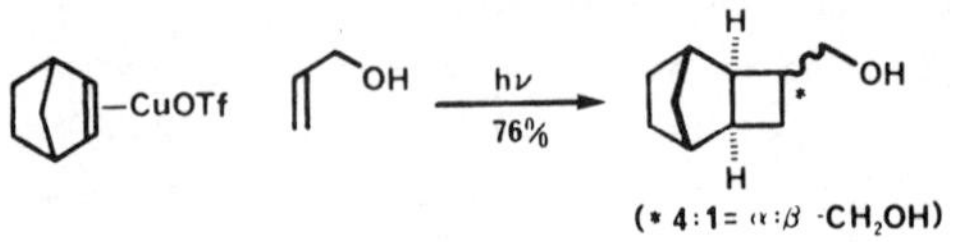

grandisol

(11)

(* 4:1 = α:β -CH_2OH)

Dienes. These reactants are included in our analysis because their cycloadditions provide the basis for a relatively unexplored but potentially efficient method for eight-membered ring synthesis. For example, sensitized excitation of butadiene gives cis- and trans-1,2-divinylcyclobutanes as major products from which cycloocta-1,5-diene is derived upon thermolysis (eq 12).[41] Alkyl substitution of the dienes provides complex mixtures of divinyl cyclobutane products; for example, isoprene gives 7 dimers and piperylene affords at least 15 products.[41a-c,42] However, as seen above for simple alkenes, intramolecular trapping of the excited state diene by a tethered alkene (eq 13)[43] can be employed to enhance efficiency and selectivity. The preferential formation of a product derived from closure to a five-membered ring biradical intermediate vis-a-vis the six-membered ring biradical is generally observed, and is sometimes referred to as the rule of five.[44] Mechanistically, this result is attributed to a kinetic preference for formation of the five-membered ring, a selectivity commonly observed in intramolecular radical additions to alkenes.[45]

hν Sens. Δ (12)

hν Ph_2CO 85% (triplet) (13)

hν (singlet) + 83% 14%

However, caution must be exercised in the application of this rule since the preference can be overridden by stereoelectronic factors. Furthermore, product ring size preference is also influenced by the relative rates of closure of the initially formed diradical(s) to cycloadduct product and its collapse to starting material.[46] It is noteworthy in connection with concerns about mode selectivity that singlet state dienes generally undergo preferential electrocyclic closure to a cyclobutene product (eq 13).[40,43]

Recently, the first examples of intramolecular diene-diene cycloadditions have been achieved in our laboratories.[47] Our interest in this area was based on the value these reactions were expected to have in the synthesis of eight-membered rings and polyquinanes. Indeed, benzophenone sensitized excitation of the readily available tetraenes efficiently provides divinylcyclobutane intermediates which upon thermolysis afford the corresponding cyclooctadiene products (eq 14). Transannular closure of the eight-membered ring in the case illustrated provides an efficient route to the antitumor, antibiotic coriolin. The stereo- and regiochemical course of this cycloaddition is in accord with initial formation of a five-membered ring. The two allylic radical groups attached to the resulting ring then close to form the observed _cis_-fused product. The other stereoisomeric intermediates, if formed, would presumably fragment to starting material since closure in these cases would ultimately involve formation of relatively strained _trans_-fused product. The mode of stereoinduction by the preexisting stereogenic center is that expected on the basis of kinetic and thermodynamic considerations since the alkoxy group occupies a less sterically encumbered position in the proposed reactive conformer and the observed product.

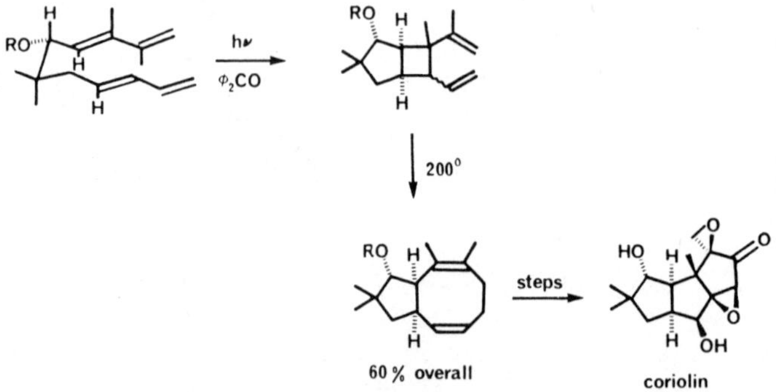

hν, ϕ_2CO

200°

steps

60 % overall

coriolin

(14)

While mechanistically unrelated to the photochemistry of dienes, it is of synthetic relevance to the foregoing analysis to note that naphthalenes, anthracenes and related aromatic compounds undergo efficient photodimerization through a 4+4 cycloaddition path to provide eight-membered ring products (eq 15).[48]

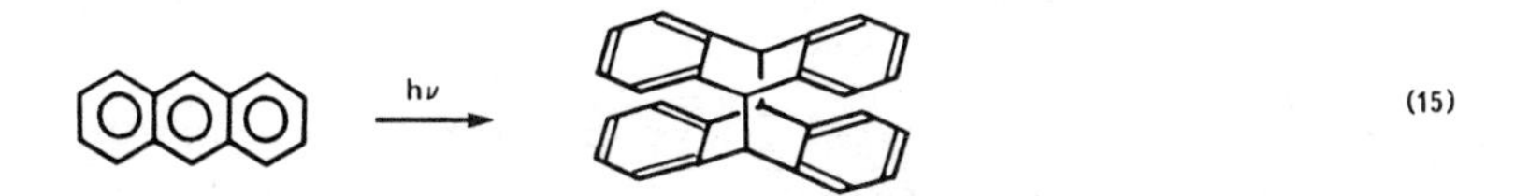

Alkene-Carbonyl. Attachment of a carbonyl group to an alkene significantly alters the photophysics and photochemistry of the latter functionality. This is particularly apparent with cyclic enones. In contrast to the simple alkenes noted above for which intersystem crossing from the singlet to the triplet excited state is relatively inefficient, cyclic enones undergo a rapid and efficient crossing thereby providing easy access to the triplet state through direct excitation. For cyclopentenones and cyclohexenones, the triplet state has a lifetime of ca. 10^{-8} - 10^{-9} seconds,[49] which is sufficiently long-lived for bimolecular reactions, including trapping by a ground state alkene to produce a cycloadduct. This cycloaddition reaction is proposed to involve a diradical intermediate formed directly or from an exciplex.[46] In either event, the diradical can fragment to starting materials or close to the cycloadduct. Because of the general value of this process in synthesis and its relevancy to photocycloadditions in general, a brief analysis of its regioselectivity and stereoselectivity is presented below.

The regioselectivity of intermolecular photo[2+2]cycloaddition is substrate dependent but generally high for reactions of excited state cyclopentenones and cyclohexenones with ground state allenes or strongly polarized alkenes. The selectivity is consistent with the intermediacy of an exciplex in which alkene and enone dipoles are aligned for optimal electrostatic interaction.[50] Based on the assumption that the enone excited state has a polarization which is complementary to its ground state, it is expected and indeed observed (eq 16)[50-52] that strongly polarized ground state alkenes react to give preferentially one cycloadduct regioisomer. In further support of this

$O^{\delta+}$ δ^- triplet; δ^- δ^+ OEt → 91% (OEt); MeO_2C δ^- δ^+ → 80% (CO_2Me, H) (16)

analysis, weakly polarized alkenes give mixtures of regioisomers (eq 17).[50,51,53] As expected from this model, solvent polarity also affects cycloaddition regioselectivity. For example, the photodimerization of cyclopentenone gives more of the head-to-tail dimer as solvent polarity is decreased (eq 18).[54] Steric

66 : 11 (17)

hν (18)

neat	43%	57%
.0 9 M/φH	15%	85%

factors also play a role in cycloaddition regioselectivity, sometimes accounting for a diminution or reversal of regioselectivity relative to that expected from the dipole-dipole model. Thus, while reaction of isobutylene with cyclopentenone gives a 2.6:1 mixture of head-to-tail and head-to-head cycloadducts, respectively, 2-trimethylsilylcyclopentenone gives only the head-to-tail adduct.[55] Finally, the stability of the diradical

R + hν → R + R (19)

R = H	2.6	1
R = $SiMe_3$	only	–

intermediates and their partitioning to cycloadduct, starting material and products derived from H-abstraction reactions (eq 20)[2,46] also contribute to the observed selectivity. While the

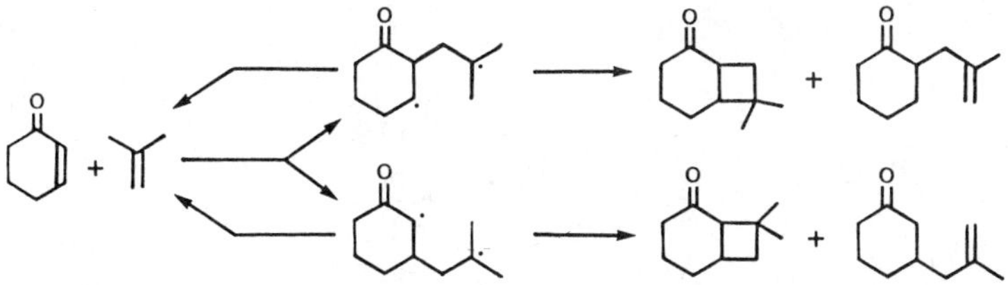

(20)

analysis of regioselectivity can obviously become quite complex, it is effectively treated in a majority of cases by application of the aforementioned excited state-ground state polarization model[50] (eq 16).

The stereochemistry observed for the photo[2+2]cycloaddition reactions of cyclopentenones and cyclohexenones is also consistent with the above model. Thus, the stereochemical integrity of the ground state acyclic alkene is lost in the cycloaddition reaction, presumably due to torsion in the relatively long-lived diradical intermediate. Consequently, acyclic E- or Z-alkene isomers generally produce the same product stereochemistry.[50] The stereochemical integrity of cyclic alkenes is preserved, however, in those cases where the ring-size inhibits E-Z isomerization. Similarly, the stereochemical integrity of the excited state alkenone is generally retained due to the torsional barrier to isomerization, although for some additions to cyclohexenones mixtures are obtained.[50] In most cases this latter result does not interfere with the synthetic utility of the reaction since *trans*-fused cycloadducts are generally efficiently converted to the *cis*-fused isomer under basic or acidic conditions.[50]

The *exo*-*endo* selectivity observed in the cycloaddition with cyclopentenones and cyclohexenones is variable but the exo cycloadduct isomer is generally favored with modest to high selectivity (*e.g.*: eqs 16, 18).

The final issue of general stereochemical importance is the influence of preexisting stereogenic centers on the stereochemical course of the cycloaddition. As exemplified in the syntheses of alpha- and beta-bourbonene[56] and acorenone[57] (eq 21 and 22), when the ground state alkene is facially

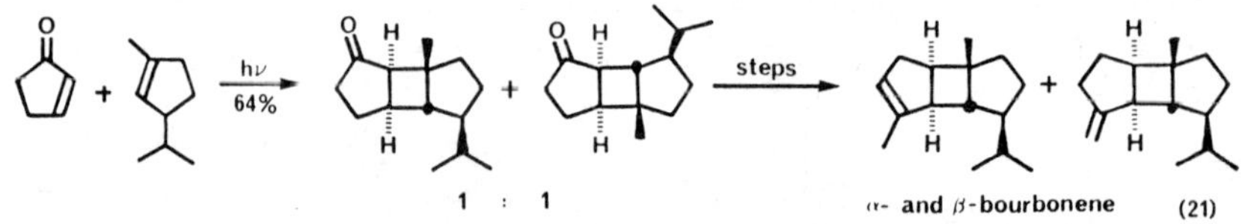

hν 64% steps

1 : 1

α- and β-bourbonene (21)

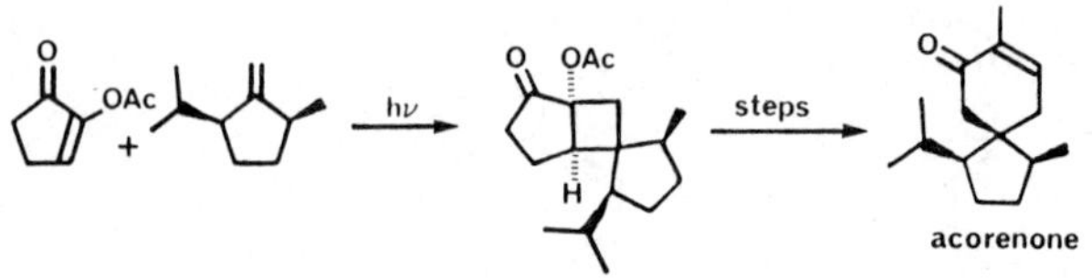

OAc hν steps

acorenone (22)

differentiated, addition of the excited state enone will occur preferentially to the less encumbered face of the alkene. The situation with the excited state component is a bit more complex but predictable in many cases. For facially differentiated enones, the stereochemical course of the cycloaddition can be analyzed in many cases by using a model empirically derived by Wiesner.[58] In essence, it is proposed that the beta-carbon of the enone excited state is pyramidalized in the stereochemistry-determining interaction with a ground state alkene. A similar situation arises in enone reductions with lithium in ammonia[59] and similar results are obtained. For example, in the reduction of octalones a hydrogen is added to the beta-carbon of the enone in a _trans_-relationship with respect to the angular group. The photoaddition of allene to these enones also occurs in a _trans_ fashion (eq 23).[58] Similarly, the conformationally more mobile monocyclic enone, 3,4-dimethylcyclohex-2-enone upon excitation in the presence of allene gives a product ratio (80:20), favoring the _trans_-dimethyl stereochemistry, not unlike that found in the metal ammonia reduction of this enone (eq 24).[61] However, results with 4-t-butylcyclohexenone[62a] and 4-i-propylcyclohexenone (eq 25)[26,62b,c] suggest that the stereochemistry

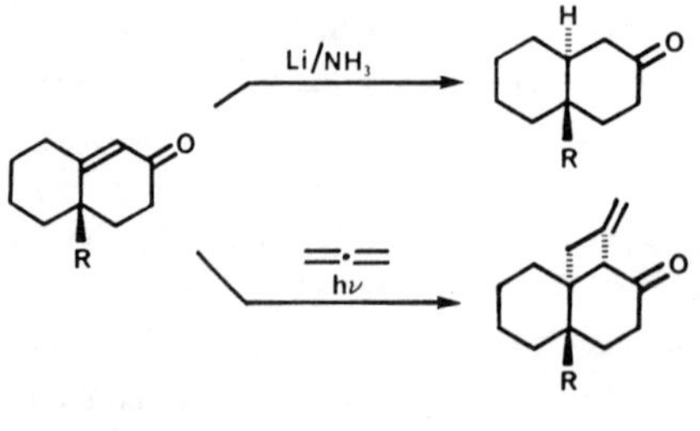

Li/NH_3

hν

(23)

(24)

T= -78° 80% 20%

T= 30 - 40° 50% 50%

(25)

ref. 62b 50% 50%

ref. 62c 34% 66%

periplanone B

of the cycloaddition might be better viewed as arising from a conformation in which the enone beta-carbon is trigonal and undergoes pyramidalization upon reaction with the ground state alkene. The planar excited state model more closely conforms to theory.[63] Limited data is available on 5-substituted cyclohexenones but mixtures are obtained and their composition consistent with a planar enone excited state (eq 26).[64] For 6-substituted cyclohexenones (eq 27),[65] again only limited data is available, but high stereoselectivity is observed which would be consistent in these cases with addition to an enone excited state conformation having a trigonal beta-carbon and a pseudoaxial C-6 group or a pyramidalized beta-carbon with a pseudoequatorial C-6 group. At present these models, which were derived primarily from studies on polycyclic systems, must be used with caution in new systems since the degree of beta-carbon pyramidalization

(26)

1. hν, =C(OEt)OEt
2. HCl, RT

CO_2Me

77% 3 : 5

steps

khusimone

(27)

hν

CO_2Me

95% at 63% conversioh

Δ 97%

steps

calameone

would be expected to vary with substrate variation. Moreover, product ratios certainly vary as a function of reaction conditions (for example, temperature: eq 24 and 25).[26,61,62b,c] Furthermore, in view of the limited information available at present, it is not clear whether product stereochemistry is controlled by approach of reactants, partitioning of isomeric intermediates or a combination of these factors.[46,2g] Further difficulties arise because of the limited data available on the sequence of bond formation involving the alpha and beta-carbons of the enone.[46]

Extension of our analysis to larger ring enones reveals, not surprisingly, that E-Z isomerization becomes a competing mode of reaction for the excited state triplet. For example, excitation of cycloheptenone or cyclooctenone in the presence of alkenes gives products derived from the E- isomer of the enone or its equivalent.[2b,50] It should be noted, however, that this competing mode of reaction is not as much related to ring size as to torsional flexibility. Excited states of large but conformationally rigid ring systems should, in principle, be sufficiently long-lived for cycloaddition. Moreover, even in somewhat more conformationally mobile rings, cycloaddition can be achieved through an intramolecular pathway. An interesting example is found in the irradiation of 3-butenyloxycyclooctenones which provide the standard cycloaddition products (eq 28).[66] It is noteworthy that the efficiency of this reaction is sensitive to alkene substitution, with phenyl and vinyl groups giving the best results.

hν (28)

The intramolecular photo[2+2]cycloaddition offers many advantages over its intermolecular counterpart.[2g] In general the excited state subunit is more efficiently trapped and constraints introduced by the intramolecular nature of the reaction frequently allow for greater control over the regio- and stereochemical course of the reaction. For example, while the reaction of isobutylene with 3-methylcyclohexenone gives a complex mixture and 2-methylcyclohexenone is a relatively poor substrate for cycloadditions, the intramolecular reaction involving these subunits (eq 29) proceeds in 77% yield and furnishes a single adduct, which is readily converted to isocomene in two further steps.[67] The regio- and stereochemical course of this reaction follows from the previous analysis.[2g,46] Thus, approach of the alkene to the excited state enone occurs on the less sterically encumbered enone face (away from the methyl group), presumably producing five-membered ring (rule of five) intermediate(s). Closure of the syn-1,4-diradical intermediate gives the observed product. The anti-1,4-diradical intermediate, if produced, could only fragment to starting material since closure would give a strained trans-fused product. Thus, the intramolecularity of the reactions serves to regulate stereoinduction, regioselectivity and endo-exo selectivity. This synthesis also illustrates one strategy for using the cyclobutane cycloadduct in the elaboration of five-membered rings.

hν, 77%; 2 steps; isocomene (29)

The versatility of this reaction is seen in other recent, representative studies (eqs 30-39)[68-77] involving syntheses of small, medium and large rings; fused, bridged and spirocyclic systems; and carbocycles and heterocycles. Further inspection of these and other cases again indicates the favored formation of the thermodynamically most stable product, suggesting that if a diradical is irreversibly formed, relative stereochemistry is set in the formation of the first bond through a transition state with product-like character. Alternatively, the diradical intermediates could be reversibly formed, and the subsequent closure reaction subject to product development control.

OR; O; hν; OR; O; H; steps; longifolene (30)

hν 80% steps (31)

*(3:1, β:α Me)

epi-bulnesene

hν 72% steps (32)

*(13:1, β:α Me)

pentalenic acid

hν 98% steps (33)

epi-precapnelladiene

hν 60% steps (34)

*1.5 : 1

hibiscone C

hν 68% steps (35)

isoamijiol

hν EtOAc 75% steps (36)

$\Delta^{9(12)}$-capnellene

hν 67% steps (37)

β-panasinsene

hν X=Cl 76% steps (38)

* 3.3:1 = β : α Me α-acoradiene

hν 78% steps (39)

phoracantholide

The evident value and potential of cycloalkenone cycloaddition reactions are enhanced considerably by the number and variety of other unsaturated carbonyl compounds and functionalized alkenes which can be used as excited state components in [2+2] cycloadditions. A partial listing of chromophores which have been shown to undergo the cycloaddition is given in Scheme V. Although limited data precludes a general analysis, the photochemistry of these systems follows the patterns seen with the cycloalkenones. Thus, ring size restrictions (7 or fewer members) are again encountered and, as noted previously, are attributable to the facile E-Z isomerization which occurs in

Scheme V

G = OH, OR, O_2CR, Cl, CN, CO_2H, CO_2R, Ph, R
n = 5 or 6 membered rings

X = O, NR

X = O, NR

G = CO_2H, CO_2R, CN, CHO, COR, NO_2

X = O, NR

R' ~ CO_2R

Ar ~ NO_2

larger rings. This limitation can be removed in many cases if the concentration of trapping agent is high or if isomerization is inhibited. For example, acyclic unsaturated esters undergo isomerization reactions quite readily but can be used in [2+2] cycloadditions when the reaction is conducted intramolecularly (eq 40).[78] Enolizable beta-dicarbonyl compounds, stabilized through H-bonding, are also effective chromophores for cycloaddition. This process, referred to as the de Mayo reaction,[2c,g,79] provides a very effective method for the addition of two differentiated carbon fragments across an unactivated double bond as illustrated in its application to the synthesis of loganin (eq 41).[30]

(40)

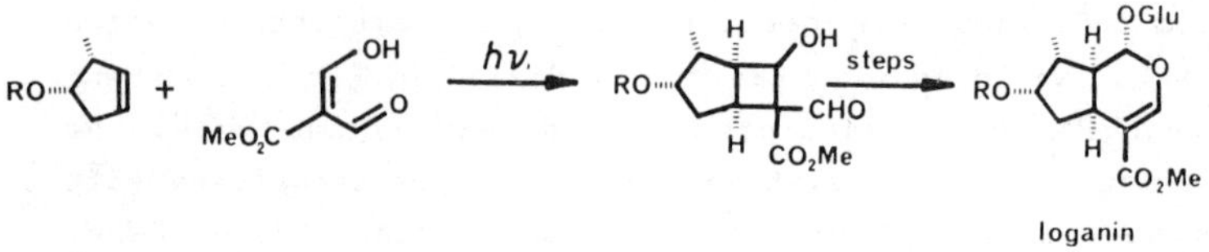

(41)

loganin

It is apparent from this overview that a wide range of excited and ground state components can be used in photo[2+2]cycloadditions. Considerable flexibility also exists for the further elaboration of the cyclobutane products. A treatment of this rich cyclobutane chemistry is beyond the scope of this chapter; however, a few general considerations are provided which bear on the options available. Elaboration of the cycloadduct of an ene-carbonyl cycloaddition is generally guided by two considerations: the cyclobutane ring is strained and at least two of its bonds are adjacent to a carbonyl group. Consequently, the carbonyl group or its derivatives can be used as an electron deficient site to induce strain-releasing migration or cleavage of an adjacent cyclobutane bond. For example, reaction of the major cycloadduct obtained from irradiation of 3-methyl-2-cyclohexenone and 4,4-dimethylcyclopentene with methyllithium gives a tertiary alcohol which upon solvolysis affords alpha-caryophyllene alcohol (eq 42).[80] Alternatively, under reducing conditions the carbonyl group or its derivatives can be used as a site of high electron density to induce fragmentation of a suitably aligned cyclobutane bond (eq 43).[65b]

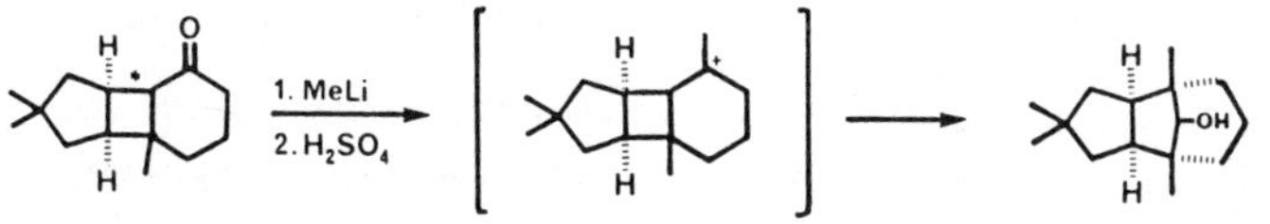

1. MeLi
2. H_2SO_4
(42)

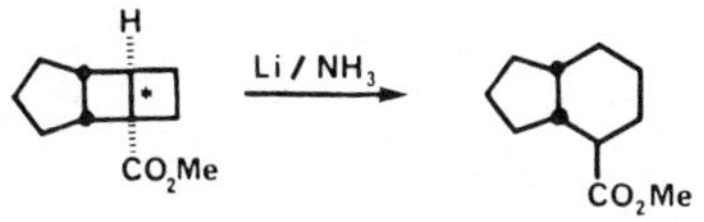

Li / NH_3
CO_2Me
CO_2Me
(43)

More heavily functionalized cycloadducts offer, when needed, greater control over the efficiency and regioselectivity of the cleavage or rearrangement reactions. This feature is operative in de Mayo reactions (eqs 30, 31, 33, 35, 39, 41) in which vicinally situated hydroxy or electron-rich groups and carbonyl groups are used to control uniquely the regiochemical course of a retro-aldol fragmentation. Reductive cleavage can also be regulated in this fashion (e.g.: eq 32, 38). Finally, the chemistry of the cyclobutane need not involve the original carbonyl group but can be based on functionality derived from the ground state alkene (eq 44)[24] or on the inherent properties of the cycloadduct (eqs 45, 46).[74,81] Alkyne cycloadducts (eq 34) or their equivalents (eq 47)[51,82] provide additional options.

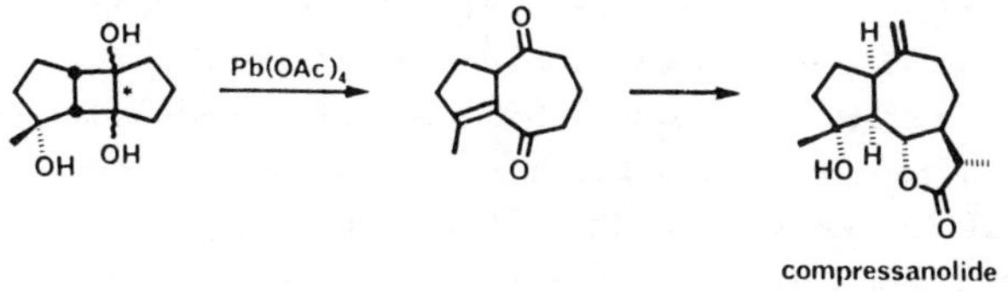

$Pb(OAc)_4$
compressanolide
(44)

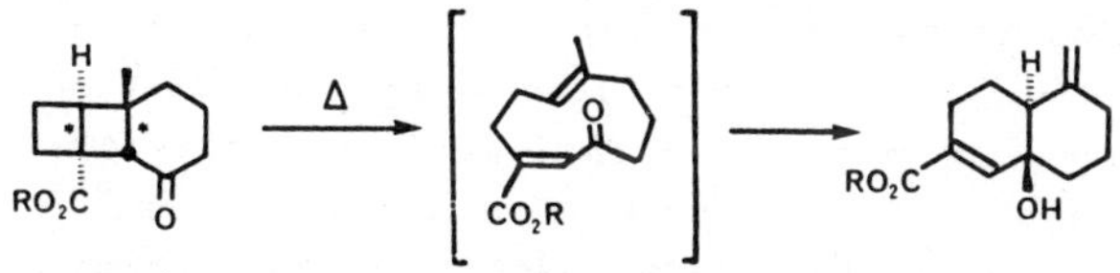

Δ
RO_2C
CO_2R
RO_2C
(45)

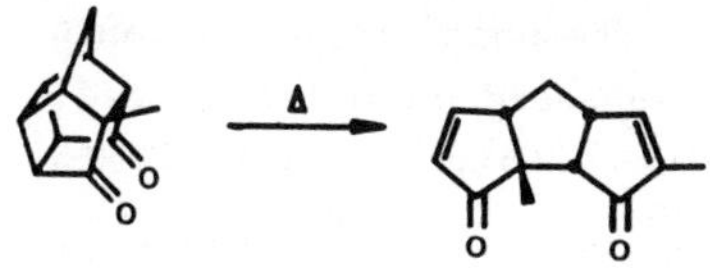

Δ
(46)

EtO OAc, EtO, CO_2R, LiOEt → [EtO, EtO, CO_2R, O] Δ → [EtO, EtO, CO_2R, O] (47)

Adding to the richness of photocycloaddition chemistry are two further recent developments. Studies on absolute stereochemical control have been reported (e.g.: eqs 48-50) in which

O, N, O; hν, $PhCOCH_3$ → O, N, O, 88% ee; steps → H, HO, grandisol (48)

CO_2R^*, RO_2C + Ar, Ar; hν → Ar, CO_2R^*, Ar, CO_2R, 94% ee (49)

R^*=bornyl, R=Me

R^*O_2C + O; hν → CO_2R^*, O, 30% ee + H CO_2R^*, H, O, 79% ee (50)

R^*= 8-phenylmenthyl

impressive selectivities have been achieved.[83-85] Based on the number of chiral auxiliaries which have been introduced for ground state reactions, it is expected that other chromophores might be suitably modified to allow for absolute stereochemical control. Finally, it is useful to note that the photocycloadditions noted above and their selectivities are subject to medium variations in addition to solvent polarity effects. For example, in the solid phase, photocycloadditions of aryl propenoate esters have been achieved with high enantiomeric selectivity resulting from the crystal lattice packing of the reactants.[86] Mixed media and organized assemblies provide a further option for modifying reaction selectivity.[87] The photocycloaddition of 1-octene to 3-butylcyclopentenone is enhanced in micellar media relative to homogeneous non-polar solvents (eq 51).[88] The hν (51)

cyclohexane	53	47
micelle	88	12

observed effects suggest that the micelle-solubilized reactants adopt the orientation of the assembly with polar surface groups and non-polar core groups. Similar effects have been noted in other systems; a recent example is found in the dimerization of acenaphthylenes for which both regioselectivity and *endo/exo* selectivity are changed when the reaction is conducted in the presence and absence of micelles.[89,90]

REFERENCES

1. G. Ciamician and P. Silber, Chem. Ber., 1908, 41, 1928.
2. a) O. L. Chapman and G. Lenz, Org. Photochem., 1967, 1, 283; O. S. Weiss and O. L. Chapman, ibid., 1973, 3, 197; b) P. E. Eaton, Acc. Chem. Res., 1968, 1, 50; c) P. de Mayo, ibid., 1971, 4, 41; d) P. G. Sammes, Quart. Rev., 1970, 24, 37; e) P. G. Bauslaugh, Synthesis, 1970, 287; f) S. W. Baldwin. "Organic Photochemistry", A. Padwa, Ed., Dekker, New York, 1981, Vol. 5, p. 123; g) W. Oppolzer, Acc. Chem. Res., 1982, 15, 135; h) W. L. Dilling, Photochem. Photobiol., 1977, 25, 605; i) J. Kossanyi, Pure Appl. Chem., 1979, 51, 181; j) G. Lenz, Rev. Chem. Intermed., 1981, 4, 369; k) R. A. Caldwell and D. Creed, Acc. Chem. Res., 1980, 13, 45; l) A. C. Weedon, "Synthetic Organic Photochemistry", W. Horspool, Ed., Plenum, New York, 1984, p. 61; m) W. L. Dilling, Chem. Rev., 1967, 67, 373.
3. G. Büchi and I. M. Goldman, J. Am. Chem. Soc., 1957, 79, 4741.
4. For recent reviews, see H. A. J. Carless in this text and "Synthetic Organic Photochemistry", W. Horspool, Ed., Plenum, New York, 1984, p. 425.
5. a)T. H. Koch and K. H. Howard, Tetrahedron Lett., 1972, 4035; b) H. Prinzbach, G. Fischer and H. Fritz, ibid., 1986, 27, 1269.
6. A. H. Lawrence, C. C. Liau, P. de Mayo and V. Ramamurthy, J. Am. Chem. Soc., 1976, 89, 2219, 3572; A. Ohno, Y. Ohnishi and Tschihashi, J. Am. Chem. Soc., 1969, 91, 5038.
7. The reader is referred to other sections of this text for a treatment of these reactions.
8. Y. Tobe, J.-I. Soto and Y. Odaira, presented at the 3rd International Kyoto Conference, November 1985. For earlier work, see Y. Tobe, S. Yamashita, T. Yamashita, K. Kakuichi and Y. Odaira, J. Chem. Soc. Chem. Commmun., 1984, 1259.
9. G. Rosini, E. Marotta, M. Petrini and R. Ballini, Tetrahedron, 1985, 41, 4633.
10. P. E. Eaton and T. W. Cole, J. Am. Chem. Soc., 1964, 86, 962, 3157; J. C. Barborak, L. Watts and R. Pettit, ibid., 1966, 88, 1328.

11. J. D. White, M. A. Avery and J. P. Carter, J. Am. Chem. Soc., 1982, 104, 5486.
12. K. Wiesner, L. Poon, I. Jirkovsky and M. Fishman, Canad. J. Chem., 1969, 47, 433.
13. A. B. Smith and D. Boschelli, J. Org. Chem., 1983, 48, 1217.
14. J. F. Bagli and T. Bogri, Tetrahedron Lett., 1969, 1639.
15. G. Mehta, V. A. Reddy, A. N. Murthy and D. S. Reddy, J. Chem. Soc. Chem. Commun., 1982, 540; K. Tatsuta, K. Akimoto and M. Kinoshita, Tetrahedron, 1981, 37, 4365.
16. M. T. Crimmins and J. A. De Loach, J. Org. Chem., 1984, 49, 2076; idem., J. Am. Chem. Soc., 1986, 108, 808.
17. F. E. Ziegler and J. A. Kloek, Tetrahedron Lett., 1974, 315; indem., Tetrahedron, 1977, 33, 373.
18. S. W. Baldwin and N. G. Landmesser, Tetrahedron Lett., 1982, 23, 4443.
19. P. A. Wender and S. L. Eck, Tetrahedron Lett., 1982, 23, 1871.
20. J. R. Williams, J. F. Callahan and C. Lin, J. Org. Chem., 1983, 48, 3162.
21. R. K. Boeckman and S. H. Cheon, J. Am. Chem. Soc., 1983, 105, 4112.
22. B. A. Pearlman, J. Am. Chem. Soc., 1979, 101, 6404.
23. B. D. Challand, H. Hikino, G. Kornis, G. Lange and P. de Mayo, J. Org. Chem., 1969, 34, 794.
24. M. Vandewalle, A. A. Devreese and P. J. de Clercq, Tetrahedron Lett., 1980, 21, 4767.
25. G. Pattenden and A. M. Buch, J. Chem. Soc. Chem. Commun., 1980, 1195.
26. E. J. Corey, R. B. Mitra and H. Uda, J. Am. Chem. Soc., 1964, 80, 485.
27. P. A. Wender and J. C. Lechleiter, J. Am. Chem. Soc., 1977, 99, 267.
28. S. L. Schreiber and C. Santini, J. Am. Chem. Soc., 1984, 106, 4038.
29. M. Ikida, K. Ohno, M. Takahashi and K.-O. Homma, Het., 1983, 20, 1005.
30. G. Büchi, J. A. Carlson, J. E. Powell and L.-F. Tretze, J. Am. Chem. Soc., 1973, 95, 540; M. R. Uskokovic, J. J. Partridge and N. K. Chada, ibid., 1973, 95, 532.
31. R. A. Whitney, Canad. J. Chem., 1983, 61, 1158.
32. R. B. Kelly, J. Zamecnik and B. A. Beckett, Canad. J. Chem., 1972, 50, 3455.
33. T. Matsumoto, K. Miyano, Y. Ohfune and S. Azuma, Tetrahedron Lett., 1974, 1545.
34. K. Yamazuki and R. J. Cvetanovic, J. Am. Chem. Soc., 1969, 91, 520.
35. D. R. Arnold, D. J. Trecker and Whipple, J. Am. Chem. Soc., 1965, 87, 2596.
36. J. S. Swenton, J. Org. Chem., 1969, 34, 3217.
37. R. Srinivasan and K. H. Carlough, J. Am. Chem. Soc., 1967, 89, 4932.
38. R. G. Salomon, Tetrahedron, 1983, 39, 485.
40. W. L. Dilling, Chem. Rev., 1969, 69, 845.
41. a) R. S. H. Liu, N. J. Turro and G. S. Hammond, J. Am. Chem. Soc., 1965, 87, 3406; b) R. S. H. Liu and G. S. Hammond, ibid., 1967, 89, 4936; c) G. S. Hammond, N. J. Turro and R. S. H. Liu, J. Org. Chem., 1963, 28, 3297.
42. G. S. Hammond and R. S. H. Liu, J. Am. Chem. Soc., 1963, 85, 477.

43. J. Saltiel and O. C. Zofiriou, Mol. Photochem., 1969, 1, 319; R. S. H. Liu and G. S. Hammond, J. Am. Chem. Soc., 1964, 86, 1892; ibid., 1967, 89, 4936; K. J. Crowley, Proc. Chem. Soc., 1962, 245, 334.
44. R. Srinivasan and K. H. Carlough, J. Am. Chem. Soc., 1967, 89, 4932; R. S. H. Liu and G. S. Hammond, ibid., 1967, 89, 4936; W. C. Agosta and S. Wolff, J. Org. Chem., 1980, 45, 3139.
45. D. J. Carlsson and K. U. Ingold, J. Am. Chem. Soc., 1968, 90, 7047; B. Maillard, D. Forrest and K. U. Ingold, ibid., 1976, 98, 7024; R. A. Sheldon and J. K. Kochi, ibid., 1970, 92, 4395; C. Walling and A. Cioffari, ibid., 1972, 94, 6059.
46. For a discussion, see a) N. J. Turro, "Modern Molecular Photochemistry", Benjamin, California, 1978, Ch. 11; b) J. Michl, Molec. Photochem., 1972, 4, 243; c) R. A. Caldwell and D. Creed, Acc. Chem. Res., 1980, 13, 45; R. O. Loutfy and P. de Mayo, J. Am. Chem. Soc., 1977, 99, 3559; D. Becker, M. Nagler, S. Hirsch and J. Ramun, J. Chem. Soc. Chem. Commun., 1983, 371.
47. P. A. Wender and C. R. D. Correia, manuscript in preparation.
48. For a review, see D. O. Cowan and R. L. Dresko, "Elements of Organic Photochemistry", Plenum, New York, 1976, p. 36.
49. P. J. Wagner and D. J. Bucheck, J. Am. Chem. Soc., 1969, 91, 5090.
50. E. J. Corey, J. D. Bass, R. Le Mahieu and R. B. Mitra, J. Am. Chem. Soc., 1964, 86, 5570.
51. T. S. Cantrell, W. S. Haller and J. C. Williams, J. Org. Chem., 1969, 34, 509.
52. P. A. Wender and J. C. Lechleiter, J. Am. Chem. Soc., 1977, 99, 267.
53. A. J. Wexler, J. A. Hyatt, P. W. Reynolds, C. Cottrell and J. S. Swenton, J. Am. Chem. Soc., 1978, 100, 512. If the ene byproducts of this reaction are included, the head-to-tail products account for 76% of the mixture and the head-to-head products 24%.
54. P. E. Eaton and W. S. Hurt, J. Am. Chem. Soc., 1966, 88, 5038; J. L. Ruhlen and P. A. Leermakers, ibid., 1966, 88, 5671; P. Wagner and D. J. Bucheck, Canad. J. Chem., 1969, 47, 713.
55. J. S. Swenton and E. L. Fritzen, Tetrahedron Lett., 1979, 22, 1951.
56. J. D. White and D. N. Gupta, J. Am. Chem. Soc., 1966, 88, 5364.
57. G. L. Lange, E. E. Neidert and W. J. Orram, Canad. J. Chem., 1978, 56, 1628.
58. a) K. Wiesner, Tetrahedron, 1975, 31, 1655; b) G. Marini-Bettolo; S. P. Sahoo, G. A. Poulton, T. Y. R. Tsai and K. Wiesner, ibid., 1980, 36, 719.
59. G. Stork and S. D. Darling, J. Am. Chem. Soc., 1960, 82, 1512.
60. R. B. Kelly, J. Zamecnik and B. A. Beckett, Canad. J. Chem., 1972, 50, 3455.
61. F. E. Ziegler, G. R. Ried, W. L. Stodt and P. A. Wender, J. Org. Chem., 1977, 42, 1991; P. A. Wender, Ph. D. Thesis, Yale University, 1974.
62. a) R. L. Cargill, G. H. Morton and J. Bordner, J. Org. Chem., 1980, 45, 3929; b) personal communication cited in 62a; c) S. L. Schreiber, Science, 1985, 227, 857.
63. C. R. Jones and D. R. Kearns, J. Am. Chem. Soc., 1977, 99, 344.
64. H.-J. Liu and W. H. Chan, Canad. J. Chem., 1982, 60, 1081.

65. a) P. A. Wender and J. C. Hubbs, J. Org. Chem., 1980, 45, 365; b) P. A. Wender and J. C. Lechleiter, J. Am. Chem. Soc., 1978, 100, 4321; c) G. L. Lange, M. A. Huggins and E. Neidert, Tetrahedron Lett., 1976, 4409; d) J. R. Williams and J. F. Callahan, J. Chem. Soc. Chem. Commun., 1979, 404, 405.
66. M. C. Pirrung and N. J. G. Webster, manuscript in preparation.
67. M. C. Pirrung, J. Am. Chem. Soc., 1981, 103, 82.
68. W. Oppolzer and T. Godel, J. Am. Chem. Soc., 1978, 100, 2583.
69. W. Oppolzer and R. D. Wylie, Helv. Chim. Acta, 1980, 63, 1198.
70. M. T. Crimmins and J. A. De Loach, J. Am. Chem. Soc., 1986, 108, 800.
71. A. M. Birch and G. Pattenden, J. Chem. Soc. Chem. Commun., 1980, 1195.
72. E. R. Koft and A. B. Smith, J. Am. Chem. Soc., 1982, 104, 5568.
73. G. Pattenden and G. M. Robertson, Tetrahedron Lett., 1986, 27, 399.
74. G. Mehta, D. S. Reddy and A. N. Murty, J. Chem. Soc. Chem. Commun., 1983, 824.
75. C. R. Johnson and N. A. Meanwell, J. Am. Chem. Soc., 1981, 103, 7667.
76. Ref. 2g; T. R. Hoye, S. J. Martin and D. R. Peck, J. Org. Chem., 1982, 47, 331; M. Fetizon, S. Lazare, C. Pascard and T. Prange, J. Chem. Soc., Perkin Trans., 1979, 1, 1407.
77. M. Ikeda, K. Ohno, M. Takahashi and K.-O. Homma, Heterocycles, 1983, 20, 1005.
78. M. C. Pirrung, Tetrahedron Lett., 1980, 4577; P. Hughes, M. Martin and J. Clardy, ibid., 1980, 4579.
79. For recent cases see references 18, 73 and H.-J. Liu and T. Dieck-Abularrach, Tetrahedron Lett., 1982, 23, 295.
80. E. J. Corey and S. Nozoe, J. Am. Chem. Soc., 1964, 86, 1652.
81. P. A. Wender and J. C. Lechleiter, J. Am. Chem. Soc., 1977, 99, 267; G. L. Lange and F. C. McCarthy, Tetrahedron Lett., 1978, 4749.
82. R. K. Boeckman, M. H. Delton, T. Nagasaka and T. Watanabe, J. Org. Chem., 1977, 42, 2946; R. K. Boeckman and S. H. Cheon, J. Am. Chem. Soc., 1983, 105, 4112.
83. A. I. Meyers and S. A. Fleming, J. Am. Chem. Soc., 1986, 108, 306.
84. L. M. Tolbert and M. B. Ali, J. Am. Chem. Soc., 1982, 104, 1742.
85. G. L. Lange, C. Decicco, S. L. Tan and G. Chamberlain, Tetrahedron Lett., 1985, 26, 4707.
86. L. Addadi and M. Lahav, J. Am. Chem. Soc., 1978, 100, 2838, and references cited therein.
87. S. Singh, G. Usha, C.-H. Tung, N. J. Turro and V. Ramamurthy, J. Org. Chem., 1986, 51, 941 and references cited therein.
88. N. Berenjian, P. De Mayo, M. E. Sturgeon, L. K. Sydnes and A. C. Weedon, Can. J. Chem., 1982, 60, 425.
89. H. Mayer, F. Schuster and J. Sauer, Tetrahedron Lett., 1986, 27, 1289.
90. For further review, see N. J. Turro, G. S. Cox and M. A. Paczkowski, "Topics in Current Chemistry" 129, Springer-Verlag, Berlin, 1985.

10
Alkenes: Photo-oxidation

By K.H. Pfoertner
F. HOFFMANN-LAROCHE AND CO. LTD., CH-4002 BASEL, SWITZERLAND

Introduction

This discourse on photo-oxidation refers to light induced oxidations with molecular oxygen, whereby electronically excited oxygen could be involved as well as ground state oxygen. Thus, we have to characterize the different states of oxygen, to describe the photochemical generation of excited oxygen, and to note the different chemical reactivities of both species.

Characterization of the oxygen states

Ground state oxygen. The 16 electrons of molecular oxygen are disposed among the available molecular orbitals. Since all orbitals up to the last are filled, attention is focussed on the half-filled highest-energy orbital with the lowest-energy configuration. According to HUND'S rule electrons enter unfilled orbitals for minimization of electron repulsion, or as spectroscopists would say, for reaching maximum multiplicity. Therefore, triplet states with their parallel electron spins have lower energies than corresponding singlet states with their antiparallel electron spins.

Electronically excited oxygen. An important and widely useful generalization when considering photophysical primary processes is the WIGNER spin conservation rule. This rule has two consequences with respect to electronically excited oxygen. The infrared electronic transitions of molecular oxygen are very highly forbidden. The transition from the lowest singlet state with its 45 min lifetime at the limit of zero pressure to the triplet ground state is quite possibly the molecular transition that is most forbidden in nature. At higher pressure, and in solution, this singlet state's lifetime is shortened by orders of magnitude. Nevertheless, a millisecond to microsecond reactive species is still of considerable chemical significance.

Spectroscopic State Characterization		Energy (kJ mol^{-1})	(kcal mol^{-1})	Spin Assignment	Type of Chemical Reactivity
Singlet	$^1\Sigma_g^+$	156.9	37.5	$\uparrow O_2 \downarrow$	two point attack in concerted reactions
Singlet	$^1\Delta_g$	94.2	22.5	$\uparrow O_2 \downarrow$	single point attack as well as concerted two point attack
Triplet	$^3\Sigma_g^-$	0	0	$\uparrow O_2 \uparrow$	single point attack in radical-like reactions

Figure 1

Compared with the long-lived first singlet state, the second one with higher energy (Fig.1) has a lifetime shorter by a factor of 400. For this reason, here we deal only with the $^1\Delta_g$ state when we use the term singlet oxygen, symbolized by $^1O_2$. That $^1O_2$ cannot be generated from triplet oxygen $^3O_2$ by direct light absorption but only by energy tranfer from so-called photosensitizers, i.e. from other light-absorbing molecules, is the second consequence of the spin conservation rule.

Generation of singlet oxygen

Fig.2 suggests that singlet and triplet excited states of the sensitizer can transfer energy to $^3O_2$. But the triplet state of the sensitizer usually has a much longer lifetime than its singlet state. Thus, with very few exceptions photo-oxidations with $^1O_2$

proceed *via* this state.

Photosensitized Generation of Singlet Oxygen

$$^{1}S_0 + h\nu \longrightarrow {}^{1}S_1 \quad \text{(light absorption)}$$

$$^{1}S_1 \longrightarrow {}^{3}S_1 \quad \text{(intersystem crossing)}$$

$$\left.\begin{array}{l} ^{1}S_1 + {}^{3}O_2 \longrightarrow {}^{3}S_1 + {}^{1}O_2 \\ ^{3}S_1 + {}^{3}O_2 \longrightarrow {}^{1}S_0 + {}^{1}O_2 \end{array}\right] \text{(energy transfer)}$$

S = sensitizer
$^{1}S_0$ = singlet ground state
$^{1}S_1$ = singlet excited state
$^{3}S_1$ = triplet excited state

O_2 = molecular oxygen
$^{3}O_2$ = triplet ground state
$^{1}O_2$ = singlet excited state

Figure 2

Some useful photosensitizers for singlet oxygenation are given in Table 1.

Sensitizer	Triplet Energy	
	kJ mol^{-1}	kcal mol^{-1}
Methylene Blue	140.2 -142.3	33.5- 34.0
Tetraphenyl Porphyrin	142.3	34.0
Hematoporphyrin	155.7	37.2
Rose Bengal	164.1 -176.6	39.2- 42.2
Eosin	180.8 -192.5	43.2- 46.0
Benzophenone	287.2 -289.2	68.6- 69.1

Table 1

Chemical reactivities of singlet oxygen

Singlet oxygen is an electrophilic species and isoelectronic with ethylene. Functionally $^{1}O_2$ seems to behave like an analogue of ethylene. Indeed, each of the three modes in which $^{1}O_2$ adds to olefins finds precedent in one of the reaction pathways of ethylene.

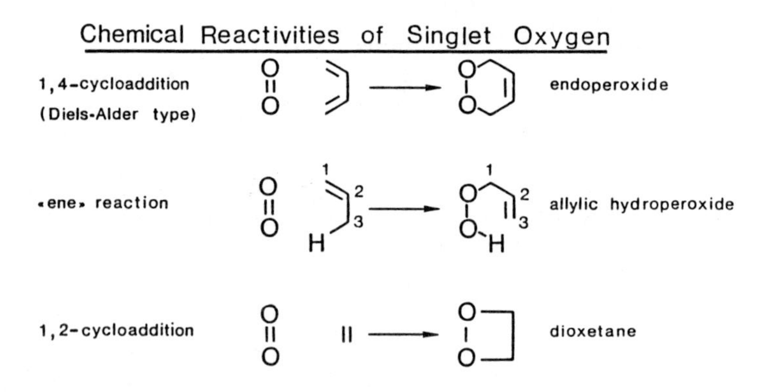

Figure 3

The addition of $^{1}O_2$ to dienes generating endoperoxides may be viewed as a DIELS-ALDER reaction with $^{1}O_2$ as dienophile. In the $^{1}O_2$ "ene" reaction, olefins containing allylic hydrogens are oxidized to the corresponding allylic hydroperoxides accompanied by a 1,2-shift of the double bond. In the final mode, the $^{1}O_2$ oxygenation of certain C,C-double bonds to yield dioxetanes is analogous to the production of cyclobutanes by a $(2_s + 2_a)$ cycloaddition of two ethylenes.

Chemical reactivity of triplet oxygen

A photosensitizer can be quenched not only by $^{3}O_2$ but also by the substrate. This is considered to be the main diversion from $^{1}O_2$ generation. Mechanistically, this process involves the production of free radicals or radical ions by interaction of the sensitizer triplet with a reducing substrate RH <u>via</u> hydrogen abstraction or electron transfer (Fig. 4). Both transients R• and $HR^{+\bullet}$ thus produced can directly react with $^3O_2$ leading to photooxidation products or can initiate free-radical chain autoxidation. The reduced sensitizer •SH or $S^{-\bullet}$ is oxidized by $^3O_2$ to generate $HO_2^\bullet$ or $O_2^{-\bullet}$ which both react further.

Chemical Reactivity of Triplet Oxygen

$$^3S_1 + RH \longrightarrow \cdot SH + R\cdot$$

$$R\cdot + ^3O_2 \longrightarrow R\text{-}O\text{-}O\cdot$$

$$\cdot SH + ^3O_2 \longrightarrow ^1S_0 + HO_2\cdot$$

(initiation of radical chains)

$$^3S_1 + RH \longrightarrow S^{\bar{\cdot}} + RH^{\dot{+}}$$

$$HR^{\dot{+}} + ^3O_2 \longrightarrow HR\text{-}O\text{-}O^{\dot{+}}$$

$$S^{\bar{\cdot}} + ^3O_2 \longrightarrow ^1S_0 + O_2^{\bar{\cdot}}$$

(initiation of radical chains)

Figure 4

Structural evidence for the different reactivity of $^1O_2$ and $^3O_2$ in analogous situations is shown in Fig.5.

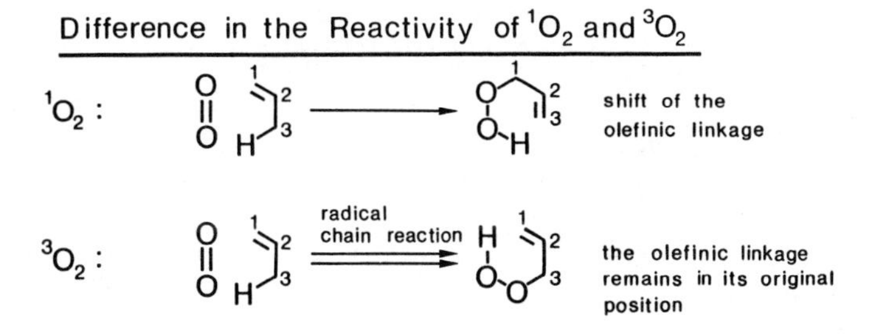

Figure 5

In general, sensitizers of ketonic structure with a triplet of the n,π^* type are powerful abstractors of hydrogen from organic molecules (e.g. benzophenone in Table 1).

Singlet oxygenations of conjugated dienes

Mechanism. The possible mechanisms for the 1,4-cycloaddition of $^{1}O_2$ to conjugated dienes are shown in Fig.6[1].

Possible Mechanisms for the 1,4-Cycloaddition of $^{1}O_2$

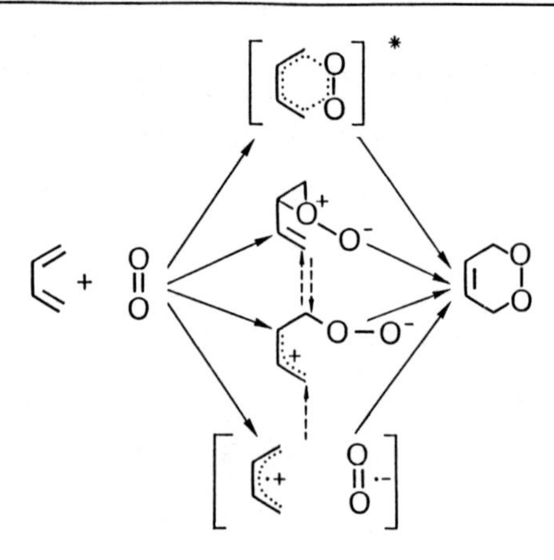

Figure 6

The mechanism of endoperoxidation seems likely to be substrate-dependent. However, for the majority of endoperoxidations investigated the concerted mechanism at the top of Fig.6 might be an adequate description.

Types of endoperoxide. Endoperoxides derived from aliphatic dienes by reaction with $^{1}O_2$ can be divided into two categories based on the nature of the conjugated diene from which they are prepared (Fig.7).

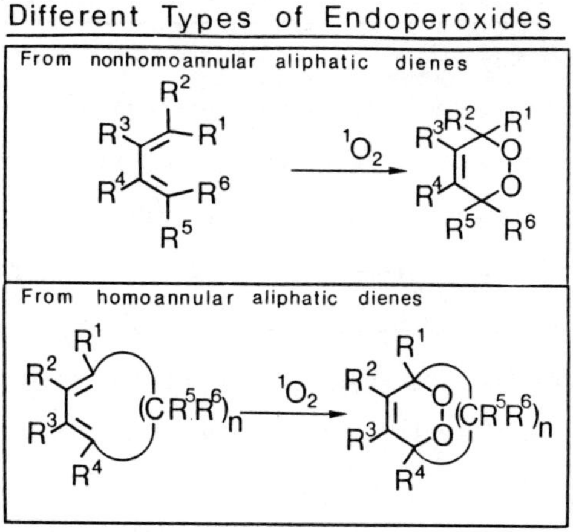

Different Types of Endoperoxides

From nonhomoannular aliphatic dienes

From homoannular aliphatic dienes

Figure 7

Synthetically useful transformations of endoperoxides

The possible transformations of endoperoxides are summarized in Fig. 8.

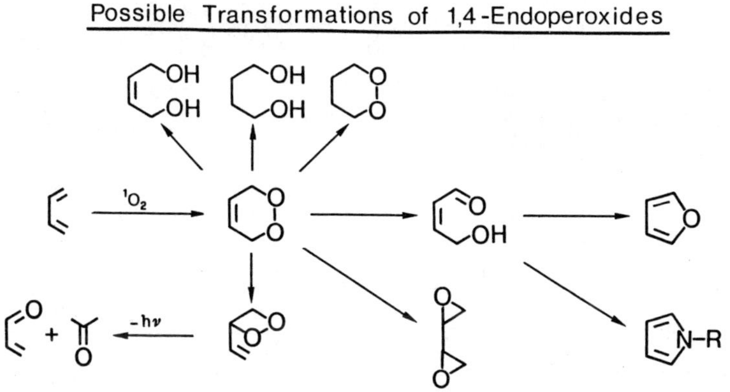

Possible Transformations of 1,4-Endoperoxides

Figure 8

Reactions of endoperoxides from nonhomoannular aliphatic dienes

The conversion of endoperoxides into furans is a reaction of considerable synthetic value as well as a process of biological interest. In general, other reactions of 1,2-dioxacyclohex-4-enes have not been studied to an extent which establishes a typical pattern of behaviour. Reduction with $LiAlH_4$ usually affords the corresponding unsaturated diol. From its configuration the configuration of the starting endoperoxide may be deduced[2]. In this way it could be demonstrated that endoperoxidation occurs stereospecifically.

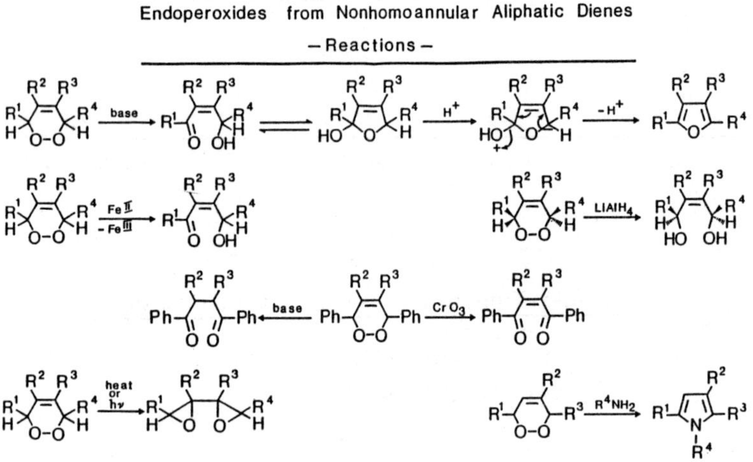

Endoperoxides from Nonhomoannular Aliphatic Dienes

– Reactions –

Figure 9

Reactions of the 3,6-diphenylendoperoxide and of the 3,4,5,6-tetraphenylendoperoxide have been investigated most thoroughly. But these endoperoxides are somewhat atypical because they give high yields of the corresponding 1,4-dione rather than the expected γ-hydroxy-α,β-unsaturated ketone which in other cases was obtained by reaction with iron(II) salts. Oxidation with CrO_3 afforded the unsaturated 1,4-diones. Thermolysis or photolysis of endoperoxides (R^1,R^2= phenyl or cross-conjugation by R^1,R^2,R^3,R^4= phenyl) yields di-epoxides and can be rationalized mechanistically on the basis of alkenedioxyl radicals formed by initial homolysis of the peroxide bond. An effective synthesis of pyrroles from endoperoxides has been reported[3] in which the base-catalysed rearrangement is carried out with KOH in EtOH in the presence of ammonia or of a primary amine.

Reactions of endoperoxides from homoannular aliphatic dienes

In this category extensive studies have been made concerning thermolysis, photolysis, and base-catalyzed isomerization of endoperoxides. For this reason characteristic patterns of behaviour are well established here.

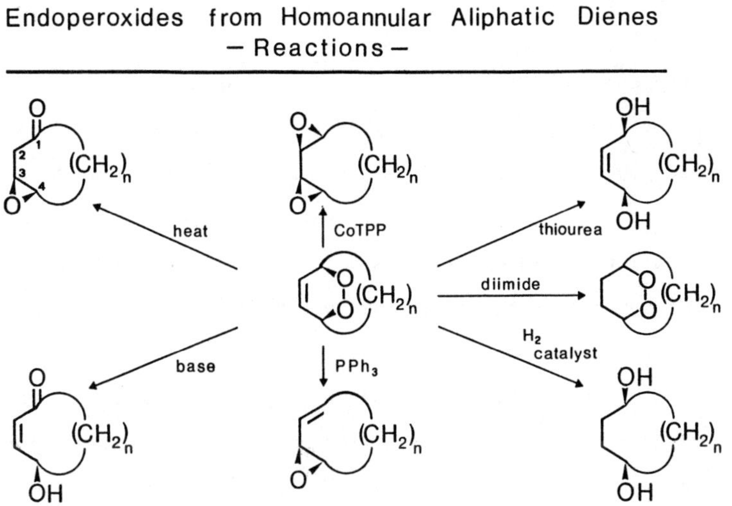

Endoperoxides from Homoannular Aliphatic Dienes
— Reactions —

Figure 10

Thermolysis of this type of endoperoxide generally leads to the formation of cis-diepoxides. The reaction is readily rationalized in terms of homolytic fission of the peroxide bond to form cycloalk-2-ene-1,4-dioxyl radicals, followed by its intramolecular addition to the adjacent double bond. The conversion of endoperoxides into 3,4-epoxycycloalkanones sometimes competes with the formation of di-epoxides, induced by thermolysis. This is a consequence of an 1,2-hydrogen shift in the 1,3-diradical formed from the initial 1,4-dioxyl radical (Fig.11).

Thermolysis of Endoperoxides to 3,4-Epoxycycloalkanones

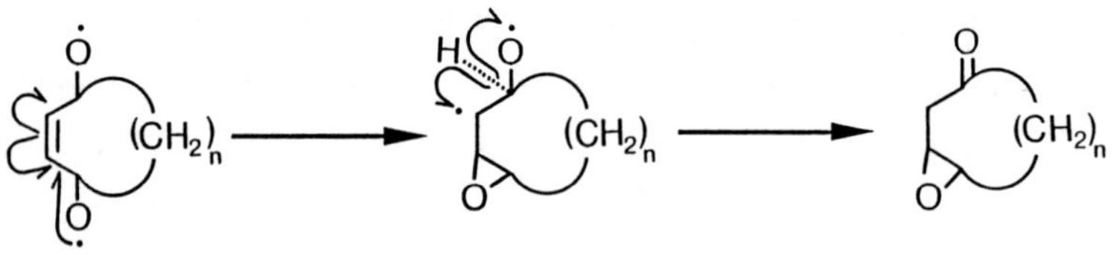

Figure 11

Catalytic amounts of transition metal compounds like cobalt(III) tetraphenylporphyrin (CoTPP) may induce transformations of endoperoxides under very mild conditions. Enough examples have been reported suggesting that the CoTPP-catalyzed rearrangement of endoperoxides into cis-diepoxides is a process superior to thermolysis[4]. The photolytic cleavage of dialkyl peroxides to form alkoxy radicals is well known. If it could be successful with endoperoxides in general, this technique would have the advantage of being able to generate the necessary diradicals at lower temperatures than those required for thermolysis. Considering this, it is disappointing to find only few examples of photolysis of that type of endoperoxide, i.e. much less than thermolysis. In addition the influence of the wavelength on the photolysis of endoperoxides has scarcely been studied[5]. Photochemical studies of this kind are in their infancy. But signs are that from future activities in this area much interesting chemistry could emerge.

Kaneko et al.[8] showed by performing photooxygenations in the presence of thiourea that cyclopentadiene and cyclohexadiene endo-peroxides can be reduced in situ to give the unsaturated diols (n=1,2 in Fig.10) in approximately 60 % yield. This procedure avoids isolation of the endoperoxide, allows the execution of $^1O_2$-oxygenation at room temperature, and is particularly attractive where the endoperoxide is very labile.

The "ene" reaction

In sharp contrast to autoxidation, $^1O_2$ ene reaction proceeds stereo-specifically in a suprafacial manner with respect to the ene unit. This means that $^1O_2$ attack and hydrogen removal occur from the same side of the olefin (Fig.12).

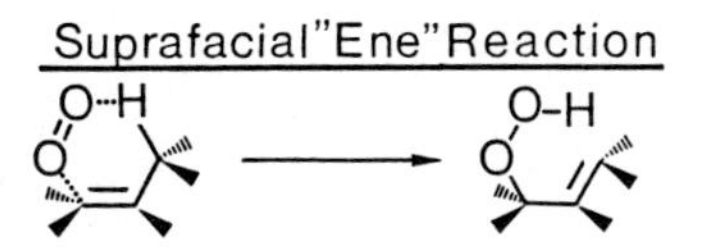

Figure 12

When applying the $^1O_2$ ene reaction to pure enantiomers of chiral olefins no racemization is observed[9]. Also E/Z-isomerization does not occur during the course of an ene reaction[10]. It is the stereo-specificity that makes an allylic functionalization by $^1O_2$ so attractive to synthetic chemists[11]. In addition its regiospecifi-city is useful. Here, trisubstituted olefins invariably react on the crowded side.

Unlike autoxidation, $^1O_2$ oxidation cannot be inhibited by adding radical-trapping reagents like 2,6-di-t-butylphenol[12]. However a sharp reduction in the rate of $^1O_2$ oxidation is observed in the presence of less than 0.005 mol l^{-1} DABCO (1,4-diazabicyclo(2.2.2)-octane) which is an efficient $^1O_2$ quencher[13]. Initiation of $^3O_2$ oxidation is delayed or inhibited at low temperatures. Little if any dependence of the reaction rate on temperature is observed with $^1O_2$. Only the type of solvent has an influence on the lifetime of $^1O_2$, but $^1O_2$ oxygenation is especially sensitive to steric factors: Singlet oxygen approaches the substrate predominantly from the sterically less hindered side. In contrast to autoxidations, the allylic H-abstraction by $^1O_2$ in an ene reaction proceeds independently of whether the H-atom is a primary, secondary or tertiary one[14]. Worthwhile too, is the lack of a strong MARKOWNIKOFF directing influence on the product distribution in the $^1O_2$ ene mode[15]. But $^1O_2$ is as an electrophilic species sensitive to the nucleophilicity of the olefinic bond. As a general rule, alkyl substitution increases the reactivity of olefins 10 to 100-fold per group[16].

Synthetically useful transformations of allylic hydroperoxides

Some of the possible transformations of allylic hydroperoxides including a silatropic reaction to give an allylic silyl peroxide are summarized in Fig.13. Because of the instability of hydroperoxides in general, and of allylic compounds in particular, it is a common procedure to reduce them to the corresponding alcohols prior to work up and isolation. The reduction can conveniently be carried out with a variety of reagents including $(PhO)_3P$, Ph_3P, $LiAlH_4$, $NaBH_4$, Na_2SO_3, MeS, and NaI/AcOH. Unstable hydroperoxides can be reduced *in situ* with excess phosphite, especially with

$(PhO)_3P$,which reacts quite sluggishly with $^3O_2$ and with $^1O_2$, in this case being the reagent of choice.

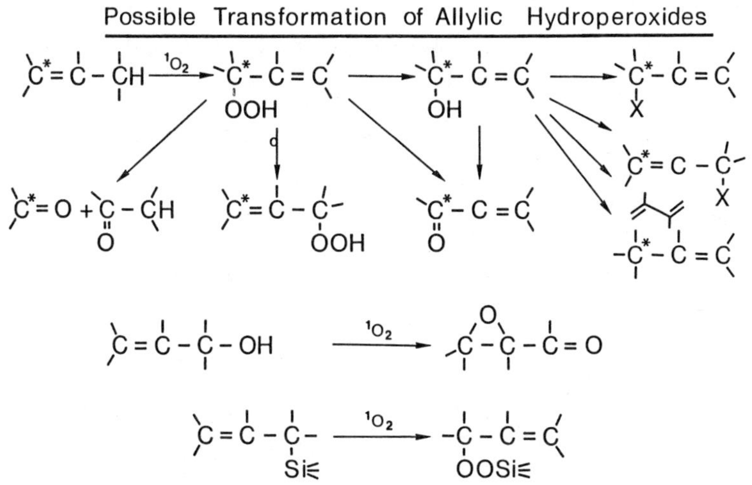

Possible Transformation of Allylic Hydroperoxides

Figure 13

"Hock-Cleavage" to carbonyl compounds

Allylic hydroperoxides undergo acid-catalyzed heterolysis of the peroxide bond generating a fragment with a positive charge on the oxygen[18] (Fig.14). The instability of such fragments induces the migration of groups to the electron-deficient oxygen with

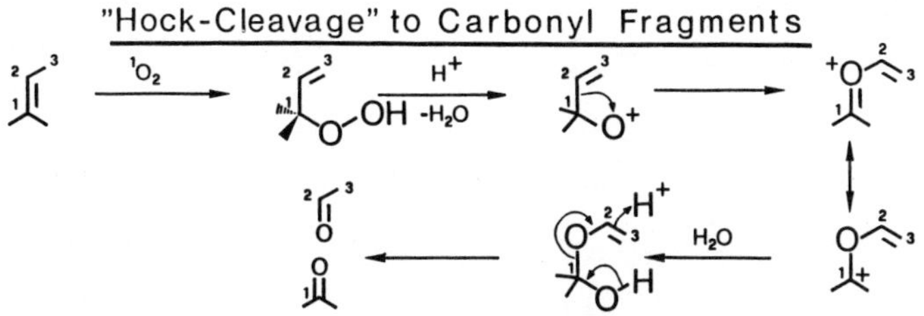

"Hock-Cleavage" to Carbonyl Fragments

The relative migratory aptitudes for this process are:

cyclobutyl > aryl > vinyl > hydrogen > cyclopentyl ≈ cyclohexyl >> alkyl

Figure 14

concomitant rearrangement of the carbon skeleton. The relative migratory aptitudes shown in Fig.14 indicate that in general the migration of the vinyl σ bond is observed. The resulting oxy-carbonium ion is nucleophilically attacked by water and the resulting hemiacetal decomposes to give two carbonyl compounds.
This process is called Hock-cleavage because HOCK found this reaction as early as in 1936 with cyclohexene hydroperoxide[19]. Although the Hock-cleavage is mainly acid-catalyzed it may also occur in the absence of any acid and is quite frequently found when a crude ene reaction solution is injected into a gas chromatograph. When the determination of a reaction mode is based on product identification only, Hock-cleavage can cause problems because decomposition of a 1,2-dioxetane may yield the same carbonyl compounds (see Fig.18).

Rearrangement of allylic hydroperoxides to epoxides[20]. Epoxides result from free-radical oxygenation processes as well as from the rearrangement of allylic hydroperoxides. For the latter two different mechanisms have to be taken into account. One starts with a homolysis of the peroxide linkage. Cyclization of the resulting alkoxy radical to an epoxide generates a radical at the adjacent carbon atom which can abstract hydrogen (path a) if it is not trapped by oxygen prior to hydrogen abstraction (path b)(Fig.15). The second mechanism is characterized by initial heterolysis which yields a positively charged oxygen. In this case formation of a peroxide generates a carbonium ion that can be trapped by the solvent (Fig.15).

Rearrangement of Allylic Hydroperoxides to Epoxides

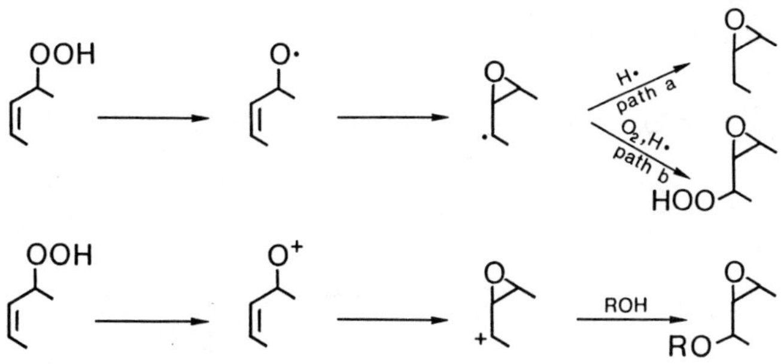

Figure 15

Base-catalyzed Kornblum-DeLaMare dehydration[21]. Only hydroperoxides with hydrogen atoms in the α-position are dehydrated in the presence of weak bases, e.g. pyridine. Unfortunately, competing reactions often lead to enol or Hock-cleavage products instead of the expected enones. Performing the dehydration e.g. in the presence of acetic anhydride converts the hydroperoxides presumably into peroxy esters which then eliminate acetic acid giving the desired enones almost exclusively.(Fig.16).

Base-Catalyzed Kornblum-DeLaMare Dehydration

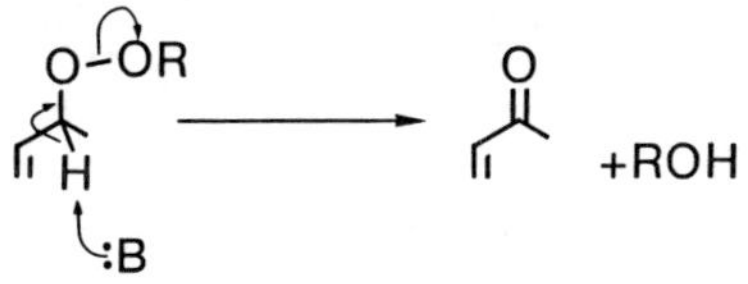

Conditions for maximum yield of enone:
:B = pyridine , R = CH_3CO from acetic anhydride

Figure 16

1,2-Dioxetanes

Cycloaddition of $^1O_2$ to enol ethers, enamines and other electron-rich alkenes is a general route to 1,2-dioxetanes[22].

1,2-Dioxetanes by Cycloaddition of Singlet Oxygen to Electron-Rich Alkenes

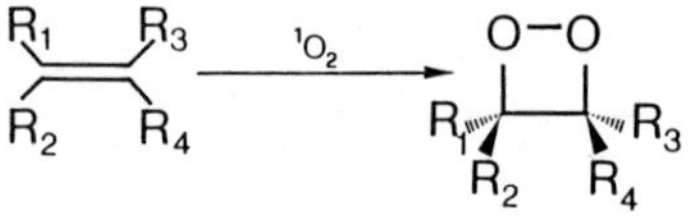

Figure 17

Decomposition of dioxetanes. To explain the production of excited triplet carbonyls a two-step diradical mechanism has been proposed as well as a concerted one. Non-luminescent reactions of isolable dioxetanes proceed via catalytic cleavage to 1,2-dicarbonyl or 2-hydroxy-carbonyl compounds. The O-O bond can be reduced by insertion, too, e.g. of trivalent phosphorus compounds yielding phosphoranes.

Possible Transformations of 1,2-Dioxetanes

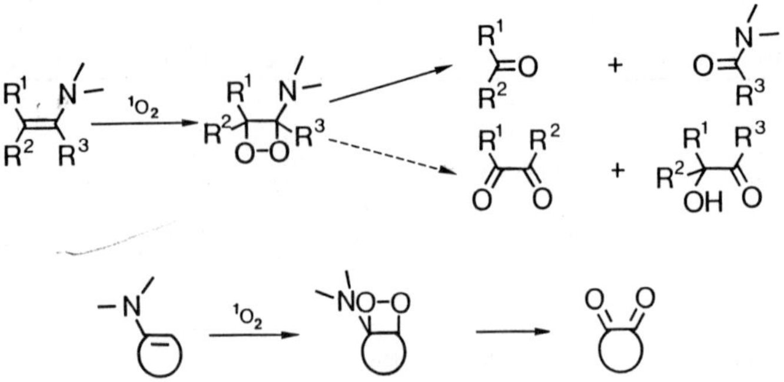

Figure 18

Synthetic applications

Singlet oxygen reactions have been utilized for the biomimetic synthesis of numerous compounds occuring in nature.

1,4-Endoperoxides as synthetic intermediates. Photochemical 1,4-endoperoxidation provides a powerful synthetic tool for the simultaneous functionalization of 1,3-dienes. The resulting endoperoxides can be transformed into diverse types of compounds as has been illustrated in Figs. 8, 9 and 10.

Reduction of the 1,4-endoperoxide from cyclopentadiene to cyclopentenediol produces an important intermediate in the synthesis of prostaglandins and jasmones[8] (Fig.19).

cis-2-Cyclopentene-1,4-diol via Photooxygenation

$^1O_2$

OH

OH

Prostaglandins

OH

OH

O

O

O

O

O

Jasmones,
Prostaglandins

Figure 19

In the synthesis of pharmacologically useful 14-hydroxymorphinanes, e.g. of naxolone and naltrexone, noroxymorphone is an essential intermediate. Its synthesis from the readily available codeine requires introduction of a 14-hydroxy group. Using $^1O_2$ this could be achieved with an overall yield of 43 %, which was never obtainable by means of metallic oxidants[23] (Fig.20).

Singlet Oxygenation in the Synthesis of 14-Hydroxymorphinans

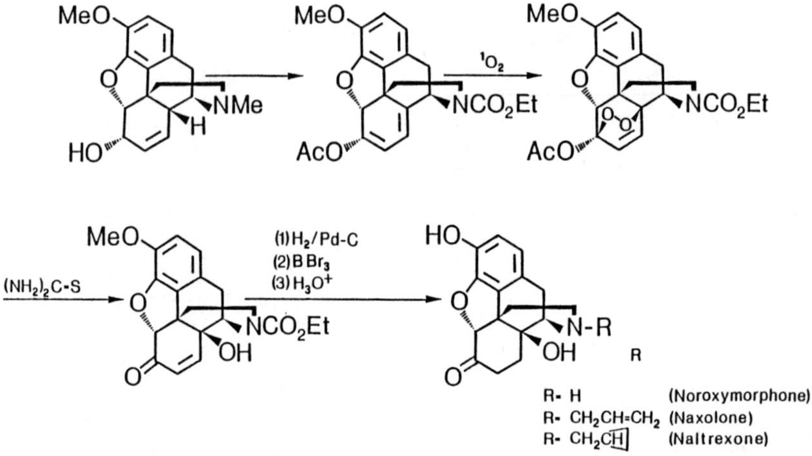
MeO
O
HO
NMe
H
MeO
O
AcO
NCO2Et
1O2
MeO
O
AcO
NCO2Et
(NH2)2C-S
MeO
O
O
NCO2Et
OH
(1)H2/Pd-C
(2)BBr3
(3)H3O+
HO
O
O
N-R
OH
R
R- H (Noroxymorphone)
R- CH2CH=CH2 (Naxolone)
R- CH2CH (Naltrexone)

Figure 20

Allylic hydroperoxides as synthetic intermediates. Figures 13 to 16 have shown the scope and limitations for the achievement of synthetic intermediates by the ene reaction. Allylic hydroperoxides themselves have been found in terrestrial and marine organisms as well as 1,4-endoperoxides. An important synthetic application of the ene reaction is the introduction of an acid-labile C(6)-benzylic hydroxy function into an antibiotically active tetracycline by its stereospecific oxygenation at a late stage of the synthetic sequence[24] (Fig.21). In the synthesis of cyclic ethers the ene reaction has been applied to the production of rose oxide and nerol oxide which are important ingredients of Bulgarian rose oil[25] (Fig.22).

Tetracyclines

A Sythetic Route Including Photooxygenation

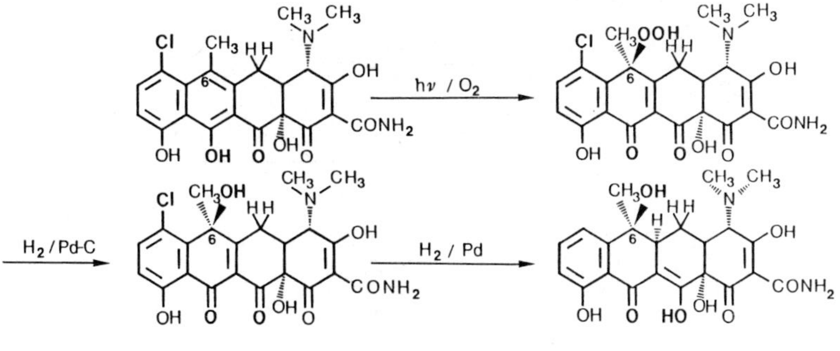

CH_3 CH_3
N
Cl CH_3 H H
6
OH
$h\nu$ / O_2
$CONH_2$
OH OH O OH O

CH_3 CH_3
N
Cl CH_3OOH H H
6
OH
$CONH_2$
OH O O OH O

CH_3 CH_3
N
Cl CH_3OH H H
H_2 / Pd-C
6
OH
H_2 / Pd
$CONH_2$
OH O O OH O

CH_3 CH_3
N
CH_3OH H H
H
6
OH
$CONH_2$
OH O HO OH O

Figure 21

Fragrant Chemicals via Photooxygenation

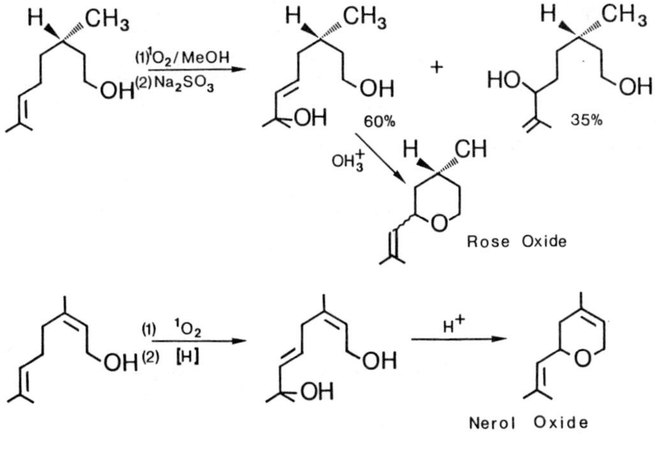

H CH_3
OH
(1) $^1O_2$/MeOH
(2) Na_2SO_3

H CH_3
OH
OH
60%

+

H CH_3
HO
OH
35%

OH_3^+

H CH
O
Rose Oxide

OH
(1) $^1O_2$
(2) [H]

OH
OH

H^+

O
Nerol Oxide

Figure 22

Dioxetanes as synthetic intermediates. The synthetic utility of 1,2-cycloaddition of $^1O_2$ is generally limited. As enamines undergo $^1O_2$ oxygenation most reliably, the useful transformations (Fig.18) are restricted to 1,2-dioxetanes derived from enamines.

References

[1]A.A. Frimer (ed.), "Singlet O_2", CRC Press, Boca Raton, Florida, 1985, Vol.2, Chapter 4.

[2]G. Rio and J. Berthelot, Bull.Soc.Chim. France , 1971, 2938.

[3]K. Kondo and M. Matsumo, Chem.Lett., 1974, 701.

[4]J.D. Boyd, C.S. Foote and D.K. Imagawa, J.Am.Chem.Soc., 1980, 102, 3641.

[5]I. Landheer and D. Ginsburg, Tetrahedon, 1981, 37, 133.

[6]J. Hamer (ed.), "1,4-Cycloaddition Reactions", Academic Press, New York, 1967, Chapter 10.

[7]D.J. Coughlin and R.G. Salomon, J.Am.Chem.Soc., 1977, 99, 655.

[8]C. Kaneko, A. Sugimoto and S. Tanaka, Synthesis, 1974, 876.

[9]C.S. Foote, S. Wexler and W. Ando, Tetrahedron Lett., 1965, 4111.

[10]F.A. Litt and A. Nickon, Adv.Chem.Ser., 1968, 77, 118.

[11]H.H. Wasserman and J.L. Ives, Tetrahedron, 1981, 37, 1825.

[12]W. Prior (ed.), "Free Radicals in Biology", Academic Press, New York, 1972, Chapter 2.

[13]C. Quannes and T. Wilson, J.Am.Chem. Soc., 1968, 90, 6527.

[14]A. Nickon et al., J.Am.Chem. Soc., 1972, 94, 5517.

[15]L.M. Stephenson, M.J. Grdina and M. Orfanopoulous, Acc.Chem.Res., 1980, 13, 419.

[16]H.H. Wasserman and R.W. Murray (eds.), "Singlet Oxygen", Academic Press, New York, 1979, Chapter 8.

[17]J.B. Plumb and C.E. Griffin, J.Org.Chem., 1963, 28, 2908.

[18]A.A. Frimer, Chem.Rev., 1979, 79, 359.

[19]H.Hock and O. Schrader, Angew.Chem., 1936, 49, 595.

[20] A.A. Frimer (ed.), "Singlet O_2", CRC Press, Boca Raton, Florida, 1985, Vol.2, Chapter 3.

[21] N. Kornblum and H.E. DeLaMare, J.Am.Chem.Soc., 1951, 73, 380.

[22] P.D. Bartlett and A.P.Schaap, J.Am.Chem.Soc., 1970, 92, 3223.

[23] M.A. Schwartz and R.A. Wallance, J.Med.Chem., 1981, 24, 1525.

[24] M.S. von Wittenau, J.Org.Chem., 1964, 29, 2746.

[25] G. Ohloff, E. Klein and G.O. Schenck, Angew.Chem.,1961, 73, 578.

11
Terpenoids

By W.M. Horspool
DEPARTMENT OF CHEMISTRY, THE UNIVERSITY, DUNDEE DD1 4HN, UK

There are many examples in the literature[1,2] of the synthesis of natural products by the photochemical addition of suitably substituted alkenes (and alkynes) to enone double bonds. Thus the main problem which faces a compiler of any lecture of this sort is the decision of which to include and which to omit. Inevitably the choice made from the many syntheses available, dependent on a photochemical step as the key process, will have a personal bias. The subjectivity shown here is one where the molecule being synthesized has been of general interest or else the actual synthesis has been especially elegant. All of the syntheses illustrated are aimed at demonstrating that the cycloaddition, while it is often the key step, is only a small part of the synthetic sequence and that the success of a total synthesis depends on ones ability as a researcher to manipulate the initial photochemical adduct to effect the desired synthesis.

In previous lectures in this series it has been highlighted that the organic photochemical route need not produce a complex mixture of products. Indeed the reactant, light, is a clean efficient reagent which in many cases can be readily controlled. Often a photochemical reaction proceeds in high yield, provided that the rules governing the reactive system are obeyed. Sadly, for a variety of reasons, photochemical reactions have received bad publicity which has served to scare away many who could have made considerable use of the technique. Hopefully these lectures will help to attract researchers to the area. For success the fundamentals controlling stereo- and regio-chemistry have to be known.

The sections 1 and 2 deal with the mechanism of the reaction and set up useful rules of thumb for the control of the photoaddition. Later sections (3-9) deal with the photochemical syntheses.

1. Mechanism

The original model for the photocycloaddition to enones affording a cyclobutane arose from the qualitative work of Corey.[3] More recently de Mayo[4] has published a detailed mechanistic treatment based on his own and his collaborators' observations. A simplified illustration of this mechanism is shown in figure 1.

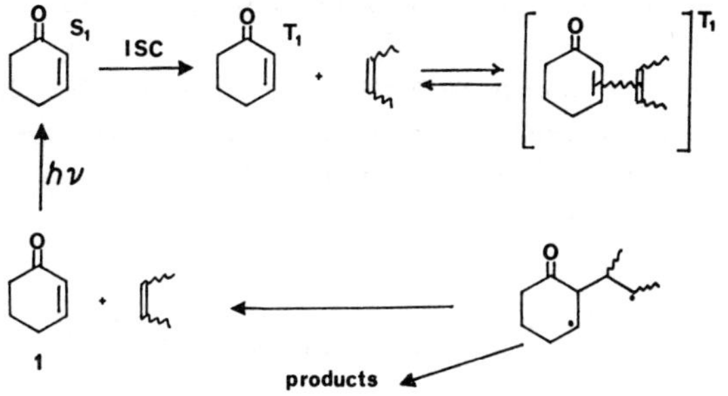

Figure 1

In this figure the enone(1) undergoes excitation to the singlet state(S_1). Once in that state a variety of options are open. Thus, with acyclic enones, relaxation of the excited state can readily occur by *trans-cis*-isomerization. This takes the enone back to the ground state without further reaction. In cyclic enones, however, because of the rigidity of the system, isomerization is often not competitive. Instead the singlet state undergoes intersystem crossing to the longer lived triplet state(T_1). Within this manifold diffusion controlled bimolecular processes

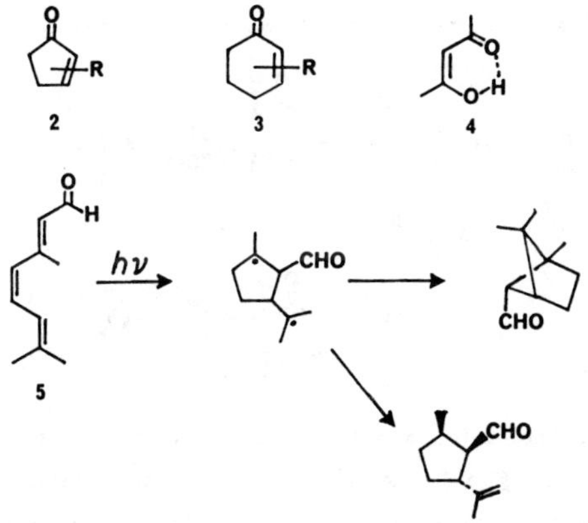

Figure 2

become important and can often compete successfully with other deactivation mechanisms. The rigidity of the enone is important and usually only cyclopentenones(2), cyclohexenones(3), and enones of reduced flexibility, such as H-bonded enols(4), will undergo cycloaddition. The addition of the enone triplet state to the alkene leads to a biradical intermediate <u>via</u> a non-quenchable exciplex. This biradical can collapse back to starting material or else go on to products by either a second bond formation, producing a cyclobutane, or else by H-abstraction yielding an alkene as seen in the intramolecular addition of the enone(5) illustrated in figure 2 where both ring closure and H-abstraction occur.

With this intramolecular example the initial bond formation occurs at the α-carbon of the enone. For intermolecular addition the problem of whether α- or β- bonding occurs first has not been resolved and often it can depend on the reactants involved.

2. Regiochemistry

Regardless of whether bonding occurs first at the α- or β-carbon regiochemistry has to be considered in addition reactions of unsymmetrical alkenes to enones. Typical of this are the two additions illustrated in figure 3.

In these none of the other regio-isomer is obtained. Corey[3], to account for this regiochemistry, was the first to propose that a complex, the exciplex in figure 1, was involved. The orientation within this was such, for the addition of electron rich alkenes, that the polar function of the alkene was distant from the carbonyl group. The reverse of this behaviour was experienced for electron deficient alkenes. Indeed later studies confirmed that the selectivity of the addition is largely due to electronic effects although sometimes steric factors can

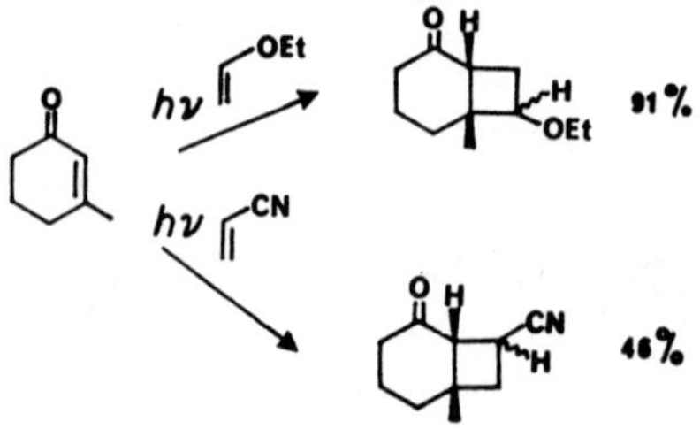

Figure 3

be important. Corey's postulates are based on the premise that enone excitation, like ketone excitation, is nπ* and thus leads to an excited state where the polarity of the carbonyl group is the reverse of that for the ground state.[5] Thus we have the usefully predictive rule outlined below in figure 4. In this figure the valence bond representation of a carbonyl group excited state is also illustrated and from this the origin of the dipole reversal can be seen.

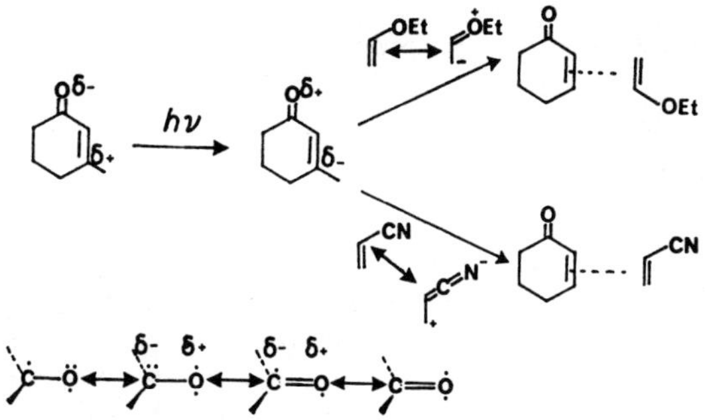

Figure 4

Occasionally the simple rules break down usually due to the steric interactions mentioned previously. Quantitative support of the use of the "complex" in predicting stereochemistry has come from work by de Mayo *et al.*[6] These results also demonstrated that such enone/alkene interactions are subject to solvent effects which in some instances can be related to a function of the dielectric constant of the medium in an approach first put forward by Kirkwood and Onsager.[7] Alternatively in situations where steric and strain effects can outweigh the "complex" approach the stability of the biradicals can be considered. Thus it is assumed that the more stable biradicals decay back to the starting material while the less stable intermediates lead to products. This approach is particularly useful in intramolecular additions where the **Rule of Five** [8] also applies i.e. in cyclizations there is a preference for the formation of five membered rings as shown in figure 5.

Figure 5

Thus there is a set of rules which is useful in the prediction of the outcome of the photocyclization. With these to hand the researcher is in a position to move on to the real syntheses. What will be obvious is that the initial step, the formation of the basic ring system is the only, or usually the only, photochemical step. Thereafter the functionalization, in the main, relies on normal ground state processes. The examples outlined throughout the rest of this review will draw on cycloadditions from both inter- and intra-molecular cyclizations.

3. Cycloadditions with Cyclopentenones and Related Compounds

There is little doubt that the photochemical method for the formation of cyclobutanes is superior to thermal paths to the same compounds. Often the ionic routes to cyclobutanes are extremely difficult to carry out. The first photochemical example chosen is probably one of the simplest total syntheses of a natural product, that of α- and β-Bourbonene(6 and 7 respectively) which are important components in *mentha piperita* and *Geranium bourbon oil*. The synthesis of these is illustrated[9,10] in figure 6 and involves the intermolecular photochemical addition of cyclopentenone(8) to 2-methyl-5-isopropyl cyclopentene(9). The utilization of the rules for prediction of regiochemistry suggests that the photochemical step will occur in the correct manner to yield the desired skeleton for the bourbonenes. The stereochemistry of the addition, due to steric interactions, will follow the *cis-anti-cis* arrangement and will also ensure that the isopropyl group will end up *cis* to the methyl group. The photo-addition turns out to be non-regiospecific, however, and affords a 1:1 mixture of the two adducts(10 and 11) as well as (2+2)-dimers of cyclopentenone. This mixture is readily separable by column chromatography. The last step, that of reaction with methylene triphenylphosphorane with isomer(10), goes in high yield and gives the desired β-bourbonene(7) which is readily converted into the α-isomer(6) by treatment with acid. This route provides the individual pure bourbonene isomers. An alternative path, addition of methyl lithium followed by dehydration with acid, yields a mixture of the two bourbonenes.

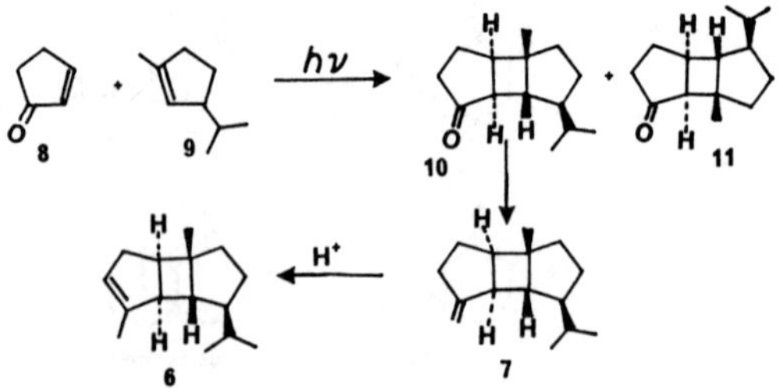

Figure 6

Another example of intermolecular cycloaddition of a cyclopentenone provides a synthetic approach to dehydrokessane(12) and 5-epikessane(13).[11] This reaction sequence (figure 7) involves the cycloaddition of 4-acetoxycyclopentenone(14) to the cyclopentene(15). The rules for regiochemistry predict that the isomer(16) should predominate. The cycloaddition confirms this and yields (16) as the major isomer along with (17) in a total yield of 80% and in a ratio of 3:1. Both isomers are of the *cis-anti-cis* configuration and both are converted by elimination of acetic acid into the cyclopentenones(18) and (19). The major isomer(19) is methylated by a 1,4-addition (yield 56%). The carbonyl group is converted to the thioacetal and removed by hydrogenolysis (yield 62%) and the two ester groups are reduced to alcohols(20) by reaction with lithium aluminium hydride. Tosylation of the primary alcohol group of (20) followed by solvolysis yields the ketone(21). In two steps this ketone is converted into the keto alcohol(22, 57%). Cyclization, using mercuric acetate followed by borohydride reduction, yields the kessane skeleton(23, 61%). Further reaction of (23) with $POCl_3$/pyridine yields dehydrokessane(12, 74%). Alternatively conversion of the alcohol (23) into the phosphoramide derivative(24) followed by reduction using lithium in ethylamine yields 5-epikessane(13, 71%).

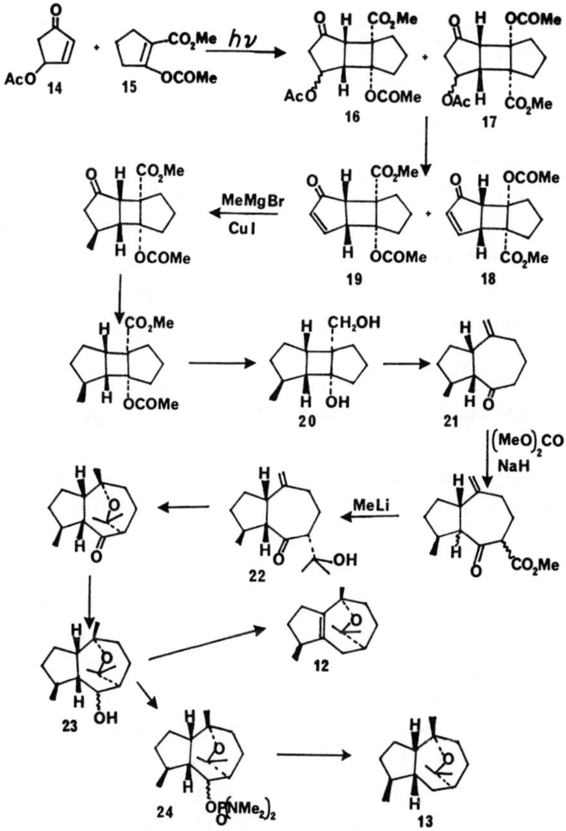

Figure 7

Intramolecular additions are also useful as can be seen in the facile synthesis of the skeleton of the natural product, hirsutic acid(25) from *stereum hirsutum* (figure 8). The photocycloaddition of enone(26) yields the unstable quadricyclic compound(27) which undergoes ready ring opening to yield a zwitterion

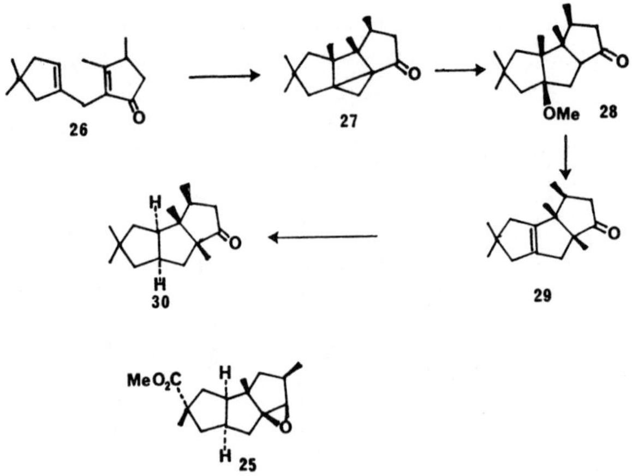

Figure 8

trapped by the solvent, methanol, used for the photolysis.[12] The result of trapping is the formation of the keto ether(28). Further chemical reaction with BBr_3 at low temperature in methylene chloride followed by ethanolic silver nitrate yields the alkene(29). Catalytic reduction produces the hirsutane skeleton(30). This approach to the hirsutanes ends at that point but it should be obvious that the route could be used to effect a total synthesis rather than the partial one described here.

4. Cycloadditions with Cyclohexenones and Related Compounds

A particularly elegant and simple synthesis by Corey[13] of the tricyclic sesquiterpene α-caryophyllene alcohol(31)(figure 9) was one of the early examples of the use of photochemistry in the synthesis of natural products. The photochemical step in this reaction is the addition of 4,4-dimethylpentene(32) to 3-methylcyclohexenone(33); the prediction of regiochemistry in this example is irrelevant since the alkene component is symmetric. The photo-addition indeed yields the required adduct(34). The development of that structure into a compound with the correct number of carbon atoms requires only the addition of a single methyl group which is achieved using methyl lithium. Treatment of the resultant alcohol(35) with acid provides the final step in the synthesis and produces α-caryophyllene alcohol(31). The acid treatment of the intermediate alcohol(35) yields an intermediate carbocation(36) within which bond "a" migrates producing the new carbocation(37). Solvation of this can only occur on the correct face to yield the desired product.

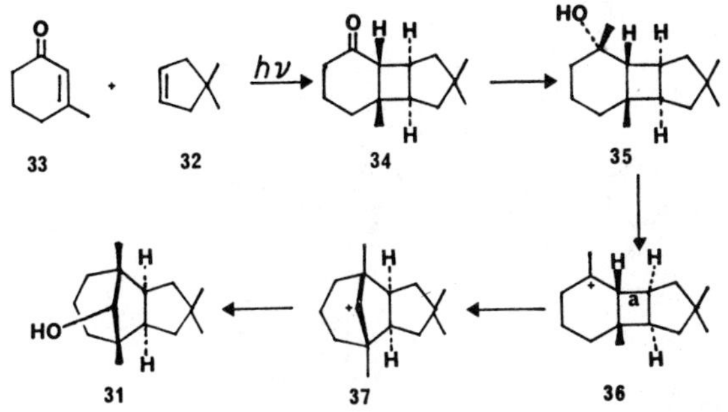

Figure 9

The above is truly a short but extremely elegant and well conceived synthesis of a fairly complex molecule. Thus we see, once again, the need for considerable skill in the manipulation of the photoadduct.

This is also seen in the many reported syntheses of Grandisol(38), one of the four pheromones[14] of the male Cotton Boll Weevil *Anthonomus grandis* Boheman. The set of syntheses illustrated[15-17] in figure 10 also helps to illustrate how relatively minor changes in the enone component can provide completely new synthetic approaches to the same compound. It is obvious from the structure of Grandisol(38) that the only photochemical step is the formation of the four membered ring. This is done using a variety of enones and irradiation in the presence of ethene as the alkene component of the reaction. This step proceeds to yield the cyclobutane component of Grandisol with the correct *cis*-arrangement of substituents. The manipulation of the adducts can be achieved by a variety of routes (figure 10). The sequences used are relatively simple and the simplest of these is the conversion of the lactone adduct(39), since this already has the appropriate oxygen functionalization for the side chain. Thus treatment with excess methyl lithium converts the adduct(39) into the diol(40) accompanied by 15% of the *trans*-isomer. Dehydration of the tertiary alcohol using acetic anhydride also converts the primary alcohol to the acetate. Grandisol(38) is then produced by reduction of the acetate using lithium aluminium hydride. As with the previous step Grandisol(38) is accompanied by the *trans*-isomer. By comparison the conversion of the adduct(42) requires the introduction of a double bond. This is achieved by bromination and elimination of HBr. Of the two adducts obtained by this path the one (43) required for the synthesis predominates. Reaction with methyl lithium followed by osmium tetroxide and sodium periodate yields the keto acid(44).

Methylene triphenyl phosphorane converts this to the adduct(45) which on reduction with $LiAlH_4$ yields Grandisol(38).

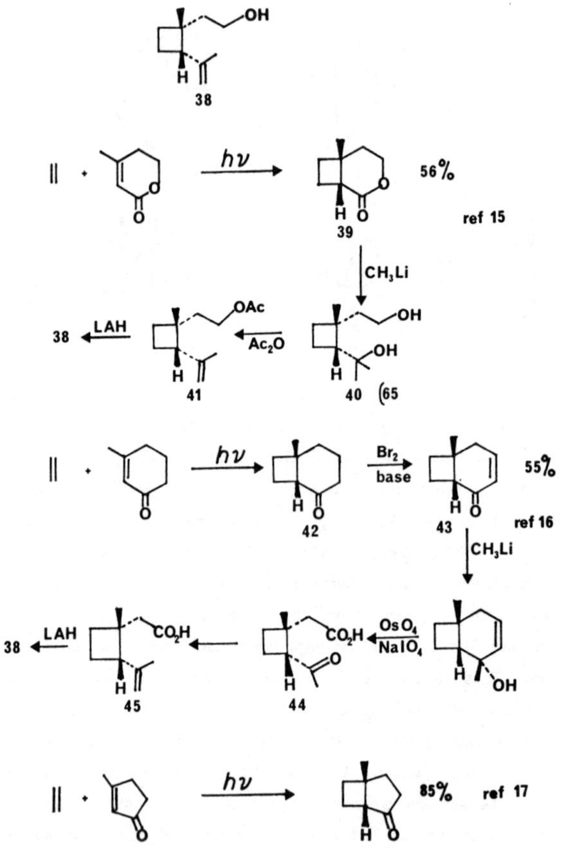

Figure 10

Considerable interest has been focussed over the years on the photochemical addition of allenes to cyclohexenone derivatives. Wiesner[18] has been one of the front runners in this area and he has used this approach as the basis for a variety of syntheses. The principal problem with this route was the interpretation of the stereo- and regio-chemistry. However, Wiesner[18] has proposed an empirical rule based on many examples to account for the outcome of the cycloaddition. This proposal hinges on the suggestion that the α-carbon of the cyclohexenone remains trigonal in the excited state while the β-carbon becomes tetrahedral. Overlap between the allene and the orbital on the β-carbon then leads to the observed product. In cyclohexenones where the two faces of the enone double bond are not identical the addition occurs to the less hindered face of the molecule. The example shown in figure 11 is a total synthesis of the phyllocladenes(46) and (47). The photochemical addition of allene to the optically active (+)-podocarpenone derivative(48)[19] gives the adduct(49) with the cyclobutane ring on the less hindered face. The adduct(49), on treatment with p-toluene sulphonic acid in benzene, rearranges by the cationic path shown yielding the bridged ketone(50, 50%). Wolff-Kishner removal of the carbonyl group yields the alkene, (+)-isophyllocladene(46) which can be readily isomerized using NBS to the (+)-phyllocladene(47).

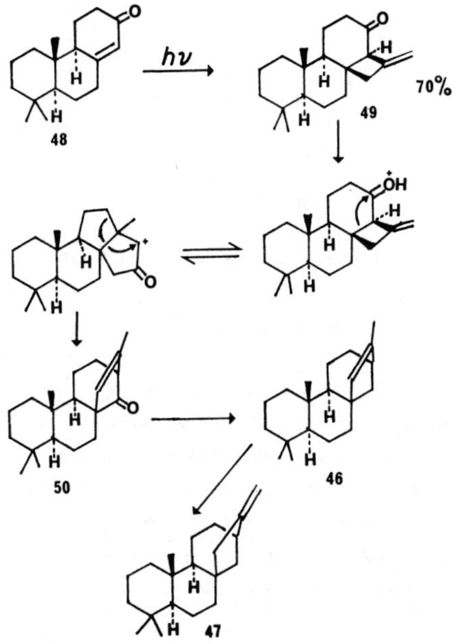

Figure 11

The intramolecular additions are also many and varied but only one example will be used in illustration. This type of cyclization has been used by Pirrung[20] to effect a total synthesis of isocomene(51). Irradiation of the enone(52) (figure 12) affords the correct adduct (53) in 77% yield. The carbonyl group in this ketone is resistant to attack by methylating agents. Thus the next step is acid catalyzed rearrangement to the isomeric ketones(54) and (55). The former, obtained in 75% yield, is reacted with methyl lithium to yield a mixture of alcohols(56) which is converted into isocomene(51) in 33% overall yield by treatment with formic acid.

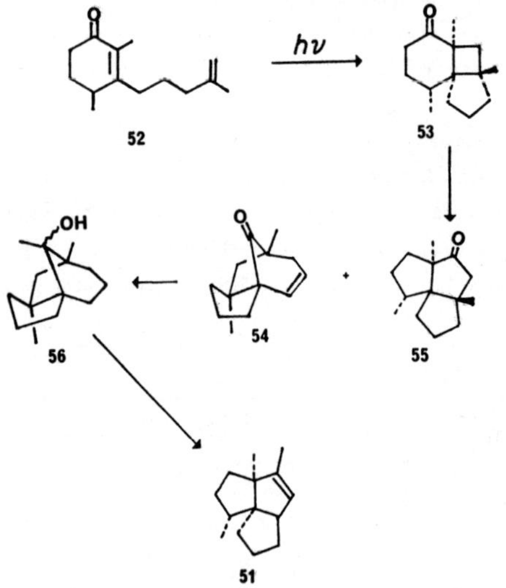

52 53 54 55 56 51

Figure 12

5. Cycloadditions with Enolized Diketones and Related Compounds

The de Mayo Reaction

Historically, if one can call a mere 20 or so years ago historical, this was the first type of (2+2)-photocycloaddition which was successfully carried out. In the first instance de Mayo demonstrated[21] that the reaction was useful as illustrated in figure 13. But, as a result of the ease by which the adduct undergoes a retro-Aldol reaction, the cyclobutane was not obtained but could only be inferred from the products. The 1,5-diketone(57) is formed and subsequently could be converted into the isomeric cyclohexenones by intramolecular Aldol condensation.

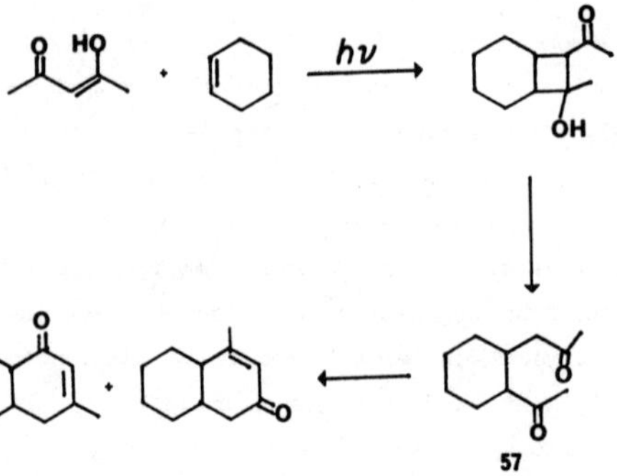

57

Figure 13

However it did not take long to demonstrate that this technique could be used to great effect in the synthesis of natural products. The variety of synthetic approaches is remarkable and only a few have been selected using both 1,3-diketones and 1,2-diketones .

6. Photocycloaddition to Enolized 1,3-Diketones. Acyclic systems

A simple approach to valerane(58) and isovalerane(59) arises from the (2+2)-addition of 1,2-dimethylcyclohexene to the enol(60). [22] This yields the unstable cyclobutane(61) which undergoes a retro-Aldol ring opening to yield the 1,5-dicarbonyl compound(62). Acid catalyzed cyclization of this affords the enone(63) which on reduction, reaction with isopropylidene triphenyl phosphorane, and reduction affords the products(58) and (59)(Figure 14).

OH hν 60 OH 61 CHO O 100% 62 O 63 58 59

Figure 14

7. Photocycloaddition to Enolized 1,3-Diketones. Cyclic Systems

A short and relatively simple synthesis of hirsutene(64), a fungal metabolite, has been reported. [23] The sequence(Figure 15) used in this was the photo-cycloaddition of 5,5-dimethylcyclohexane-1,3-dione as its enol to 2-methylcyclopent-2-enol to afford the two adducts(65) and (66) in a total yield of 86%. Because of the instability of the adducts, the problem of a second retro-Aldol reaction, they were separated by chromatography of the corresponding t-butyldimethyl silyl derivatives. This gave three of the four possible silyl derivatives in a ratio of 1:1:1. Two of these were shown to be isomers of the structure(67) where the only difference is in the stereochemistry of the silyl group while the third product was the other regio-isomer(67). The isomers were taken through to the next stage using low valent titanium to bring about coupling between the carbonyl groups. This yielded the alkenes(69) and (70) in yields of 17 and 38%. The alkene(69) was catalytically reduced with reduction occurring on the more hindered face presumably as a result of assistance by the oxygen function. Desilylation and oxidation is the final step and this regenerates the carbonyl group which is readily converted into a methylene by reaction with the appropriate Wittig reagent. The product of this sequence is hirsutene(64).

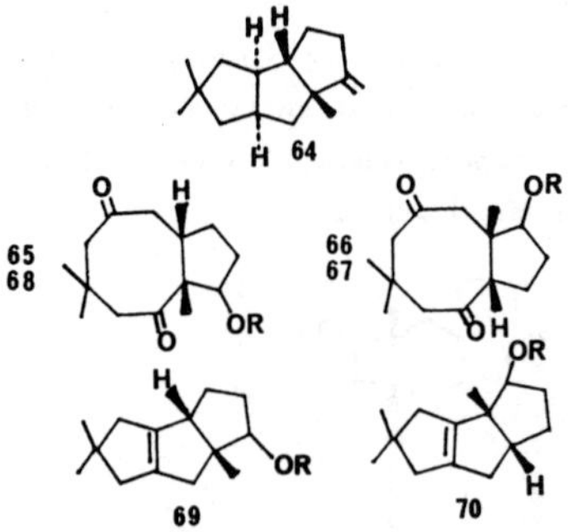

65, 66 R = OH
67, 68 R = $OSiMe_2t$-Bu
69, 70 R = $OSiMe_2t$-Bu

Figure 15

Another example of addition to a cyclohexanedione is aimed at the synthesis of the taxane skeleton(71), taxanes are unusually hydroxylated diterpenes found, for example, in the yew tree *Taxus baccata L* This involves the (2+2)-cycloaddition of the diketone(72) in its enol form to cyclohexene and affords the adduct(73, 56%) which can be ring opened by treatment with base to afford the desired structure(71)(Figure 16).[24]

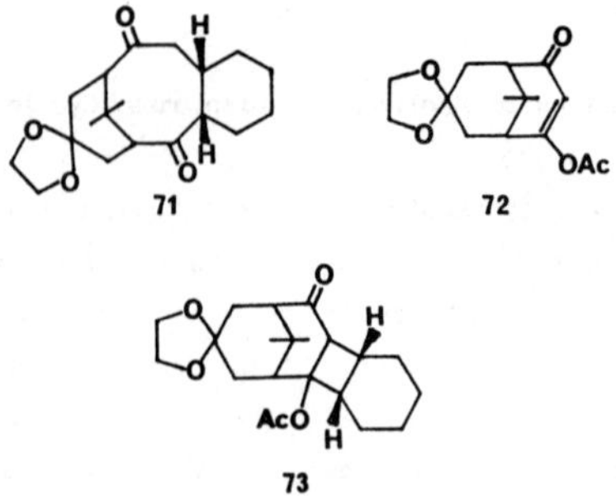

Figure 16

8. Cycloadditions with Enolized 1,2-Diketones

Cyclohexane-1,2-diones in the enolized form do not undergo (2+2)-cycloadditions with alkenes. This is also the case with 2-substituted cyclohexenones. However, the cyclopentane-1,2-diones do undergo addition. Their use in this respect is illustrated for the synthesis of acorenone(74). The cycloaddition of the enol acetate(75) to the methylene cyclopentane(76) affords the adduct (77).[25] The key step in the use of the 1,2-enol diketones is the base catalyzed rearrangement to the bicyclo(2,2,1)heptanone system. The hydrolysis and rearrangement in this affords the ketone(78) which is readily ring opened oxidatively with lead tetra acetate to yield the ketoester(79). This can be elaborated to the desired product, acorenone(74)[25](Figure 17).

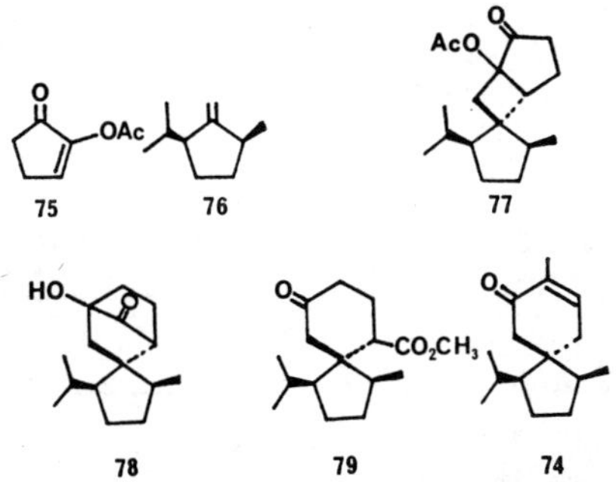

Figure 17

9. Intramolecular Cycloadditions to Enolized Diketones

The cycloadditions within this class are also useful for the synthesis of structures which can be converted into natural products. The one example chosen is the synthesis of β-bulnesene(80)(figure 18). The photocycloaddition of the enone(81) yields the adduct(82). Treatment of this with methyl magnesium bromide

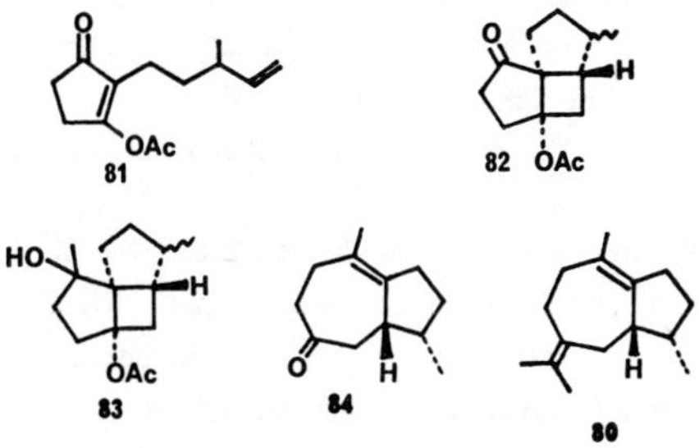

Figure 18

affords the tertiary alcohol(83). Hydrolysis of the acetate group followed by mesylation and solvolysis yields the enone(84). This enone is readily converted to β-bulnesene(80).[26]

A review of this sort can never hope to be exhaustive. It is hoped that the foregoing will have given a taste for the photochemical approach. Hopefully within the several examples given there are sufficient to illustrate the power of the method. But as was said at the outset, the photochemistry is only the start. The elaboration of the initial adduct is often the key to a successful synthesis.

References

1. A.C. Weedon in "Synthetic Organic Photochemistry", Ed. W.M. Horspool, Plenum 1984, p.61.
2. J. Kossanyi, Pure Appl. Chem., 1979, 51, 181-202; S.W. Baldwin, Org. Photochem.,1981, 5, 123-225.
3. E.J. Corey, J.D. Bass, R. Le Mahieu, and R.B. Mitra, J. Am. Chem. Soc., 1964, 86, 5570-5583.
4. P.de Mayo, Acc. Chem. Res., 1970, 4, 41-47.
5. J.C. Dalton and N.J. Turro , Ann. Rev. Chem. Phys., 1970, 499-560.
6. B.D. Challand and P. de Mayo, J. Chem. Soc. Chem. Commun., 1968, 982-983.
7. N. Berenjian, P. de Mayo, M.E. Sturgeon, L.K. Sydnes, and A.C. Weedon, Can. J. Chem., 1982, 60, 425-436.
8. N.J. Turro in "Modern Molecular Photochemistry", Benjamin Cummings Publishing Co Inc., Menlo Park, California, 1978, pp 429, 458 and references therein.
9. J.D. White and D.N. Gupta, J. Am. Chem. Soc., 1966, 88, 5364-5365.
10. J.D. White and D.N. Gupta, J. Am. Chem. Soc.,, 1968, 90, 6171-6177.
11. H.J. Liu and S.P. Lee, Tetrahedron Lett., 1977, 3699-3702.
12. J.S.H.Kueh, M. Mellor, and G. Pattenden, J. Chem. Soc. Chem. Commun., 1978, 5-6.
13. E.J. Corey and S. Nozoe, J. Am. Chem. Soc., 1965, 87, 5733-5735.
14. J.A. Katzenellenbogen, Insect Pheromone Synthesis: A New Methodology, Science, 1976, 194, 139-148.
15. R.C. Gueldner, A.C. Thompson, and P.A. Hedin, J. Org. Chem., 1972, 37, 1854-1856; H. Kosugi, S. Sekiguchi, R.I. Sekita, and H. Uda, Bull. Chem. Soc. Jpn., 1976, 49, 520-528.
16. R.L. Zurfluh, L. Dunham, V.L. Spain, and J.B. Sidall, J. Am. Chem. Soc., 1970, 92, 425.
17. R.L. Cargill and B.W. Wright, J. Org. Chem., 1975, 40, 120-122.

18. K. Wiesner, Tetrahedron, 1975, 31, 1655-1658; G. Marini-Bettolo, S.P. Sahoo, G.A. Poulton, T.Y.R. Tsai, and K. Wiesner, Tetrahedron, 1980, 36, 719-721; J.F. Blount, G.D. Gray, K.S. Atwal, T.Y.R. Tsai, and K. Wiesner, Tetrahedron Lett., 1980, 21, 4413-4416.
19. D.K.M. Duc, M. Fetizon, and S. Lazaire, J. Chem. Soc. Chem. Commun., 1975, 282; D.K.M. Duc, M. Fetizon, and S. Lazaire, J. Chem. Res. M, 1978, 167-177.
20. M.C. Pirrung, J. Am. Chem. Soc., 1981, 103, 82-87.
21. P. de Mayo, H. Takeshita, and A.B.M.A. Sattar, Proc. Chem. Soc., 1962, 119.
22. S.W. Baldwin and R.E. Gawley, Tetrahedron Lett., 1975, 3969-3972.
23. B.W. Disanayaka and A.C. Weedon, J. Chem. Soc. Chem. Commun., 1985, 1282.
24. H.Neh, S. Blechert, W. Schnick, and M. Jansen, Angew. Chem. Int. Ed. Engl., 1984, 23, 905-906.
25. G.L. Lange, E.E. Niedert, W.J. Orrom, and D.J. Walker, Can. J. Chem., 1978, 56, 1628-1633.
26. W. Oppolzer and R.D. Wylie, Helv. Chim. Acta, 1980, 63, 1198-1203.

12
Aromatic Compounds: Isomerisation and Cycloaddition

By P.A. Wender and T.W. von Geldern
DEPARTMENT OF CHEMISTRY, STANFORD UNIVERSITY, STANFORD, CALIFORNIA 94305, USA

Aromatic photochemistry is a relatively new field, tracing its origin back less than 30 years. Prior to this time benzene and its derivatives had been considered to be photochemically inert, and in fact were often chosen as solvents for photochemical reactions. In 1956, as a control experiment for a study of radical generation and trapping[1], Bryce-Smith and coworkers irradiated isopropylbenzene with UV light and observed the formation of a yellow compound in solution. Subsequent experiments[2] with benzene led to the identification of fulvene as product, and a new area of research was opened.

Since these early studies numerous advances have been made in the field of aromatic photochemistry. The photoexcited arene has been shown to undergo various isomerization reactions, and to react with a range of substrates to give addition products. These processes have been subject to frequent review[3]; it is the intent of this chapter to provide the reader with a brief, general overview of the field, highlighting its synthetic merit.

ARENE PHOTOISOMERIZATIONS

Early work in the field understandably concentrated on the isomerization of aromatic rings. Photophysical studies on the system indicated a primary photoproduct, arising from the S_1 singlet[4], which underwent rapid return to the ground state. Clues to the nature of this intermediate came from the photolysis of benzene in the presence of acids; the products were shown[5] to have the unusual bicyclic structure **1**. It was postulated that the non-

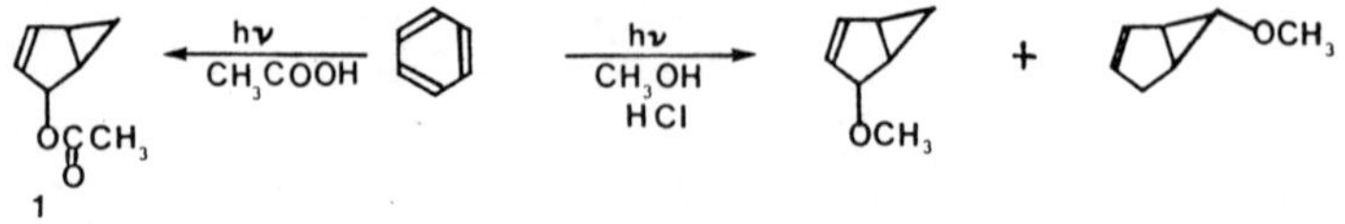

1

aromatic valence isomer **2**, termed "benzvalene," is produced in the photoreaction by way of a "prefulvene" 1,3-diradical; these intermediates react further to provide the observed products. The formation of benzvalene was confirmed by its subsequent isolation;[6] the addition of acid was demonstrated to be purely a thermal process.[4]

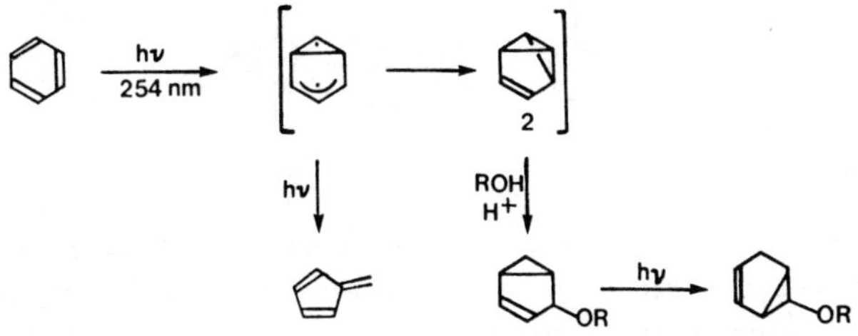

As more substituted benzenes began to be studied, a new mode of photoreactivity was recognized. Photolysis of o-xylene at 254 nm results in the formation of m-xylene as a main product.[7]

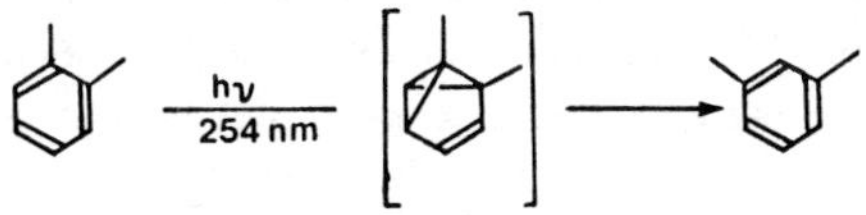

Experiments with triplet quenching agents indicated that the conversion occurs from a singlet[8]; the specificity of the isomerization suggested benzvalene as a logical intermediate. Further evidence for the mechanism came from studies on trisubstituted arenes[9]; in the case of tri-t-butyl benzene the intermediate benzvalene has been detected.[10] Benzvalene intermediates have also been postulated for the interesting interconversion[11] of 2,2-cyclophanes. The isomerization has synthetic utility for the preparation of bicyclo[3.1.0]-hexanes; photolysis of phenol in the presence of acid gives a 20% yield of **3**.[12]

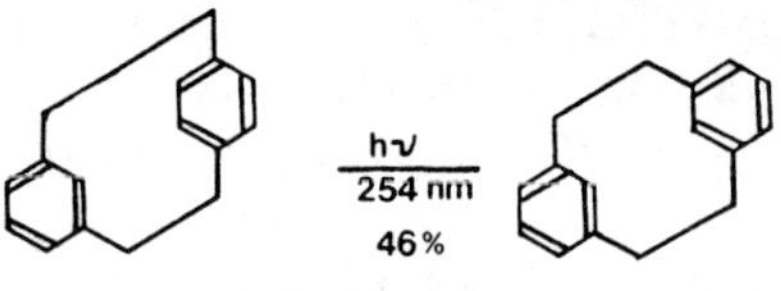

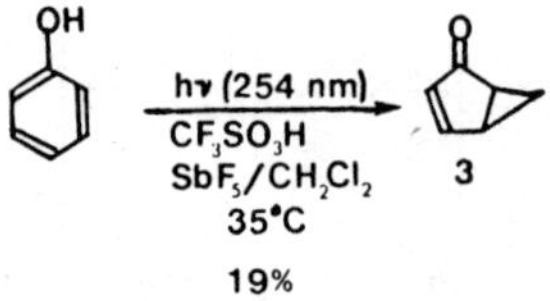

Not all photoisomerizations of arenes implicate benzvalene or the prefulvene diradical. When benzene is photolysed using 180 nm light a new product is produced in addition to those previously described. This product has been shown to be the Dewar isomer of benzene,[13] and has been suggested to arise via a non-planar 1,4-diradical. Such a species is predicted to correlate with the S_2 singlet of benzene[14]; selective excitation into S_2 gives primarily para-bonded products.[15] Irradiation of hexafluorobenzene yields the Dewar isomer selectively,[16] and the perfluoroalkyl benzenes exhibit a similar tendency.[17] More importantly, tri-t-butylbenzene gives a single, isolable para-bonded product,[18] highlighting the fact that substitution by bulky groups stabilizes the valence isomers.

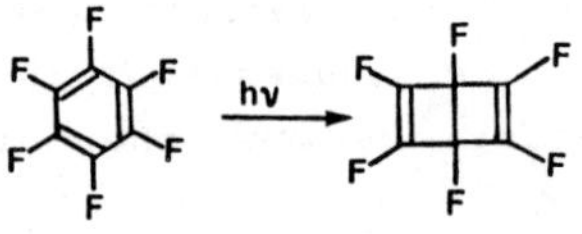

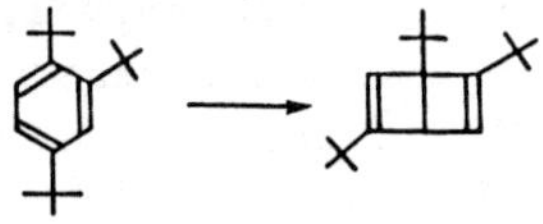

Cyclophanes exhibit an enhanced tendency toward valence isomerization. The conversion of [6]-paracyclophane into the Dewar isomer upon mercury-lamp irradiation is rapid and efficient; the thermal reversion of this product is quite facile.[19] When the aromatic ring is substituted with two carboxylic ester groups the single product has the prismane skeleton **4** ![20] Prismanes are also produced during photolysis of other arenes, for example hexakis-(trifluoromethyl)benzene,[17] and are generally assumed to derive from the corresponding Dewar isomers, although there are exceptions.[20]

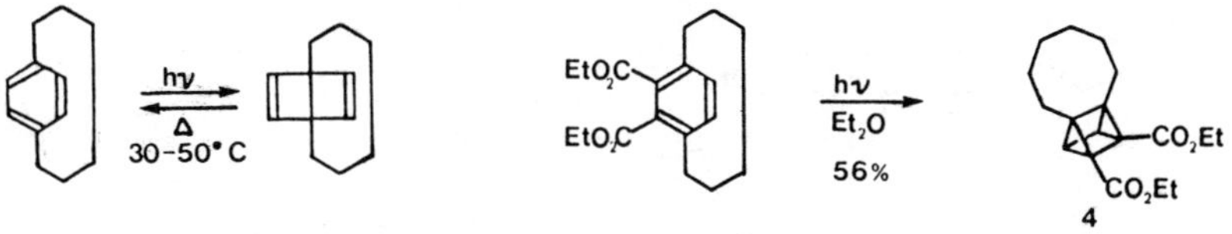

Naphthalenes convert photochemically to both the valene and Dewar isomers;[21] the choice of product depends on substitution, and seems to be related to steric factors, with congestion minimized in the latter. In a similar manner 9-t-butylanthracene is converted into the 9,10-bonded product[22]. The isomerizations of hetero-aromatics are also interesting. Photolysis of pyridine produces the Dewar isomer **5**,[23] which has a relatively short lifetime, but may be trapped by reduction with $NaBH_4$. Perfluoroalkyl substitution serves to stabilize the valence isomers, and a wavelength-dependent conversion to either the Dewar or prismane is observed.[24]

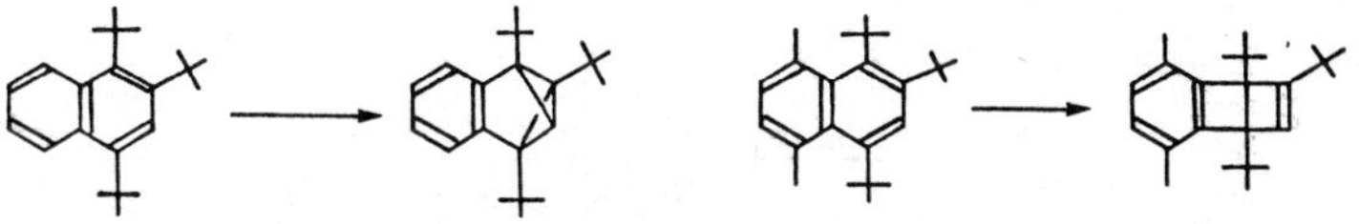

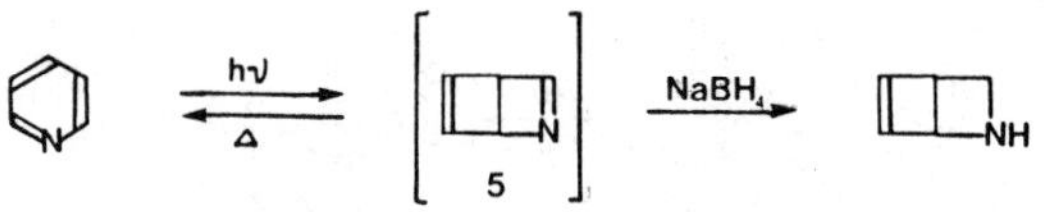

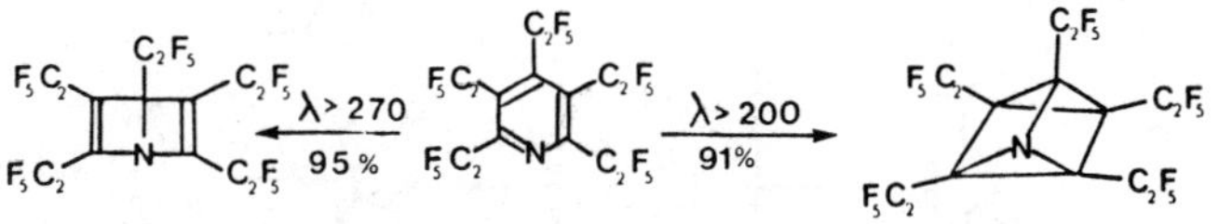

Photolysis of pyridylacetonitrile at 254 nm leads to anthranilo-nitrile by way of the isolable alternative Dewar isomer **6**.[25] A similar process converts halogenated pyridazines into the corresponding pyrazines;[26] isolation of intermediates **7** and **8** is believed to define the reaction pathway. On the other hand, pyridinium ions apparently react <u>via</u> the heteroniabenzvalene,[23,27] as do pyrylium ions.[28]

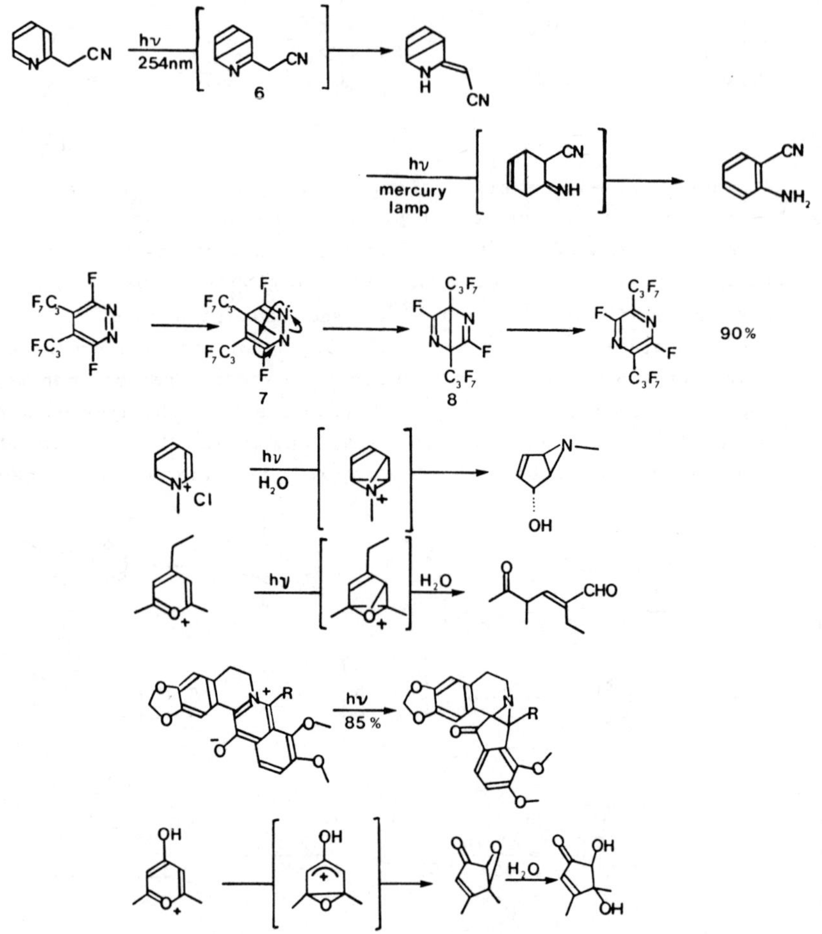

CN hν 254nm N CN 6 N H CN

hν mercury lamp CN NH CN NH_2

F F_7C_3 N N F_7C_3 F → F_7C_3 F N N F_7C_3 F 7 → C_3F_7 F N N F C_3F_7 8 → C_3F_7 F N N F C_3F_7 90%

N+ Cl hν H_2O N+ N— OH

+ hν H_2O + O CHO

N+ R O hν 85% N R O

OH + OH + O O H_2O O OH OH

Photoisomerizations also occur in five-membered heterocycles. Permutation of substituents has been observed in furans[29], pyrroles[30] and thiophenes[31]; Dewar isomer **9** and cyclopropenecarbonyl compound **10** (and sometimes both) have been postulated as intermediates, but the mechanistic details have not been completely marked out. Oxazoles and isoxazoles interconvert,[32] as do pyrazoles and imidazoles.[33] For the former transformation an acylazirene has been demonstrated as an intermediate in some cases, while the latter is postulated to occur via internal cyclization/1,3-shift. Photoisomerization of both starting materials and products is possible, and the actual processes are quite complex.[34]

hν 1,3 1,3

9

hν

10

hν

254 nm

> 300nm

hν

ARENE PHOTOADDITIONS

Efforts to trap the proposed photoisomerization intermediates with alkenes led to the discovery of several new reactions. Irradiation of benzene in the presence of maleic anhydride gave an intermediate cycloadduct in which the alkene carbons attached to ortho (1,2-) related carbons of the arene; subsequent interest in this process led in the sixties to the discovery of two alternative modes of cycloaddition, one involving meta- or 1,3-bonding and the other para- or 1,4-bonding. Other functional groups, including alkynes and amines, were found to add to excited arenes, and in the late sixties an intramolecular version of the meta-addition was reported. The eighties heralded the first application of this process to complex molecule synthesis, along with a delineation of the factors influencing stereochemistry and stereoinduction in this reaction class.

MECHANISTIC CONSIDERATIONS

The work of Houk,[35] Gilbert,[36] and Bryce-Smith and coworkers[37] has provided a qualitative theoretical basis for understanding the partitioning between ortho-, meta-, and para-cycloaddition. For the simple case of benzene and ethylene, the relevant electronic configurations and their corresponding energies are as shown below.

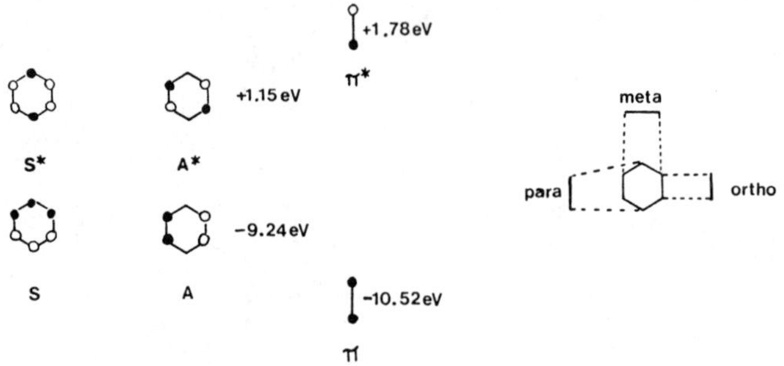

+1.78eV
π^*
S* A* +1.15eV
S A −9.24eV
−10.52eV
π
meta para ortho

The actual molecular orbitals for benzene are combinations of these; the lowest excited singlet is $^1B_{2u}$ (SA*-AS*). The ethylene π-orbital interacts with benzene S to stabilize meta-, and with A to stabilize ortho-addition. π^* and S* cannot interact; with A* both ortho- and meta-addition are supported. Consequently, only meta-approach of ethylene completely stabilizes S_1, specifically the S→A* component. In fact meta-addition is predicted to predominate whenever the alkene HOMO and LUMO are close in energy to the SOMOs of the excited arene; this is analogous to Bryce-Smith and Gilbert's generalization[38] that alkenes with IP's similar to that of benzene give meta-products.

Substituent effects can change this picture. The placement of a donor group on the olefin, or an acceptor on the aromatic, raises the energy of π and π^* relative to the benzene orbitals. When the filled π pushes above half-filled S or A, charge transfer may occur. $\pi \rightarrow$A charge transfer selectively stabilizes ortho-approach, while $\pi \rightarrow$S favors meta-addition. Because overlap of the relevant orbitals is most efficient with the ortho carbons of the arene, the former process is expected to be favored. On the other hand, acceptor substitution on the olefin (or donor on the arene) lowers the relative energies of π and π^*. As the empty π^* falls below half-filled S* or A*, charge transfer again becomes a major process; the "allowed" A*$\rightarrow \pi^*$ stabilizes both ortho- and meta-addition. In this case two additional factors stabilize ortho-approach: 1) interaction between S and the new half-filled orbital (A*+π^*) resulting from A*$\rightarrow \pi^*$ charge transfer, and 2) more efficient ortho-overlap. Para-addition is unlikely due to poor orbital overlap and because the only stabilizing interactions, S $\rightarrow \pi^*$ and S*$\rightarrow \pi$, are clearly minor.

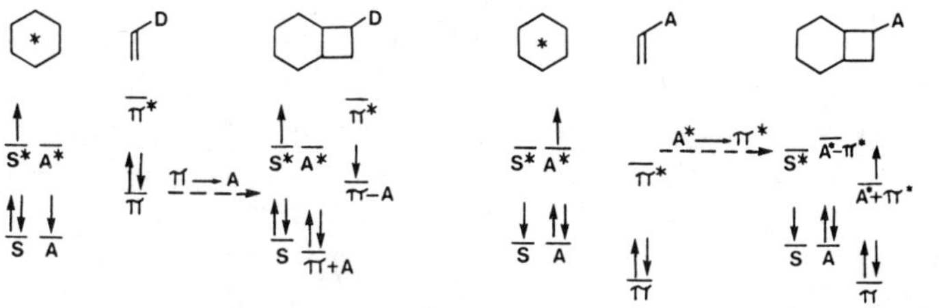

In summary, the theory predicts that the ortho-cycloadduct will predominate whenever charge transfer is an important process. This premise has found strong support in the recent work of Mattay[39], who has correlated product distribution with ΔG(charge transfer) for a large number of photoadditions. His empirical rules also predict alternate modes of reaction, e.g. substitution. Exceptions, and apparent exceptions, to these simple selection rules have been observed, however. Several examples in particular highlight the complexity of the process and the difficulties of drawing conclusions based on work performed under differing conditions. The cycloaddition of vinylene carbonate,[40] a strongly donor-substituted olefin, with benzene was initially reported to give 1,3-products, in contravention of the selection principles; however, as Gilbert has pointed out, many 1,2-products are photolabile, and the conditions of the reaction (high arene concentration) are conducive to triplet-sensitized photodecomposition of these products. More thorough studies, of the dichlorovinylene carbonate/benzene system,[41] indicate that 1,2-products actually predominate, and implicate charge transfer. Conversely, reaction of benzonitrile with acceptor-substituted olefins,[42] predicted to proceed *via* *meta*-addition, often occurs instead through 1,2-closure; it has been shown that it is the olefin which is excited in these cases. This emphasizes that our theoretical model refers specifically to the interaction of S_1 arene with S_0 olefin; the reverse case of S_1 olefin $\rightarrow S_0$ arene can give *ortho*-cycloaddition *via* a concerted process.[14]

1,2-PHOTOADDITIONS

In an early attempt to confirm the formation of fulvene, benzene was irradiated in the presence of maleic anhydride. In this case the reaction took an entirely different course, forming the 2:1 adduct **12**.[43] *Ortho*-adduct **11** was later demonstrated conclusively to be an intermediate[44], and the initial cyclization was shown to involve charge-transfer[45] from the S_1 singlet of the arene.[38]

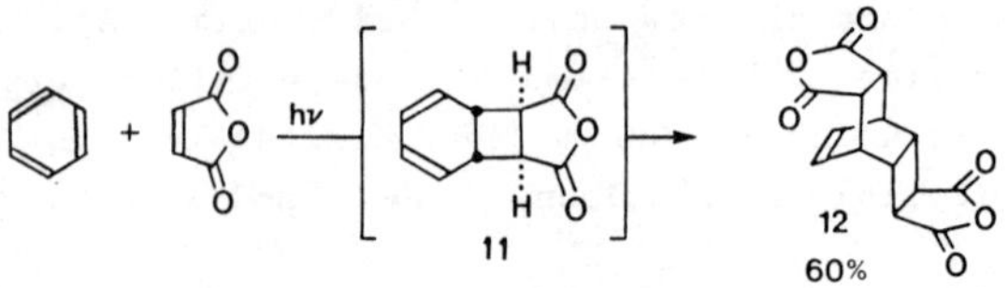

Ortho-cycloaddition is general for dienophilic olefins; for example benzene and acrylonitrile give the main photoadduct **13**.[46] The products usually have the substituent placed in an *exo*-orientation;[46,47] this has been rationalized as resulting from an *ortho*-exciplex, in which the acceptor group is expected to interact with the aromatic S* in an antibonding sense.[35] Significantly, weaker withdrawing groups are poorer *exo*-directors; for example, methyl acrylate gives a mixture of *exo* and *endo* isomers with benzene.[47] When the aromatic ring is substituted, there is also a question of regioselectivity. Stronger donor substituents lead to greater regiocontrol; the toluene-acrylonitrile cyclization gives about a 5:2 ratio of 1,2- *vs*. 3,4-addition[36] (with the *exo*-orientation preferred), while anisole is completely selective for the 1,2-*exo*-product. This is consistent with the overall M.O. picture for the reaction. A donor substituent at position 1 increases the orbital coefficients at carbons 2 and 6 (and less so

CN hν CN

13

CN hν CN CN CN

4 : 1 : 2 (2:1)

OMe CN hν CH_3CN OMe CN

95% of product

at 4); this encourages bond formation involving the 2-carbon (i.e. 1,2-addition) and also increases the antibonding character of the secondary orbital interaction (leading to exo-product). As expected from the charged nature of the transition state, solvent effects on selectivity[36] and rate[38,48] have also been observed. Further transformations of the photoproduct, particularly those which restore aromatic character, are possible[49]; in one especially interesting case, ring expansion of the intermediate cyclobutane leads to heterospirocycles in good yields.[50]

$N^{\oplus}$ H CO_2Me hν CH_3CN $NH^{\oplus}$ CO_2Me NH CO_2Me H

50%

Ortho-addition with the olefin as donor is also common; benzene and 1,4-dioxene react to give the 1,2-adduct **14**,[51] and enol ethers often give similar results.[47,52] The ortho-cycloaddition of

O O hν H O O H 14 MeO OMe hν OMe OMe H^+ O

O hν (254) Φ=0.7 H O H

dimethoxyethylene has been employed in a convenient synthesis of cyclooctatrienone.[52] Charge transfer is encouraged by electron-withdrawing groups on the arene, and a number of examples of donor-acceptor pairs have been reported.[36,53] When the aromatic substituents are sufficiently good acceptors (viz. benzonitrile[53] or perfluorobenzene[54]), even simple olefins can act as donors. Photolysis of trimesic ester and cyclooctene provides the interesting triscyclobutane **15**[55], apparently by way of an ortho-adduct. Competition from oxetane (or azetine) formation often complicates these processes,[56,57] and it has been suggested that the

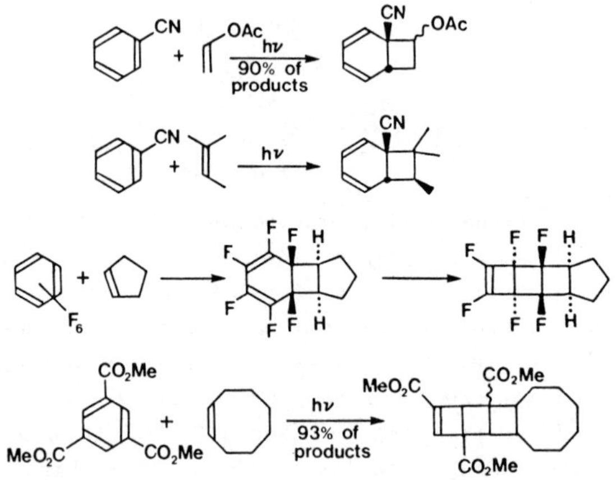

cyclization onto the aromatic nucleus arises from a triplet ($\pi \rightarrow \pi^*$) state.[57]

The stereochemistry of the donor substituent was originally assumed to be endo-[38]; subsequent reinvestigations have demonstrated that the products are in fact exo-,[47] a result more in line with the theoretical prediction that secondary orbital interactions between substituent and aromatic should be antibonding.[35] These inconsistencies may be related to the observation that the exo-adducts in particular are photolabile,[52] and are selectively destroyed upon continued irradiation. Such distinctions between "kinetic" and "thermodynamic" product distributions have been considered all too rarely during studies of photochemical processes in general.

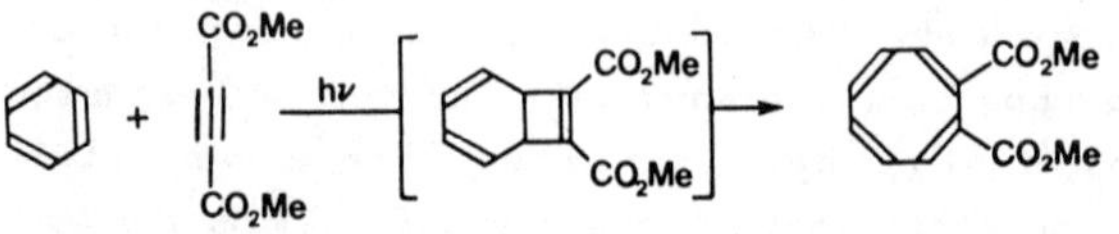

Ortho-cycloaddition is also a major photoprocess when the olefinic partner is replaced by an acetylene. In this case the intermediate reacts further, providing cyclooctatetraenes.[58] The acetylenic partners generally have conjugating groups attached; it has been demonstrated that often the alkyne, not the arene, is excited.[59] This is in line with the prediction that the interaction of S_1 olefin and S_0 arene selectively stabilizes 1,2-cycloaddition.[37] While cyclization of simple non-conjugated alkynes does occur, the process is generally quite inefficient.[37,60] With cyclooctyne the corresponding cyclooctatetraene is formed in high

hν, Φ< 0.001 (no semibullvalene forms)

hν, ØH — mixture of isomers — stops at 1.9:1 product : octyne

yield, but the reaction stops at partial completion, presumably due to autoinhibition.[61] The reaction has some synthetic utility[62,63]; phenylacetylene and hexafluorobenzene give 86% of a mixture of two products **17** and **18**, which may be separated.[64] More importantly, it is possible to isolate the putative bicyclic 1,2-intermediate **16** under certain conditions. An intramolecular variant, occurring <u>via</u> the arene singlet, has been explored[61].

16 — 17 + 18 86%

hν, Φ=0.0033 — mixture of isomers

Substituted naphthalenes add 1,2- acting as donors or acceptors, or in charge-balanced systems[65]. 1,2-Addition can also represent a

hν, CN, 36% — OH⁻, 76%

hν, Φ=0.68

significant event in the photochemistry of heteroaromatics. A variety of olefins[66] and acetylenes[67] add across the 3 and 4 positions of perfluoropyridine; the exo product predominates. With five-membered heterocycles ortho-addition is often the major process, particularly in the case of acetylenes; however, many other photoreactions can compete, and the actual course of the reaction is quite substrate-dependent.

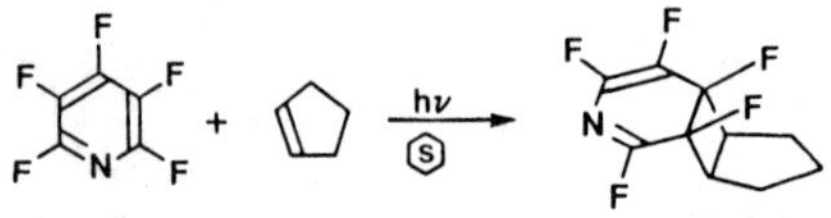

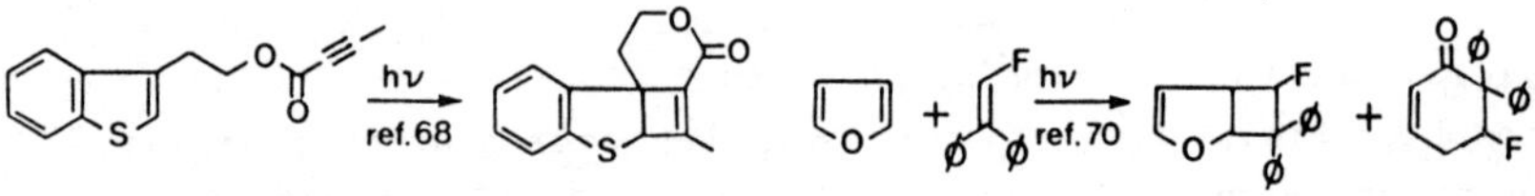

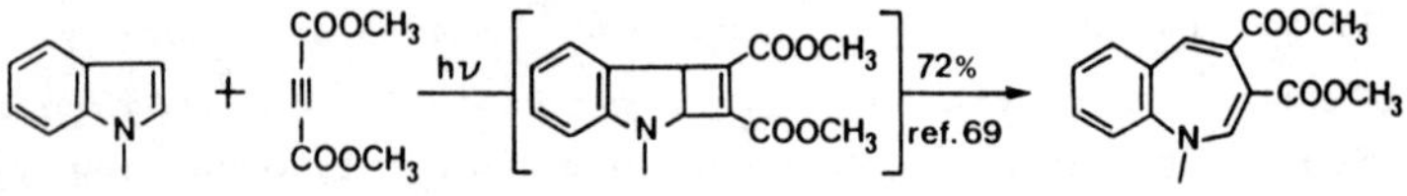

1,4-CYCLOADDITIONS

As predicted, poor orbital overlap between excited arene and ground-state olefin makes para-cycloaddition a relatively uncommon event. Not surprisingly, the first observed and most prevalent examples of this process involve the cyclization of dienes, where these difficulties are minimized. Reaction of benzene and butadiene is complex, but the major products derive from the labile para-adduct **19**.[71] Isoprene has been employed in an efficient synthesis of **20**.[72] Several 1,4 adducts have been suggested and employed as precursors to novel benzene dimers.[73]

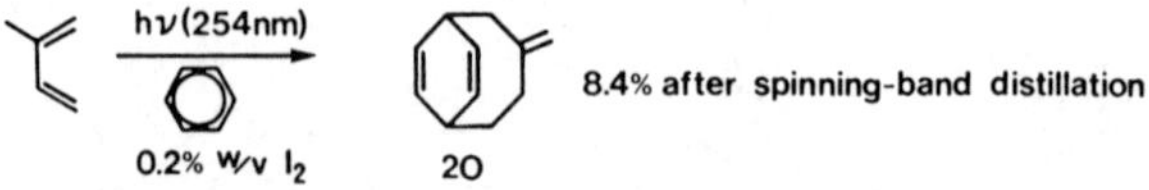

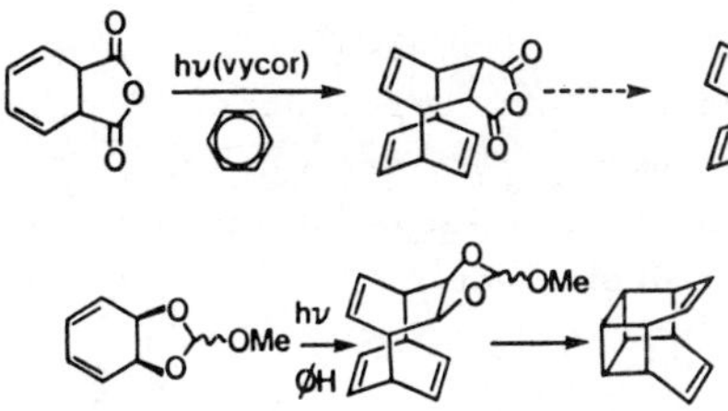

Isolated double bonds only occasionally enter into 1,4-cycloadditions; the most notable examples are allenes.[74]

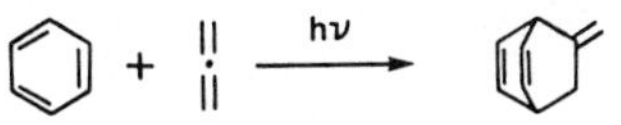

Intramolecular cyclization of phenethyl vinyl ether leads primarily to the 2,5-addition product; this is interesting in light of the fact that 5-phenylpentene and allyl benzyl ether both give 1,3-products under similar conditions,[75] but no satisfactory explanation has been provided.

Extension of the aromatic system tends to encourage 1,4-addition. A number of examples of additions of 1,3-dienes and simple olefins to napthalene have been reported[76]; for anthracene this often represents the main photoprocess, and has been employed in the synthesis of the 1,2- and 1,4-benzene/anthracene adducts.[77]

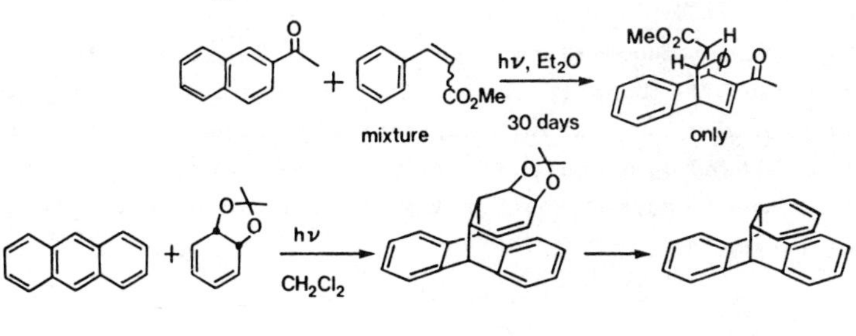

A final important example involves the 1,4-addition of primary and secondary amines. This quite general process[78] has been suggested to involve charge transfer from the amine to the S_1 arene. Tertiary amines also undergo para-addition, but have the amine α-carbon as the site of attachment.

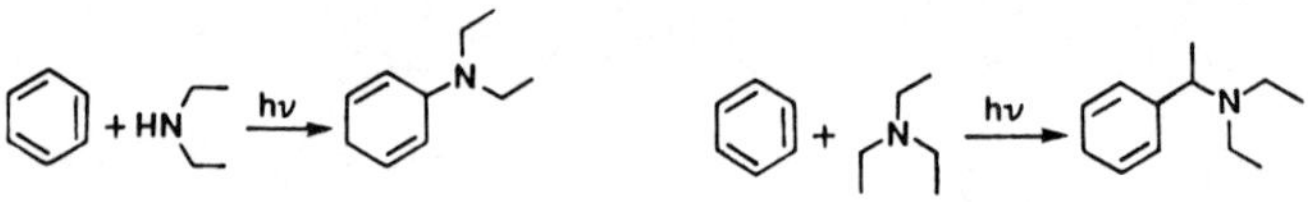

1,3-CYCLOADDITIONS

The "strange process"[79] of meta-cycloaddition was first reported independently by two groups in 1966.[80] Despite the massive bond reorganization involved in this transformation, it appears to be quite facile and general for olefins having simple, non-polar substituents. Additions to cis- and trans-2-butene proceed with retention of olefin geometry,[81] suggesting a singlet intermediate. Careful studies of quantum yields for several systems indicate an interaction between S_1 arene and S_0 olefin.[82] The endo adduct is generally preferred over the exo, and is often the exclusive product. A number of apparent exceptions to this "endo rule" were later found to have been misassigned.[81]

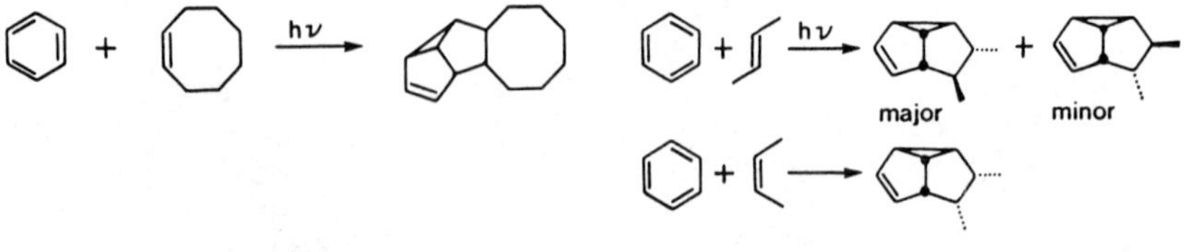

major

minor

Three basic suggestions have been made regarding the course of the reaction.[36] Bond reorganization in the aromatic may precede interaction with the olefin; alternatively, arene-olefin association may be followed by promotion to an exciplex[83], or the process may be entirely synchronous. Most of the mechanistic data at the present time are in line with the exciplex model. Toluene and the various xylenes react with cyclic olefins to give <u>meta</u> products in which the 1-position is always substituted.[84,85] Results are similar with

anisole;[85,86] in both cases only <u>endo</u> products are observed. This regiochemistry is not expected for addition to the most stable

arene, but <u>is</u> consistent with the formation of a stabilized exciplex. To test the possibility of a diradical intermediate, Sheridan has prepared diazenes **21** and **22**.[87] Photolysis of each of these compounds gives the same product ratio observed in the photochemical process. In addition, Scharf and coworkers have recently observed what they believe is exciplex fluorescence in a benzene/1,3-dioxole system.[88] These results support a concerted but

asynchronous process. The <u>endo</u> orientation is predicted in the theoretical treatment, which makes no distinction as to the timing of bond formation;[35] interaction between arene S and olefin π (or

between A* and π*) occurs in a favorable (bonding) sense. The steric bulk of a 1-substituent also serves to favor the _endo_-orientation.

The further transformations of these _meta_-cycloadducts were first studied by Srinivasan. He explored cyclopropane cleavage _via_ suprafacial 1,5-hydrogen shift as a method for identifying the photolysis products; thus, the minor compound from a benzene/_trans_-2-butene cycloaddition thermolyzes at 300°C to give **23**, while the major is inert at this temperature.[89] Vinylcyclopropane rearrangement, which interconverts the two photoproducts, is also possible at higher temperatures. The conversion of benzene/bicycloheptadiene adduct **24** to **25** upon thermolysis at 230°C, while formally similar, is mechanistically quite distinct.[90]

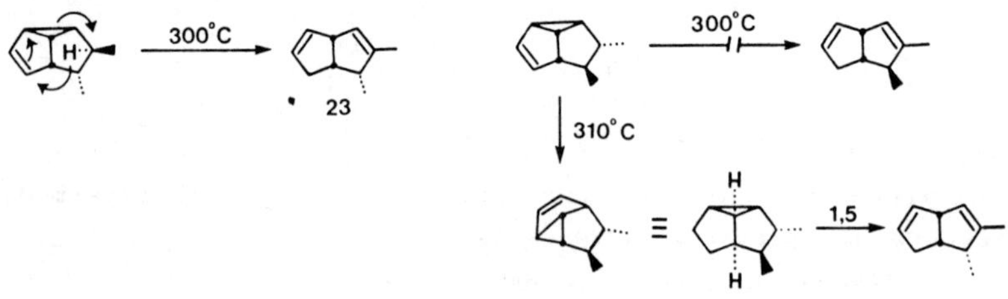

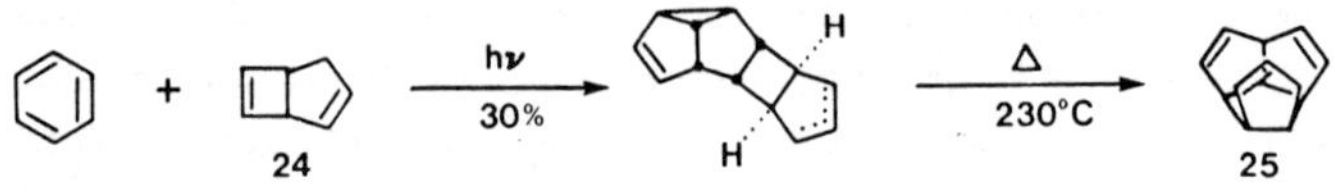

The same net transformation results from treatment of the photoadducts with anhydrous acid, and has been used to make [4.3.3]-propellanes.[91] When anisole is the aromatic partner, acid treatment provides a bridged ketone, which has been cleaved again to the perhydroazulene skeleton.[92] A number of unsuccessful attempts have been made to prepare semibullvalene from 1,3-adducts[93,94].

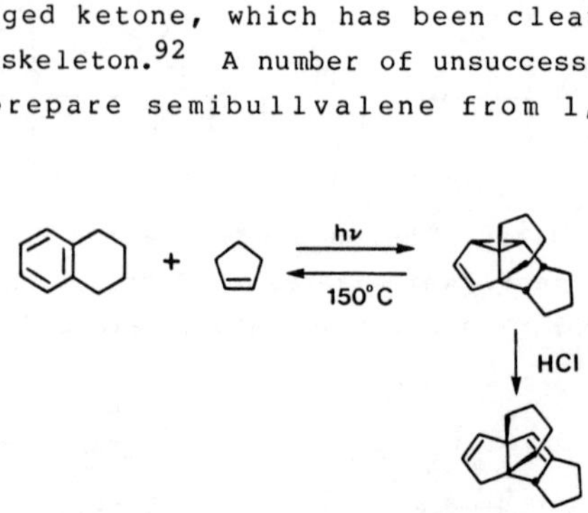

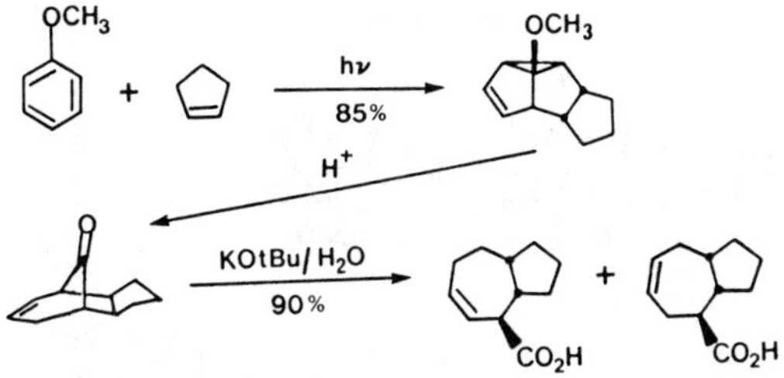

The *meta*-cycloaddition, which results in the greatest increase in molecular complexity[95] of any general reaction, is the only one whose use in natural products synthesis has been explored to any great extent. This work, largely developed in our laboratories, was based on the recognition that the reaction provides a general and powerful approach to a number of commonly encountered skeleta. At the outset of our studies, only Morrison's work[83] provided any information on intramolecular *meta*-additions. He found that 6-phenyl-hex-2-ene formed a singlet exciplex efficiently and gave two major products upon photolysis; *cis*/*trans* isomerization of the olefin was a minor process. The interesting regiochemical course of the reaction was not speculated upon.

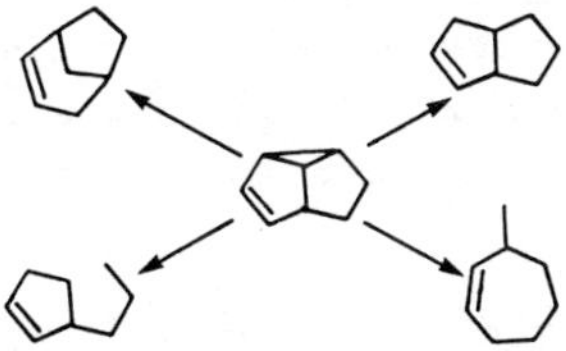

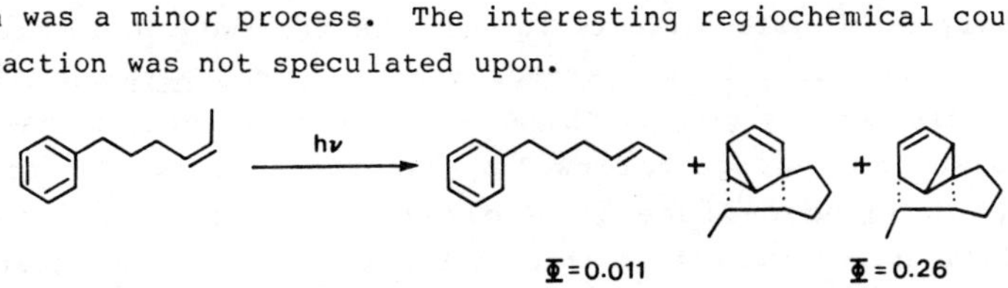

$\Phi = 0.011$ $\Phi = 0.26$

Our earliest study, which led to the synthesis of (±)-α-cedrene[96], was designed to test mode, regio-, and *endo*/*exo* selectivities along with the as-yet unexplored question of stereo-induction attending the process. *Meta*-addition of **26** was expected on the basis of previous theoretical and experimental studies; it was felt that the low efficiencies observed for intermolecular addition to *ortho*-substituted anisoles would not jeopardize the intramolecular case. While addition across both alkyl and methoxy groups is electronically favorable, non-bonded interactions in the

Δ

hν 65%

26

1:1

1. Br_2
2. nBu_3SnH

α-cedrene

exciplex leading to *ipso*-addition was expected to disfavor the former process. Overlap between the olefinic carbons and the aromatic *meta*-positions would appear to be most efficient when the sidechain is placed *exo*-, in contrast to the *endo*-orientation which prevails in intermolecular cases. Stereoinduction was expected to arise *via* conformation **27**, in which the interaction between the methyl and aryl methoxy are minimized. In practice, these

ipso exo endo 27

expectations were all realized, and stereoinduction was complete; the only photolysis products formed were the two vinylcyclopropane isomers resulting from stereocontrolled *exo*-, *meta*-cycloaddition across the methoxy group. These isomers conveniently converge upon brominative cleavage followed by dehalogenation. Our four-step route to (±)-α-cedrene gave clear indication of the design simplification inherent in this powerful reaction; modifications to the route allowed for an 8-step synthesis of pipitzol.[97]

I_2 THF/H_2O

pipitzol

These factors were explored further in the synthesis of isocomene.[98] Again stereoinduction is complete, and olefin geometry is retained; the cyclopropane ring is cleaved *via* the known homo-1,5-shift, which can occur only in isomer **28**. The two photoisomers converge as before, this time via photochemical (or thermal) vinylcyclopropane rearrangement.

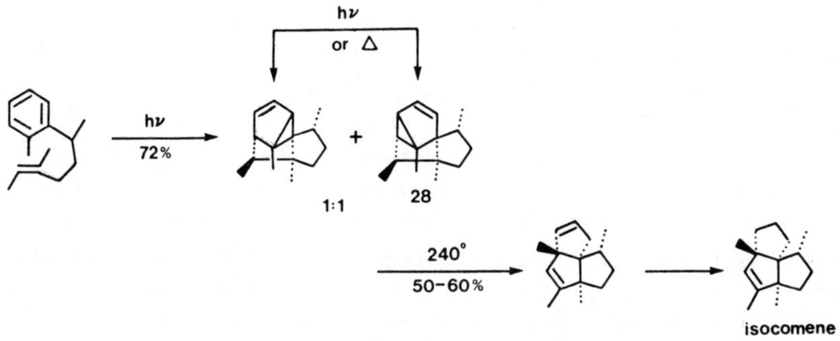

The synthesis of isocomene is also significant in that it opens up access to the angularly-fused triquinane skeleton. A number of natural products have this skeleton, and their synthesis has allowed for further study of the reaction. In our studies on silphinene[99] it was determined that the cyclopropane bond most nearly aligned with the π-system of the olefin may be cleaved by addition of

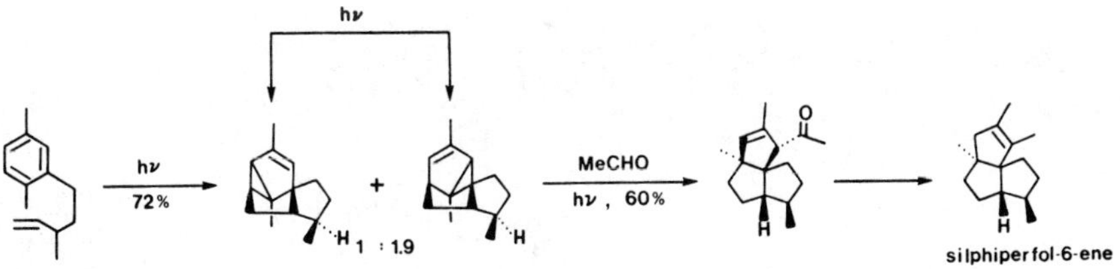

sulfide radical or an electron.[100] While in this case only one of the two isomers may be converted to product, the three-step synthesis is quite efficient overall. The synthesis of the silphiperfolenes[101] demonstrated the use of photogenerated radicals in the cyclopropane cleavage, and also highlighted a new mode of asymmetric induction. Based on the parallel-planes model it was expected and observed that products arise via exciplex **29** in which the interactions of the allylic substituent with both the olefin and the approaching aromatic are minimized. Again induction is efficient, and only the two interconvertible vinylcyclopropane isomers are isolated; these converge in the next step. This same route is presently being exploited in an approach to retigeranic acid;[102] in an approach to laurenene[103] restriction of the side chain into the orientation predicted by the parallel-planes model results in efficient cycloaddition to a single vinylcyclopropane isomer.

29

retigeranic acid

laurenene

The linearly-fused triquinanes are also approachable by cleavage of the same cyclopropane bond in the other photoisomer. In our synthesis of hirsutene[104] this "linear" isomer was conveniently produced as the major product. Note that the acetoxy group provides only a 3:1 induction of the five new asymmetric centers, and the overall yield is lower, presumably due to the steric stress involved

AcO hν 2% + 7% + 23%

1) KOH 2) H^+ 71%

hirsutene

in formation of three contiguous quartenary centers. These new centers are, however, absolutely determined with respect to one another. Acid-induced cleavage of the weak cyclopropane bond of the major isomer generates the linear skeleton. A modification of this route also allowed for the synthesis of coriolin.[105] Although cycloaddition of an enol ether would in principle provide the C8-hydroxyl directly, substrates of this type proved ineffective presumably due to an inappropriate matching of ionization potentials. The acetal **30** gave a dramatic improvement in the cyclization, and was easily converted into the desired functionality by Baeyer-Villiger oxidation.

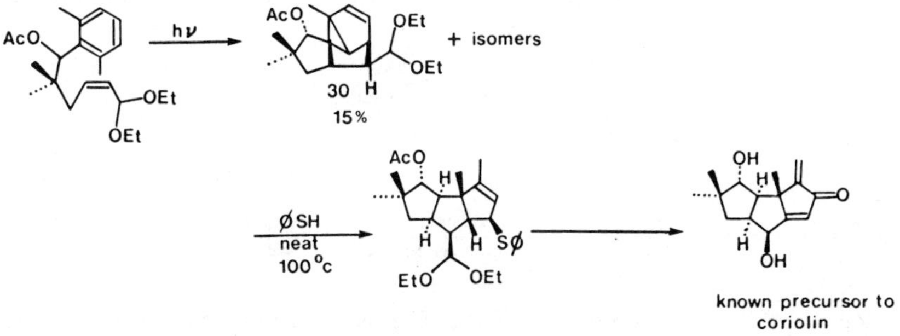

known precursor to coriolin

Although ipso-addition is disfavored due to non-bonded interactions in the sidechain, this process may be encouraged, and has been employed in a synthesis of des-dimethylquadrone[106]. Thus, photolysis of ester **31** produces a mixture which includes **32**. Not surprisingly, this cyclization is much less efficient than those explored previously. Homo-1,5-shift to cleave the cyclopropane generates the quadrone skeleton in three steps from commercial materials; **33** may be further elaborated to the target. One major product from a similar photocyclization has been employed in a synthesis of [5.5.5.5]-fenestranes[107].

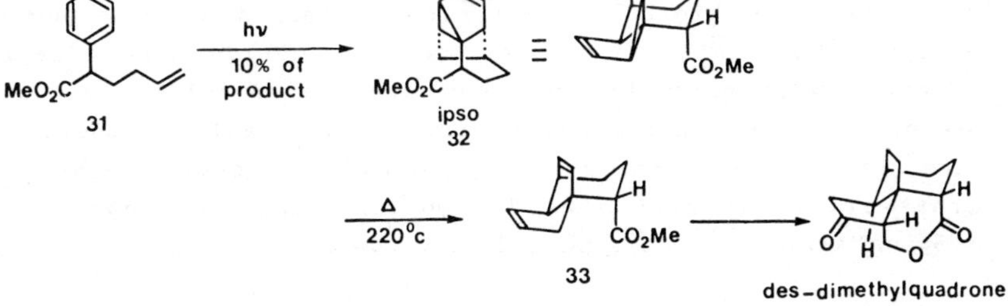

des-dimethylquadrone

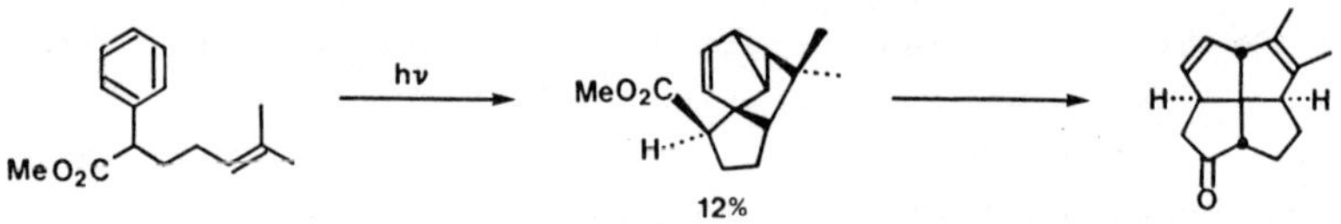

The *meta*-cycloaddition has also been employed synthetically in an intermolecular sense. The synthesis of the propellane modhephene begins with irradiation of a solution of indane and vinyl acetate to provide *endo*-cycloadduct **34** as the major product.[108] Deprotonation of the corresponding ketone results in generation of a semibullvalene enolate, whose complete alkylation provides three of the four methyl groups required for the natural product. Cleavage of the vinylcyclopropane moiety is accomplished by organocuprate addition, which occurs homo-1,4 rather than homo-1,6 to avoid quaternization of a carbon adjacent to the bulky *gem*-dimethyl group. Starting from the known adduct of benzene and vinyl acetate, a similar sequence provides isoiridomyrmecin.[109]

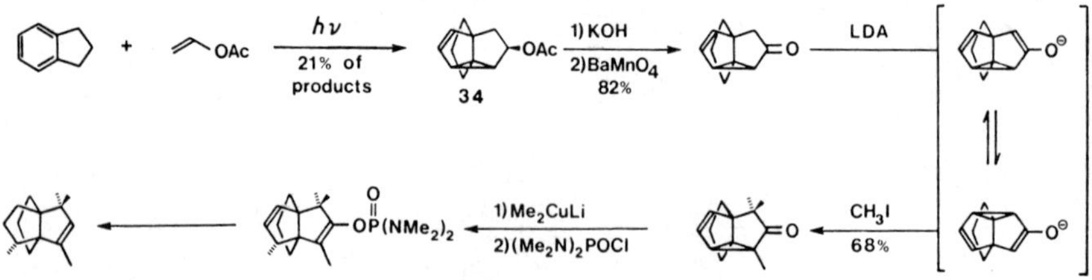

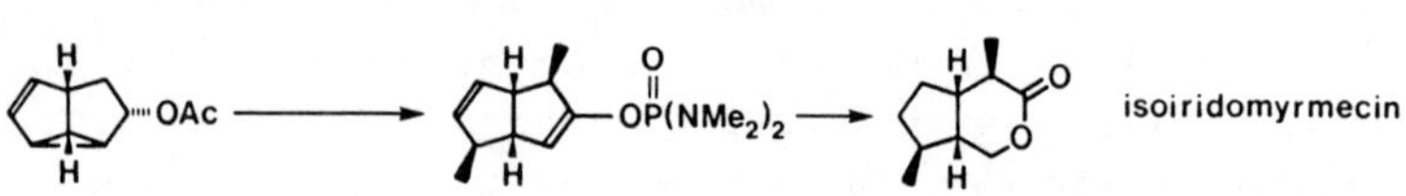

Finally, the application of two-bond cleavage to cycloheptane synthesis has also been explored. For the synthesis of the pseudo-guaianolide rudmollin, *meta*-cycloaddition of *cis* olefin **35** leads to two major products;[110] note that the silyl ether seems to be a much more effective stereoinducing element than the corresponding acetate. The vinylcyclopropane isomers converge upon oxidative cleavage with mercuric acetate to the cedrene-like tricyclic skeleton. After manipulation of functionality and stereoselective side chain introduction, a Grob fragmentation cleaves the bicyclic framework to generate the 7-membered ring. An approach to grayanotoxin-II[111] is expected to proceed in a similar fashion.

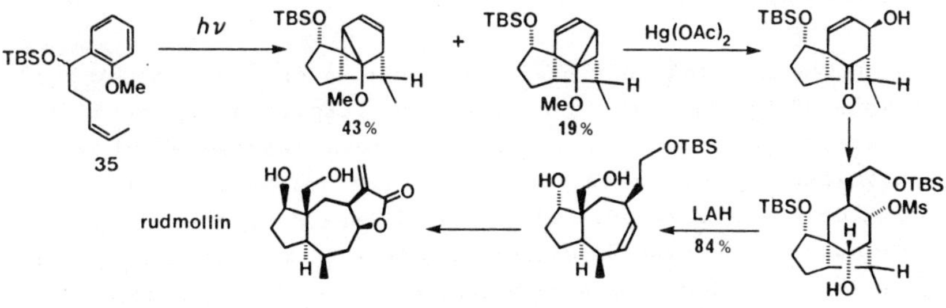

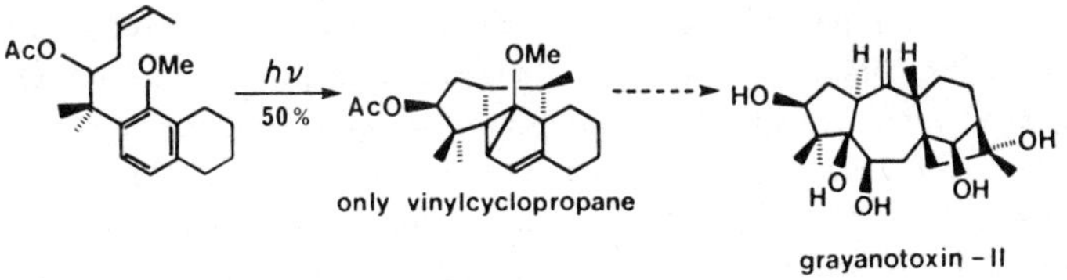

This synthesis provides the first example of the one remaining mode of sidechain stereoinduction; the substituent apparently places itself pseudoequatorial on the less sterically hindered convex face of the new 5-membered ring. The extent of this stereoinduction (only a single cycloadduct detected) is rather remarkable.

An important and as-yet-unaddressed point is with regard to the partitioning between the two possible vinylcyclopropane isomers. After exciplex formation, the choice of product depends only on what is formally a direction of electron shift. MM2 calculations on the

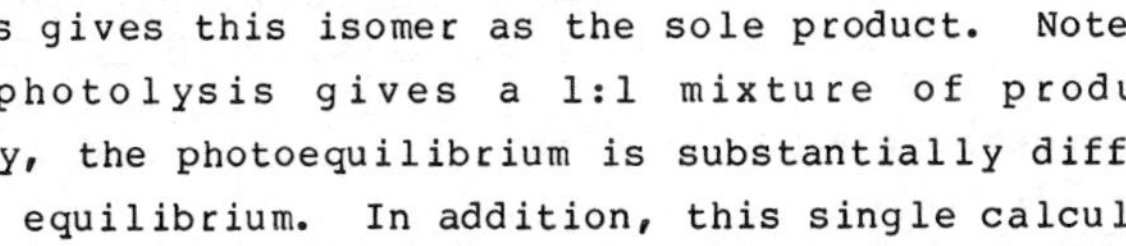

ΔH_f(MM2) = 44.84 kcal/mole

ΔH_f(MM2) = 41.26 kcal/mole

simplest intramolecular reaction indicate that the "linear" isomer is favored over the "angular" by 3.6 kcal/mole.[97] This explains nicely the fact that thermal equilibration of the cedrene photoadducts gives this isomer as the sole product. Note, however, that the photolysis gives a 1:1 mixture of products; not surprisingly, the photoequilibrium is substantially different from the thermal equilibrium. In addition, this single calculation does not represent a completely general trend; in the case of the laurenene study, the same method predicts that the angular product is much more stable[103]. Relatively little work has been done in this area, and the factors controlling the isomeric ratio are yet to be understood.

CONCLUSIONS

It is apparent that the identification of Bryce-Smith's "yellow compound" has opened an enormous and largely unexplored area of research. A tremendous amount has been learned about aromatic photochemistry in a short period of time; a tremendous amount remains to be learned. It is hoped that future studies will be cognizant of the crucial differences between kinetic and equilibrium results, so that their results will be more quantifiable and comparable. In particular [we feel that] more applied studies are in order; the example of 1,3-addition has demonstrated that such investigations inevitably lead to a deeper understanding of the overall process. In this regard it seems that the problem of photolability of 1,2-adducts has been overdramatized; with the proper care this can be a useful synthetic procedure. The factors governing 1,4-addition still need to be explored. Finally, the enormous synthetic potential of the 1,3-addition in all its manifestations makes its continued study worthwhile.

REFERENCES

1. D. Bryce-Smith, J. Chem. Soc. 1956, 1603.
2. J.M. Blair and D. Bryce-Smith, Proc. Chem. Soc. 1957, 287.
3. see in particular "Photochemistry: A Specialist Periodical Report", vol. 1-17, D. Bryce-Smith, ed.; Royal Society of Chemistry.
4. L. Kaplan, D.J. Rausch and K.E. Wilzbach, J. Am. Chem. Soc. 1972, **94**, 8638.
5. E. Farenhorst and A.F. Bickel, Tetrahedron Lett. 1966, 5911; L. Kaplan, J.S. Ritscher and K.E. Wilzbach, J. Am. Chem. Soc. 1966, **88**, 2881.
6. K.E. Wilzbach, J.S. Ritscher and L. Kaplan, J. Am. Chem. Soc. 1967, **89**, 1031.
7. K.E. Wilzbach and L. Kaplan, J. Am. Chem. Soc. 1964, **86**, 2307.
8. R.B. Cundall and A.J.R. Voss, J. Chem. Soc., Chem. Commun. 1968, 902.
9. K.E. Wilzbach, A.L. Harkness and L. Kaplan, J. Am. Chem. Soc. 1968, **90**, 1116.
10. I.E. Den Besten, L. Kaplan and K.E. Wilzbach, J. Am. Chem. Soc. 1968, **90**, 5868.
11. M.H. Delton, R.E. Gilman and D.J. Cram, J. Am. Chem. Soc. 1971, **93**, 2329; D.J. Cram, R.C. Helgeson, D. Lock and L.A. Singer, J. Am. Chem. Soc. 1966, **88**, 1324.
12. R.F. Childs, G.S. Shaw and A. Varadarajan, Synthesis 1982, 198.

13. H.R. Ward and J.S. Wishnok, J. Am. Chem. Soc. 1968, **90**, 5353.
14. D. Bryce-Smith and A. Gilbert, Tetrahedron 1976, **32**, 1309.
15. D. Bryce-Smith, A. Gilbert and D.A. Robinson, Angew. Chem. Int. Ed. 1971, **10**, 745.
16. I. Haller, J. Am. Chem. Soc. 1966, **88**, 2070; J. Chem. Phys. 1967, **47**, 1117; M.G. Barlow, R.N. Haszeldine and J.G. Dingwall, J. Chem. Soc., Perkin Trans. I 1973, 1542.
17. M.G. Barlow, R.N. Haszeldine and M.J. Kershaw, J. Chem. Soc., Perkin Trans. I 1975, 2005.
18. E.E. van Tamelen, S.P. Pappas and K.L. Kirk, J. Am. Chem. Soc. 1971, **93**, 6092.
19. S.L. Kammula, L.D. Iroff, M. Jones, Jr., J.W. van Straten, W.H. deWolf and F. Bickelhaupt, J. Am. Chem. Soc. 1977, **99**, 5815.
20. J. Liebe, C. Wolff and W. Tochtermann, Tetrahedron Lett. 1982, 2439.
21. Z. Yoshida, S. Miki, F. Kawamoto, T. Hijiwa and H. Ogoshi, Chem. Abs. 1980, **92**, 163323.
22. H. Gusten, M. Mintas and L. Klasinc, J. Am. Chem. Soc. 1980, **102** , 7936.
23. K.E. Wilzbach and D.J. Rausch, J. Am. Chem. Soc. 1970, **92**, 2178.
24. M.G. Barlow, R.N. Haszeldine and J.G. Dingwall, J. Chem. Soc., Perkin Trans. I 1973, 1542.
25. Y. Ogata and K. Takagi, J. Am. Chem. Soc. 1974, **96**, 5933; J. Org. Chem. 1978, **43**, 944.
26. R.D. Chambers, J.A.H. MacBride, J.R. Maslakiewicz and K.C. Srivastava, J. Chem. Soc., Perkin Trans. I 1975, 396, and references therein.
27. M. Hanaoka, S. Yasuda, K. Nagami, K. Okajima and T. Imanishi, Tetrahedron Lett. 1979, 3749.
28. J.A. Barltrop, K. Dawes, A.C. Day, S.J. Nuttall and A.J.H. Summers, J. Chem. Soc., Chem. Commun. 1973, 410; J.A. Barltrop, J.C. Barrett, R.W. Carder, A.C. Day, J.R. Harding, W.E. Long and C.J. Samuel, J. Am. Chem. Soc. 1979, **101**, 7510; J.A. Barltrop, A.C. Day and C.J. Samuel, J. Am. Chem. Soc. 1979, **101**, 7521.
29. H. Hiraoka and R. Srinivasan, J. Am. Chem. Soc. 1968, **90**, 2720; E.E. van Tamelen and T.H. Whitesides, J. Am. Chem. Soc. 1968, **90**, 3894; S. Boue and R. Srinivasan, J. Am. Chem. Soc. 1970, **92**, 1824; E.E. van Tamelen and T.H. Whitesides, J. Am. Chem. Soc. 1971, **93**, 6129.
30. H. Hiraoka, J. Chem. Soc., Chem. Commun. 1971, 1610; J. Barltrop, A.C. Day, P.D. Moxon and R.W. Ward, J. Chem. Soc., Chem. Commun. 1975, 786.
31. H. Wynberg and H. van Driel, J. Am. Chem. Soc. 1965, **87**, 3998; H. Wynberg, R.M. Kellogg, H. van Driel and G.E. Beekhuis, J. Am. Chem. Soc. 1966, **88**, 5047.

32. E.F. Ullman and B. Singh, J. Am. Chem. Soc. 1966, **88**, 1844; 1967, **89**, 6911; S. Albanesi, A. Marchesini and B. Gioia, Tetrahedron Lett. 1979, 1875.
33. H. Tiefenthaler, H. Dorscheln, H. Goth and H.Schmid, Tetrahedron Lett. 1964, 2999; P. Beak, J.L. Miesel and W.R. Messer, Tetrahedron Lett. 1967, 5315; P. Beak and W. Messer, Tetrahedron 1969, **25**, 3287.
34. A. Padwa, in "Rearrangements in Ground and Excited States", P. deMayo, ed., Academic Press (1980); vol. 3, p 501 provides a good general review of photorearrangements of five-membered heterocycles.
35. K.N. Houk, Pure & Appl. Chem. 1982, **54**, 1633.
36. A. Gilbert, Pure & Appl. Chem. 1980, **52**, 2669; A. Gilbert and P. Yianni, Tetrahedron 1981, **37**, 3275.
37. D. Bryce-Smith and H.C. Longuet-Higgins, J. Chem. Soc., Chem. Commun. 1966, 593; D. Bryce-Smith, J. Chem. Soc., Chem. Commun. 1969, 806; D. Bryce-Smith, B. Foulger, J. Forrester, A. Gilbert, B.H. Orger and H.M. Tyrrell, J. Chem. Soc., Perkin Trans. I 1980, 55.
38. D. Bryce-Smith and A. Gilbert, Tetrahedron 1977, **33**, 2459; D. Bryce-Smith, A. Gilbert, B. Orger and H. Tyrrell, J. Chem. Soc., Chem. Commun. 1974, 334.
39. J. Mattay, Tetrahedron 1985, **41**, 2392; ibid., 2405.
40. H.-G. Heine and W. Hartmann, Angew. Chem. Int. Ed. 1975, **14**, 698.
41. G. Hesse and P. Lechtken, Angew. Chem. 1971, **83**, 143; H.-D. Scharf and R. Klar, Tetrahedron Lett. 1971, 517.
42. A. Gilbert and P. Yianni, Tetrahedron Lett. 1982, 4611.
43. H.J.F. Angus and D. Bryce-Smith, Proc. Chem. Soc. 1959, 326; J. Chem. Soc. 1960, 4791; D. Bryce-Smith, B. Vickery and G.I. Fray, J. Chem. Soc. (C) 1967, 390.
44. W. Hartmann, H.-G. Heine and L. Schrader, Tetrahedron Lett. 1974, 3101.
45. D. Bryce-Smith and J.E. Lodge, J. Chem. Soc. 1962, 2675; G.S. Hammond and W.M. Hardham, Proc. Chem. Soc. 1963, 63.
46. B.E. Job and J.D. Littlehailes, J. Chem. Soc. (C) 1968, 886; R.J. Atkins, G.I. Fray and A. Gilbert, Tetrahedron Lett. 1975, 3087.
47. R.J. Atkins, G.I. Fray, A. Gilbert, M.W. bin Samsudin, A.J.K. Stewart and G.N. Taylor, J. Chem. Soc., Perkin Trans. I 1979, 3196.
48. A. Gilbert and G. Taylor, Tetrahedron Lett. 1977, 469.
49. see, for example, C. Pac, K. Mizuno, H. Okamoto and H. Sakurai, Synthesis 1978, 589.
50. P.S. Mariano, J.L. Stavinoha, G. Pepe and E.F. Meyer, Jr., J. Am. Chem. Soc. 1978, **100**, 7114.
51. J. Mattay, H. Leismann and H.D. Scharf, Mol. Photochem. 1979, **9**, 119.

52. A. Gilbert, G.N. Taylor and M.W. bin Samsudin, J. Chem. Soc., Perkin Trans I 1980, 869.
53. J.G. Atkinson, D.E. Ayer, G. Buchi and E.W. Robb, J. Am. Chem. Soc. 1963, **85**, 2257.
54. B. Sket and M. Zupan, J. Chem. Soc., Chem. Commun. 1976, 1053; 1977, 365.
55. Y. Katsuhara, T. Nakamura, A. Shimizu, Y. Shigemitsu and Y. Odaira, Chem. Lett. 1972, 1215.
56. D.R. Arnold, Adv. Photochem. 1968, **6**, 301; N.C. Yang and W. Eisenhardt, J. Am. Chem. Soc. 1971, **93**, 1277.
57. T.S. Cantrell, J. Am. Chem. Soc. 1972, **94**, 5929.
58. E. Grovenstein, Jr. and D.V. Rao, Tetrahedron Lett. 1961, 148.
59. D. Bryce-Smith, A. Gilbert and J. Grzonka, J. Chem. Soc., Chem. Commun. 1970, 498.
60. D. Bryce-Smith and J.E. Lodge, Proc. Chem. Soc. 1961, 333; J. Chem. Soc. 1963, 695.
61. W. Lippke, W.I. Ferree, Jr., and H. Morrison, J. Am. Chem. Soc. 1974, **96**, 2134.
62. carbomethoxycyclooctatetraene: G.L. Grunewald and J.M. Grindel, Org. Photochem. Syn. 1976, **2**, 20 (30 g prepared).
63. 1,2-dicarbomethoxycyclooctatetraene: L.A. Paquette and R.S. Beckley, Org. Photochem. Syn. 1976, **2**, 45 (4 g prepared).
64. B. Sket and M. Zupan, J. Am. Chem. Soc. 1977, **99**, 3504.
65. W.H.F. Sasse, P.J. Collin and G. Sagowdz, Tetrahedron Lett. 1965, 3373; I.A. Akhtar and J.J. McCullough, J. Org. Chem. 1981, **46**, 1447; J.J. McCullough, W.F. MacInnis, C.J.L. Lock and R. Faggiani, J. Am. Chem. Soc. 1980, **102**, 7780.
66. M.G. Barlow, D.E. Brown and R.N. Haszeldine, J. Chem. Soc., Chem. Commun. 1977, 669; J. Chem Soc., Perkin Trans. I 1978, 363; B. Sket, N. Zupancic and M. Zupan, J. Org. Chem. 1982, **47**, 4462.
67. M.G. Barlow, D.E. Brown and R.N. Haszeldine, J. Fluorine Chem. 1982, **20**, 745.
68. A.H.A. Tinnemans and D.C. Neckers, J. Org. Chem. 1978, **43**, 2493.
69. P.D. Davis and D.C. Neckers, J. Org. Chem. 1980, **45**, 456.
70. B. Sket and M. Zupan, Tetrahedron Lett. 1978, 2607.
71. K. Kraft and G. Koltzenburg, Tetrahedron Lett. 1967, 4357.
72. J. Ipaktschi and M.N. Iqbal, Synthesis 1977, 633.
73. N.C. Yang, C.V. Neywick and K. Srinivasachar, Tetrahedron Lett. 1975, 4313; N.C. Yang and M.G. Horner, Tetrahedron Lett. 1986, 543.
74. D. Bryce-Smith, B.E. Foulger and A. Gilbert, J. Chem. Soc., Chem. Commun. 1972, 664; J.C. Berridge, J. Forrester, B.E. Foulger and A. Gilbert, J. Chem. Soc., Perkin Trans. I 1980, 2425.
75. A. Gilbert and G.N. Taylor, J. Chem. Soc., Perkin Trans. I 1980, 1761.

76. for example: D.R. Arnold, L.B. Gillis and E.B. Whipple, J. Chem. Soc., Chem. Commun. 1969, 918; N.C. Yang and J. Libman, J. Am. Chem. Soc. 1972, **94**, 9228.
77. N.C. Yang, M.-J. Chen, P. Chen and K.T. Mak, J. Am. Chem. Soc. 1982, **104**, 853.
78. M. Bellas, D. Bryce-Smith and A. Gilbert, J. Chem. Soc., Chem. Commun. 1967, 862; M. Bellas, D. Bryce-Smith, M.T. Clarke, A. Gilbert, G. Klunkin, S. Krestonosich, C. Manning and S. Wilson, J. Chem. Soc., Perkin Trans. I 1977, 2571; A. Gilbert, S. Krestonosich and S. Wilson, Tetrahedron Lett. 1982, 4061.
79. D. Bryce-Smith, Pure & Appl. Chem. 1968, **16**, 47.
80. K.E. Wilzbach and L. Kaplan, J. Am. Chem. Soc. 1966, **88**, 2066; D. Bryce-Smith, A. Gilbert and B.H. Orger, J. Chem. Soc., Chem. Commun. 1966, 512.
81. A. Morikawa, S. Brownstein and R.J. Cvetanovic, J. Am. Chem. Soc. 1970, **92**, 1471; K.E. Wilzbach and L. Kaplan, J. Am. Chem. Soc. 1971, **93**, 2073.
82. J. Cornelisse and R. Srinivasan, Chem. Phys. Lett. 1973, **20**, 278; V.Y. Merritt, J. Cornelisse and R. Srinivasan, J. Am. Chem. Soc. 1973, **95**, 8250.
83. H. Morrison and W.I. Ferree, Jr., J. Chem. Soc., Chem. Commun. 1969, 268; W.I. Ferree, Jr., J.B. Grutzner and H. Morrison, J. Am. Chem. Soc. 1971, **93**, 5502.
84. R. Srinivasan, J. Am. Chem. Soc. 1971, **93**, 3555; J. Cornelisse, V.Y. Merritt and R. Srinivasan, J. Am. Chem. Soc. 1973, **95**, 6197.
85. R.S. Sheridan, Tetrahedron Lett. 1982, 267.
86. R. Srinivasan and J.A. Ors, Chem. Phys. Lett. 1976, **42**, 506; J.A. Ors and R. Srinivasan, J. Org. Chem. 1977, **42**, 1321.
87. R.S. Sheridan, J. Am. Chem. Soc. 1983, **105**, 5140; D.E. Reedich and R.S. Sheridan, J. Am. Chem Soc. 1985, **107**, 3360.
88. J. Mattay, J. Runsink, H. Leismann and H.-D. Scharf, Tetrahedron Lett. 1982, 4919.
89. R. Srinivasan, Tetrahedron Lett. 1971, 4551.
90. R. Srinivasan, J. Am. Chem. Soc. 1972, **94**, 8117.
91. C.S. Angadiyavar, J. Cornelisse, V.Y. Merritt and R. Srinivasan, Tetrahedron Lett. 1973, 4407.
92. R. Srinivasan, V.Y. Merritt and G. Subrahmanyam, Tetrahedron Lett. 1974, 2715.
93. A. Gilbert and M.W. bin Samsudin, Angew. Chem. Int. Ed. 1975, **14**, 552; J. Chem. Soc., Perkin Trans. I 1980, 1118; A. Gilbert and P. Heath, Tetrahedron Lett. 1979, 2831.
94. A. Gilbert, M.W. bin Samsudin, G.N. Taylor and S. Wilson, J. Chem. Soc., Perkin Trans. I 1980, 1225.
95. S.H. Bertz, J. Am. Chem. Soc. 1981, **103**, 3599.
96. P.A. Wender and J.J. Howbert, J. Am. Chem. Soc. 1981, **103**, 688.
97. J.J. Howbert, Ph.D. dissertation, Harvard University.
98. P.A. Wender and G.B. Dreyer, Tetrahedron 1981, **37**, 4445.

99. P.A. Wender and R.J. Ternansky, Tetrahedron Lett. 1985, 2625.
100. these two methods for cyclopropane cleavage had previously been employed in the synthesis of coriolin; see ref. 97.
101. P.A. Wender and S.K. Singh, Tetrahedron Lett. 1985, 5987.
102. P.A. Wender and S.K. Singh, unpublished results.
103. P.A. Wender and T.W. von Geldern, unpublished results.
104. P.A. Wender and J.J. Howbert, Tetrahedron Lett. 1982, 3983.
105. P.A. Wender and J.J. Howbert, Tetrahedron Lett. 1983, 5325.
106. D.J. Wolanin, Ph.D. dissertation, Harvard University.
107. J. Mani, J.H. Cho, R. Astik, E. Stamm, P. Bigler, V. Meyer and R. Keese, Helv. Chim. Acta 1984, **67**, 1930; J. Mani and R. Keese, Tetrahedron 1985, **41**, 5697.
108. P.A. Wender and G.B. Dreyer, J. Am. Chem. Soc. 1982, **104**, 5805.
109. P.A. Wender and G.B. Dreyer, Tetrahedron Lett. 1983, 4543.
110. P.A. Wender and K.J. Fisher, Tetrahedron Lett., in press.
111. P.A. Wender and A.G. Olivero, unpublished results.

13

Practical Photochemistry: Scale-up

By K.H. Pfoertner
F. HOFFMANN-LAROCHE AND CO. LTD., CH-4002, BASEL, SWITZERLAND

Introduction

In this treatment of problems which always accompany the scale-up of a photoreaction no chain reactions are considered. Chain reactions can be started not only photochemically but also by thermal production of radicals, e.g. by thermal decomposition of peroxides. In addition we confine ourselves to photoreactions in the liquid phase.

Decision on scaling up a photoreaction

The photoreaction must be economically feasible. This is true for the following types of photoreactions: no competitive reaction sequence without any photochemical step exists. Figure 1 shows the production of vitamin D and of its hydroxylated derivatives. The latter do not act as vitamins but as drugs against bone deseases etc. Estimated worldwide output of vitamin D: 50 ton year^{-1}.

$\xrightarrow{h\nu}$ $\xrightarrow{\Delta}$

Vitamins

R^1= H , R^2= $CH(CH_3)CH{=}CHCH(CH_3)CH(CH_3)_2$ Vitamin D_2

R^1= H , R^2= $CH(CH_3)CH_2CH_2CH_2CH(CH_3)_2$ Vitamin D_3

Drugs

R^1= H , R^2= $CH(CH_3)CH_2CH_2CH_2C(CH_3)_2OH$ 25-Hydroxy vitamin D_3

R^1= OH, R^2= $CH(CH_3)CH_2CH_2CH_2C(CH_3)_2OH$ 1α,25-Dihydroxy vitamin D_3

Figure 1

The introduction of a photoreaction reduces the production cost, e.g. by considerable shortening of comparable conventional syntheses. Some examples are given in Fig.2: production of ε-caprolactam for Nylon 6 via photonitrosation of cyclohexane with an estimated output of 10 000 ton year^{-1}, production of 2,3-dichloro-tetrahydrofuran by stoichiometric photochlorination of tetrahydrofuran with an estimated output of 2 ton year^{-1}, and production of the fragrant chemical rose oxide by photosensitized oxidation of citronellol with an estimated output of 9.6 ton year^{-1}.

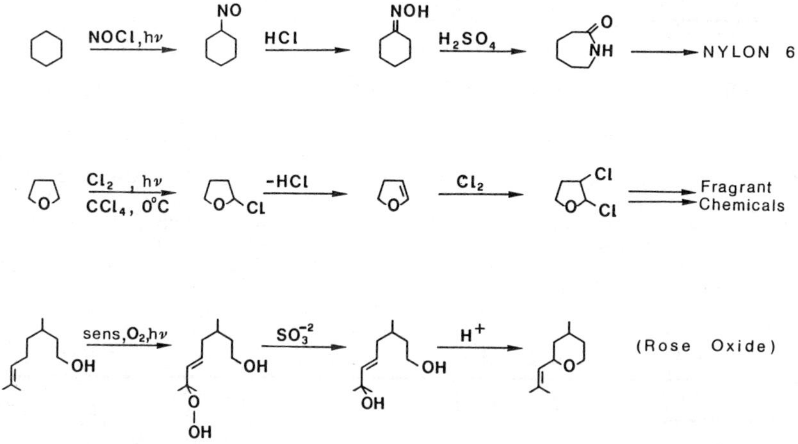

Figure 2

Unusual structures are produced by photochemical syntheses including light-induced skeletal rearrangements. In Fig.3 it is shown that 2,2,5,5a-tetramethyl-5H-1-benzopyran photochemically derived from cheap β-ionone is an important starting material for the production of various fragant chemicals, and the fragrant chemical chrysanthenone can be produced by a light-induced one-step skeletal rearrangement.

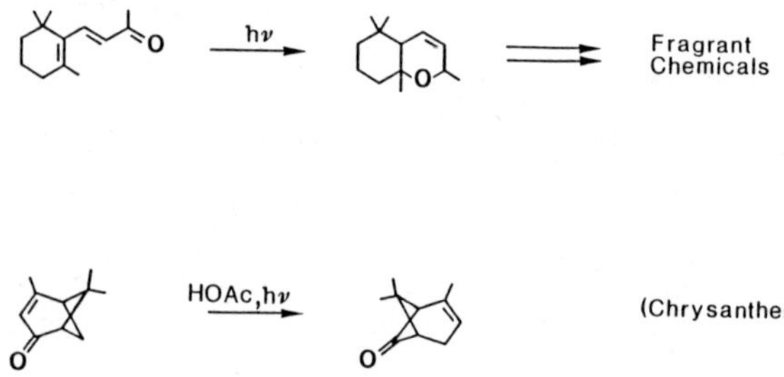

hν

Fragrant Chemicals

HOAc,hν

(Chrysanthenone)

Figure 3

Light absorption

The first law of photochemistry, formulated by GROTTHUS (1817) and DRAPER (1843), states: Only light which is absorbed by a molecule can be effective in producing a reaction which changes the molecule. Although this law is a truism today, it implies the fact that there must be some overlap between the wavelengths of the light entering the photoreactor and those absorbed by the starting material or sensitizer if a photochemical reaction is to occur.

Determination of the light absorbed by one component at a given wavelength. To avoid confusing mol of substance with mol of photons the latter are converted into "Einstein"[1]. The light absorption during its penetration of the path length L of an absorbing solution is given in Fig.4 using the Beer-Lambert law.

Light Absorption

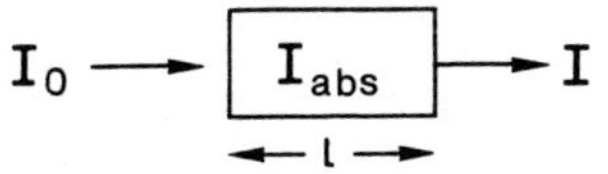

$$I_0 \longrightarrow \boxed{I_{abs}} \longrightarrow I$$

$\longleftarrow l \longrightarrow$

$$I_{abs} = I_0 - I \qquad (1)$$

Beer-Lambert Law: $I = I_0 \cdot 10^{-\varepsilon C l}$ (2)

$$I_{abs} = I_0 (1 - 10^{-\varepsilon C l}) \qquad (3)$$

Definitions

$6.023 \cdot 10^{23}$ photons = 1 mol photons = 1 Einstein

Incident radiant flux	I_0 (Einstein h^{-1})
Absorbed radiant flux	I_{abs} (Einstein h^{-1})
Penetrating radiant flux	I (Einstein h^{-1})
Decadic molar absorption coefficient	ε ($l\ mol^{-1}\ cm^{-1}$)
Concentration of absorbing molecules	C ($mol\ l^{-1}$)
Absorbing path length	l (cm)

Figure 4

Industrial photochemists frequently are not equipped for an exact measurement of the incident radiant flux I_0 (Einstein h^{-1}). But this is no reason to give up. In many cases it is sufficient to express I_0 as 100 %. Then, eq.3 can be replaced by

$$I'_{abs} = 100(1 - 10^{-\varepsilon cl})\ \% \qquad (3a)$$

Refined mathematical treatment of the light absorption, considering the inner filter effect[2]. When more than one light-absorbing component is present at the beginning of a photoreaction or developed in its course, eq.3 and eq.3a are too simple to characterize the real situation. In this case one has to consider the concentrations c_i of every light absorbing component i present in the reaction mixture as well as its molar absorption coefficient ε_i (i=1,2,3...). By an appropriate choice of the parameters the exponential term of eq.4 becomes neglegible. Then eq.4 reduces to the much simpler eq.5.

Light Absorption of Multi-Component Systems

$$(4)\quad I_{abs,i} = I_0 \frac{\varepsilon_i C_i}{\sum_{n=1}^{m} \varepsilon_n C_n}\left(1-10^{-l \sum_{n=1}^{m} \varepsilon_n C_n}\right) \qquad i = 1, 2, 3 \ldots n$$

The exponential term becomes negligible when $10^{-l \sum_{n=1}^{m} \varepsilon_n C_n} \ll 1$

$$(5)\quad I_{abs,i} = I_0 \frac{\varepsilon_i C_i}{\sum_{n=1}^{m} \varepsilon_n C_n}$$

Figure 5

Again expressing the incident radiant flux I_0 as 100 % instead of Einstein h^{-1} might be helpful. Another method to overcome the problems which are correlated with an exact knowledge of the incident radiation flux is to utilize such extremely high values of the concentration c_0 in combination with the accompanying molar absorption coefficients ε that even after an extensive irradiation virtually all incident photochemically effective light is absorbed. This procedure is quite simple from the experimental standpoint and eliminates the time-dependence of the absorbance, but it limits the extent to which the measured yield may be put. Another restriction in strongly absorbing solutions is the geometric inhomogeneity of light absorption which will be discussed later on in connection with photoreactor design.

A generalized description of photoreactions[3]

Photochemical reactions are brought about by the promotion of a molecule M to an excited state M* either by absorption of a quantum of light (eq.6) or by transfer of energy from a sensitizer in its excited state (eq.7), followed by the transformation of M* into a product or products (eq.8)

$$M + h\nu \longrightarrow M^* \quad (6)$$

$$M + sens.^* \longrightarrow M^* + sens. \quad (7)$$

$$M^* \longrightarrow product(s) \quad (8)$$

$$\text{Quantum yield } \varphi = \frac{\text{Number of moles converted or produced}}{\text{Number of Einsteins absorbed}} \quad (9)$$

Figure 6

To define the consumption of M or the quantity of product formed the term "quantum yield" was introduced. As in eq.9 the dimension of the nominator equals that of the denominator φ is a dimensionless factor of efficiency.

Industrial photochemists have to consider that production managers do not calculate in mol but in kg or ton. Therefore, the molecular weight of the desired photoproduct must be taken into account (Fig.7[4]) as it determines the amount of material produced by photoreactions with the same molar conversion in the same irradiation time. Example: In photonitrosations substitution of cyclohexanone oxime for Nylon 6 (Fig.2) by cyclododecanone for Nylon 12 produces 1.7 times the amount by weight[5].

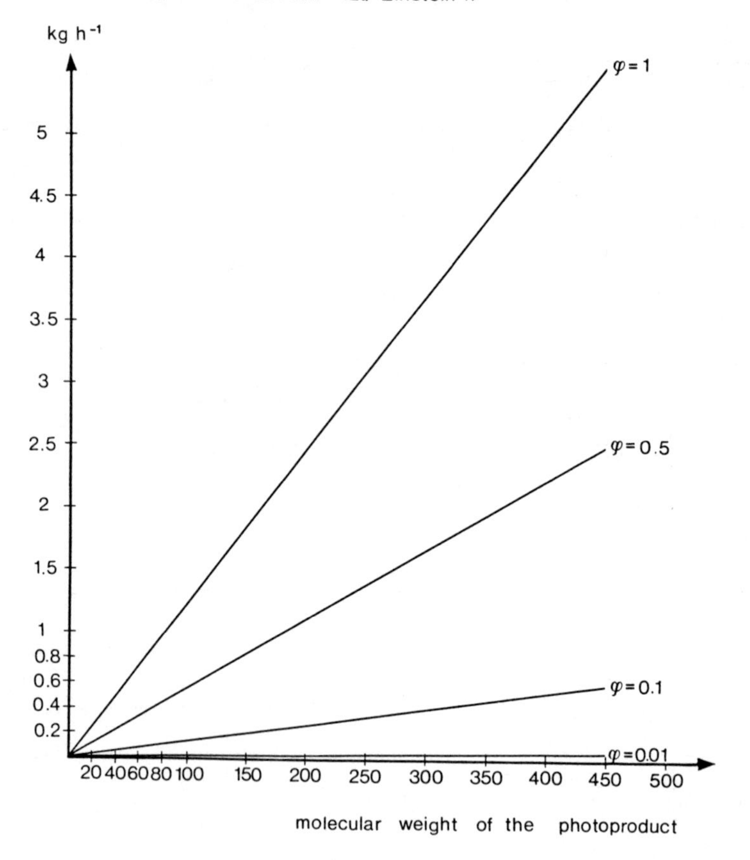

Figure 7

Refined approach to photoreactions. Quantum yields (excluding chain reactions) can vary over the range $0<\varphi\leqslant 1$. To explain this effect it is necessary to refine our first approximation. The atoms of most compounds with non-ionic bonds are held together by homeopolar bonds consisting of electron pairs with antiparallel spins.

Spectroscopists call states with antiparallel electron spins singlet states, symbolized by ^{x}S. Triplet states have parallel electron spins and are symbolized by ^{x}T. Electronic excitation processes and their corresponding lifetimes in solution are summarized in Fig.8[7] and Fig.9[7], respectively.

Step	Process
1. Excitation	$^{0}S + h\nu \longrightarrow {}^{1}S$
2. Excited singlet state ^{1}S decays by the internal conversion (IC) process to the vibronic state $^{0}S^*$.	$^{1}S \longrightarrow {}^{0}S^*$
Vibronic state $^{0}S^*$ rapidly dissipates by vibrational relaxation to the ground state ^{0}S .	$^{0}S^* \longrightarrow {}^{0}S$
3. Fluorescent emission (F).	$^{1}S \longrightarrow {}^{0}S + h\nu_F$
4. Excited singlet state ^{1}S converts by intersystem crossing (ISC(S)) process to the vibronic state $^{1}T^*$.	$^{1}S \longrightarrow {}^{1}T^*$
Vibronic state $^{1}T^*$ rapidly dissipates by vibrational relaxation to the triplet state ^{1}T	$^{1}T^* \longrightarrow {}^{1}T$
5. Phosphorescent emission (P)	$^{1}T \longrightarrow {}^{0}S + h\nu_P$
6. Excited triplet state ^{1}T decays by intersystem crossing (ISC(T)) process to the vibronic state $^{0}S^*$	$^{1}T \longrightarrow {}^{0}S^*$
Vibronic state $^{0}S^*$ rapidly dissipates by vibrational relaxation to the ground state ^{0}S	$^{0}S^* \longrightarrow {}^{0}S$

Figure 8

Step	Process	Lifetime (s)
1. Excitation	$^{0}S + h\nu \longrightarrow {}^{1}S$	10^{-15}
2. Internal conversion (IC)	$^{1}S \longrightarrow {}^{0}S + \Delta$	$10^{-9} - 10^{-7}$
3. Fluorescent emission (F)	$^{1}S \longrightarrow {}^{0}S + h\nu_F$	$10^{-11} - 10^{-6}$
4. Intersystem crossing (ISC(S))	$^{1}S \longrightarrow {}^{1}T + \Delta$	$10^{-11} - 10^{-8}$
5. Phosphorescent emission (P)	$^{1}T \longrightarrow {}^{0}S + h\nu_P$	$10^{-3} - 10^{2}$
6. Intersystem crossing (ISC(T))	$^{1}T \longrightarrow {}^{0}S + \Delta$	$10^{-3} - 10^{2}$

Figure 9

Concerning Fig.8 and Fig.9 it has to be mentioned that heavy atoms and certain paramagnetic species increase the probability of intersystem crossing (ISC) even in low concentrations.

The necessary irradiation time is a crucial factor in scale-up.

The suitability of a photoreaction for scaling up is based on the excitation of a sufficient number of molecules to their reactive states in a reasonable time to obtain the required yield of product. The efficacy of eq.8 (Fig.6) determines the amount of product. Since there is no direct way to measure the concentration of excited molecules M*, the usual measure of photochemical efficiency is the overall quantum yield φ , which is defined by eq.8 in Fig.6 and includes all events given in Fig.8. If M* is produced by photosensitization (eq.7 in Fig.6) the quantum yield is based on the number of Einstein absorbed by the senitizer. Determination of the overall quantum yield φ requires the measurement of the radiant flux I_0 (Einstein h^{-1}) from the light source within the limits of the absorption which is characteristic for the substrate M, the fraction Q_M (Einstein $L^{-1}h^{-1}$) of the source output absorbed by the quantity of M per volume unit, and the quantity of the product(s) formed or of the substrate consumed during the photoreaction (mol h^{-1}).

Measurement of the radiant flux from the radiation source. To determine the radiant flux from the source within the limits of the substrates characteristic absorption is a problem of serious experimental complexity. Research chemists prefer chemical actinometers for light-intensity measurements. Several systems show an exactly known product quantum yield which is rather insensitive to changes in temperature, concentration of the reactant, and wavelength of the absorbed light. Quantitative analysis of the product is in most cases simple and precise. An advantage is gained by choosing a liquid-phase actinometer for photoreactions involving liquids. This procedure eliminates corrections for differences between the fraction of incident light reflected from the front window of the photoreactor and the fraction causing the actual photochemical reaction. For technical purposes using strong lamps with an electrical input from 10 to 100 kW, chemical actinometers are in general too sensitive if not useless.

Measurement of the light absorption by M within the volume of the irradiated solution. The evaluation of Q_M (Einstein l^{-1} h^{-1}) can cause severe difficulties. Only two aspects of this problem can be considered here. If the photoreaction occurs with appreciable conversion of M to product(s), it is possible that the absorbance of the solution will change significantly during the course of irradiation. In that case, assuming Beer-Lambert's law to be valid, the radiant flux absorbed in the considered volume has to be integrated over the duration of the experiment in which the concentration of M varies with time[6]. If concentration gradients are produced in the photoreactor (see photoreactor design) the situation may be further complicated. In that case one cannot assume adherence to Beer-Lambert's law, since these gradients will exhibit a time dependence.

Photochemical kinetics

The law of Beer-Lambert hitherto used considers only the attenuation of light penetrating the path length ℓ (cm) of an absorbing solution. On the other hand the quantum yield is a dimensionless factor of efficiency. As the mathematical treatment of kinetics works with concentrations of reactants and products (mol l^{-1})

we have to convert the radiant flux (Einstein h^{-1}) absorbed by the educt M into a volumetric radiant energy absorption Q_M (Fig.10).

The Rate of Light Absorption by M

Definitions:

Volumetric energy absorption by M $$Q_M = \frac{I_{abs,M}}{V} \text{ (Einstein } l^{-1}h^{-1}) \quad (10)$$

From Beer-Lamberts law:

Dimensionless factor for light attenuation by its penetration of the distance l of a solution containing absorbing molecules M

Absorbance $$A_\lambda = \log \frac{I_0}{I} = \varepsilon_{M,\lambda} C_M l \quad (11)$$

Figure 10

Differences in thermal and photochemical kinetics using the differential form of a hypothetical first order reaction are demonstrated in Fig.11.

Differences in Thermal and Photochemical Kinetics	
$M_1 \xrightarrow{\Delta} M_2$	$M_1 \xrightarrow{h\nu} M_2$
$-\frac{dC_1}{dt} = k_1 C_1$ (12a)	$-\frac{dC_1}{dt} = \varphi \cdot Q_1$ (12 b)
Dimension of the rate constant: k (t^{-1})	Dimensionless factor of efficiency: φ = quantum yield (eq.9)

Figure 11

Mathematical treatment of a complex conventional reaction.

Often reaction intermediates are present in very small concentrations which for a certain time remain constant in the course of the reaction. The intermediate is then said to be in a steady state, and to a good approximation its time derivative can be set equal to zero. This procedure is called the BODENSTEIN principle. Its result is usually a vast simplification of the rate equation as seen from eq.13 to eq.18 in Fig.12. But simplification of a rate equation <u>via</u> the steady-state approximation should be examined critically to see if it is valid in the case considered.

Complex Thermal Reaction

$$M_1 \underset{k_{21}}{\overset{k_{12}}{\rightleftharpoons}} M_2 \underset{k_{32}}{\overset{k_{23}}{\rightleftharpoons}} M_3$$

$$-\frac{dC_1}{dt} = k_{12}C_1 - k_{21}C_2 \qquad (13)$$

$$-\frac{dC_2}{dt} = (k_{21}+k_{23})\,C_2 - k_{12}C_1 - k_{32}C_3 \qquad (14)$$

Steady state of M_2 assumed : $\frac{dC_2}{dt} = 0$

$$C_2 = \frac{k_{12}C_1 + k_{32}C_3}{k_{21}+k_{23}} \qquad (15)$$

$$-\frac{dC_1}{dt} = \frac{k_{12}\cdot k_{23}}{k_{21}+k_{23}}\,C_1 - \frac{k_{21}\cdot k_{32}}{k_{21}+k_{23}}\,C_3 \qquad (16)$$

eq.17 is formally the same as that for a first order reversible reaction $M_1 \underset{k_r}{\overset{k_f}{\rightleftharpoons}} M_3$

$$-\frac{dC_1}{dt} = k_f C_1 - k_r C_3 \qquad (17)$$

and can easily be integrated :

$$k_f\,t = \frac{C_{3,e}}{C_{1,t=0}} \ln \frac{C_{3,e}}{C_{3,e} - C_{3,t}} \qquad (18)$$

$C_{3,e}$ = concentration of M_3 in the equilibrium

$C_{3,t}$ = concentration of M_3 after the time t

$C_{1,t=0}$ = concentration of M_1 at the beginning of the reaction

Figure 12[8,9]

Provided that the exponential term of eq.4 (Fig.5) is negligible[10], Fig.13 presents a photochemical reaction which is formally analogous to the thermal reaction shown in Fig.12. Again it is evident that photochemical reactions have to be treated mathematically in another way than thermal reactions.

Complex Photochemical Reaction

$$M_1 \underset{\varphi_{21}}{\overset{\varphi_{12}}{\rightleftharpoons}} M_2 \underset{\varphi_{32}}{\overset{\varphi_{23}}{\rightleftharpoons}} M_3$$

$$-\frac{dC_1}{dt} = \varphi_{21}Q_2 - \varphi_{12}Q_1 \quad (19)$$

$$-\frac{dC_2}{dt} = \varphi_{12}Q_1 + \varphi_{32}Q_3 - (\varphi_{21}+\varphi_{23})Q_2 \quad (20)$$

$$-\frac{dC_3}{dt} = \varphi_{23}Q_2 - \varphi_{32}Q_3 \quad (21)$$

From the equations 4and 4a one comes to

$$-\frac{dC_1}{dt} = \frac{Q_0}{\sum_{n=1}^{m=3}\varepsilon_n C_n}(\varphi_{21}\varepsilon_2C_2 - \varphi_{12}\varepsilon_1C_1) \quad (22)$$

$$-\frac{dC_2}{dt} = \frac{Q_0}{\sum_{n=1}^{m=3}\varepsilon_n C_n}(\varphi_{12}\varepsilon_1C_1 + \varphi_{32}\varepsilon_3C_3 - \varphi_{21}\varepsilon_2C_2 - \varphi_{23}\varepsilon_2C_2) \quad (23)$$

Treatment of this system of differential equations is only possible by standard integration methods (e.g. Runge-Kutta) using a computer.

$$-\frac{dC_3}{dt} = \frac{Q_0}{\sum_{n=1}^{m=3}\varepsilon_n C_n}(\varphi_{23}\varepsilon_2C_2 - \varphi_{32}\varepsilon_3C_3) \quad (24)$$

Figure 13

Light sources

The choice of the light source is dictated by a suitable combination of spectral energy distribution, radiant flux and filter for the special case considered. If the reactant absorbs in more than one region of the spectrum - nearly all organic molecules do so - irradiation in different absorption bands may induce different photochemical reactions. Using polychromatic instead of monochromatic light (which is unusual in technical photochemistry) the resulting effect is not predictable. This is especially the case when the absorption bands are quite broad and their absorbances differ significantly. Thus, mainly the irradiation is restricted to an absorption in the longest wavelength region. Competitive absorption caused by more than one reactant or by products formed has already been considered in eq.4 (Fig.5).

Available strong light sources. High pressure discharges up to a power input of 100 kW are available. For economic reasons the industrial photochemist must know that the radiant flux of mercury medium pressure arcs increases strongly up to 10 kW, then remains nearly proportional to the power input above this limit[5]. For technical purposes it is also useful to know the conversion of the radiant flux ϕ in Watt emitted by a lamp into I_0 in Einstein h^{-1} as a function of the wavelength λ in nm[11]

$$I_0 = 3.01 \times 10^{-5} \times \phi \times \lambda \qquad (25)$$

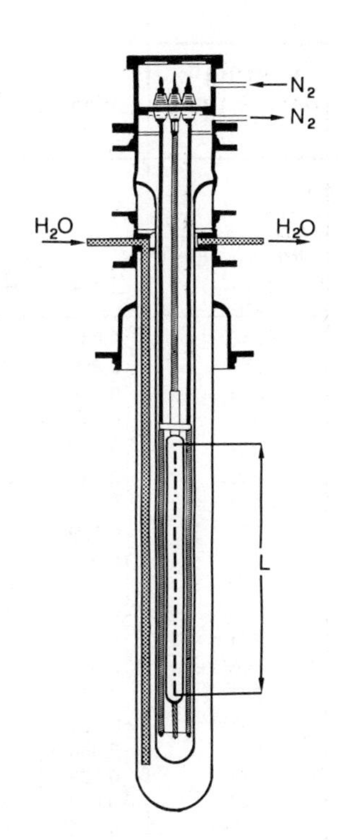

Figure 14

The majority of technical light sources are of the immersion lamp type which is schematically represented by Fig.14 for a power input from 2 to 60 kW. The spectral energy distribution of the widely used mercury medium pressure lamps is given in Table 1.

radiating part L [mm]	1044	1044	1535	1520	1520	2000
	Einstein h^{-1}					
Wave length λ [nm]	2 kW	4 kW	10 kW	20 kW	40 kW	60 kW
248	0,045	0,140	0,76	1,40	2,24	3,36
254	0,280	0,730	3,06	7,90	17,58	26,22
265	0,096	0,420	1,44	2,10	3,59	4,23
270	0,023	0,059	0,37	0,59	1,46	1,46
275	0,020	0,053	0,25	0,56	0,75	1,16
280	0,040	0,140	0,96	1,50	2,53	3,71
289	0,015	0,045	0,37	0,52	1,48	1,65
292	0,009	0,023	0,16	0,15	0,44	0,53
296	0,070	0,200	1,63	2,30	4,83	7,15
302	0,150	0,480	2,07	3,20	6,64	8,71
313	0,430	1,050	3,72	6,30	15,83	18,65
334	0,020	0,073	0,58	1,00	2,11	2,71
366	0,730	1,820	7,66	12,70	27,98	34,92
391	0,006	0,016	0,12	0,26	0,36	0,59
405/8	0,320	0,800	2,52	4,70	14,70	15,42
436	0,530	1,490	4,36	8,40	23,23	27,82
492	0,009	0,025	0,21	0,07	0,44	0,44
546	0,850	2,100	5,95	13,10	37,14	45,85
577/9	0,330	1,180	9,64	19,80	39,49	56,72

Table 1

Composition of stable liquid cut–off filters (1 l)					
Absorbance A = 90% for a path length l = 1 cm					
λ nm	$CuSO_4 \cdot 5H_2O$ g	$Fe_2(SO_4)_3 5H_2O$ g	$FeSO_4 \cdot 1(2)H_2O$ g	H_2SO_4 conc. ml	H_2O ml
282	8.5	——	——	1.0	995
297	35.0	——	——	1.0	988
304*	36.8	0.13	0.04	1.0	990
344	38.8	0.53	——	3.9	1028
366	33.0	1.0	——	7.7	994
395	24.5	5.3	1.8	38.5	1015

* corresponds to the light attenuation of 0.5 cm pyrex glass

Table 2

Most of the known light filters are not stable when they are combined with strong lamps. Therefore, liquid cut-off filters have been developed[12] which do not change their absorption characteristics even under strong irradiation and are given in Table 2.

Reactor design

According to the Beer-Lambert law (eq.2) photoreactions are characterized by their spatial inhomogeneity. High molar absorption coefficients and/or high concentrations of the light-absorbing substances may reduce the thickness of the photoreaction zone l', defined by eq.26 in Fig.15, to several micrometers, *i.e.* to a region where even the turbulent flow of a liquid becomes laminar by adhesive forces from the wall. The result is a decreasing contribution of macroscopic homogenization processes to the exchange of reactants out of and into the reaction zone. As an example from the industry Fig.15 shows irradiation of provitamin D_3 with light of different wavelengths. From this we learn that it is not always favourable to use light which corresponds to the substrate's absorption maximum. Irradiation into less absorbing parts of the chosen absorption band may elongate the reaction zone l' significantly. Thicknesses of photoreaction zones for l' (90 % absorption and 99 % absorption) are given in Fig.15. In industrial photochemistry high substrate concentrations are preferred because of the cost of solvent losses and recycling. The disadvantages that arise concerning the reaction zone should then be compensated by an adequate photoreactor design. There are three technically suitable possibilities which are outlined in Fig.16.

Photoreaction Zone

From Absorbance $A = \log \frac{I_0}{I} = \varepsilon C L$ (see eq. 11)

Definition of a Photoreaction Zone $L' = \frac{A}{\varepsilon C}$ (cm) (26)

Example: 7-Dehydrocholesterol, $C = 0{,}13$ (mol l^{-1})

λ nm	ε l mol^{-1}cm^{-1}	A	Absorption %	Zone L' μ
282	11 770	1	90	65
		2	99	131
254	4 720	1	90	163
		2	99	326

Figure 15

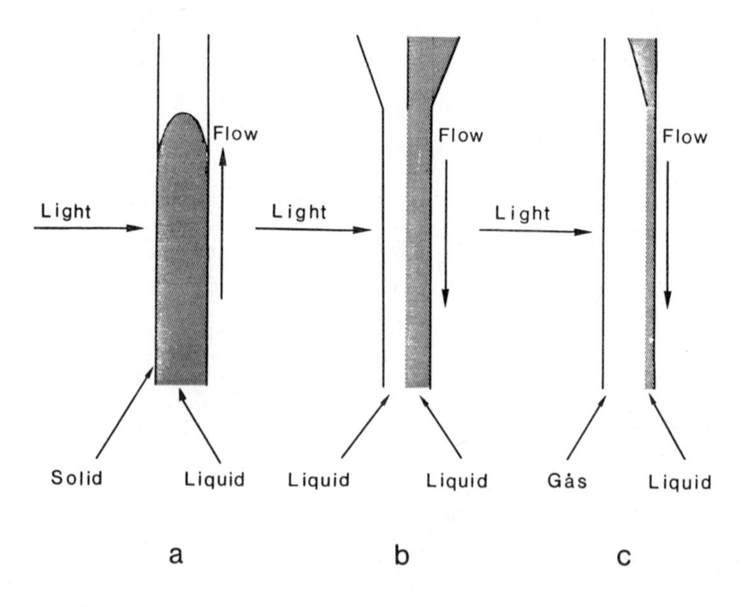

Figure 16

Fig.16, example a: The above defined photoreaction zone l' is located in the contact region solid-liquid. Here, in the case of $A/\varepsilon\cdot c \ll 1$ (for A = 1 or 2) the zone l' can be torn off from the wall only by a flow speed of twice the velocity of sound. Fig.16, example b: The reaction zone l' is located in the contact region liquid-liquid, provided that one of the non-miscible liquids is permeable for the photoreactive light and the substrate is only soluble in the other one. Then the motion of the reaction zone may be varied within certain speed limits[13]. Fig.16, example c: The reaction zone is located in the contact region gas-liquid. That means irradiation of a falling film[14]. The schematic representations Fig.16 a to c are simplified to demonstrate the matter of importance. In reality photoreactors consist of two or - when filter solutions are used - of three concentric tubes surrounding a tube-like light source. Every system shown in Fig.16 has its advantages and disadvantages. Utilization of case a depends on the magnitude of l'. Case b requires a suitable pair of non-miscible liquids, and in case c the vapor phase over the irradiated solution should neither react photochemically nor attenuate the radiation.

Materials

As the light transmission of pyrex glass decreases drastically below 320 nm, the appropriate light-transmitting material is quartz in its different qualities. This does not only concern the transmission of the strong mercury line 254 nm (Fig.17) but also its excitation by the latter, causing fluorescence which is not negligible for strong lamps. Quality 1 gives a blue-violet and quartz glass of quality 2 a pale green fluorescence.

When we express the fluorescence in arbitrary units, attaching to quality 1 the factor 100, then 2.05 is the factor for quality 2, and 0.12 the factor for quality 3.

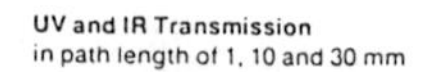

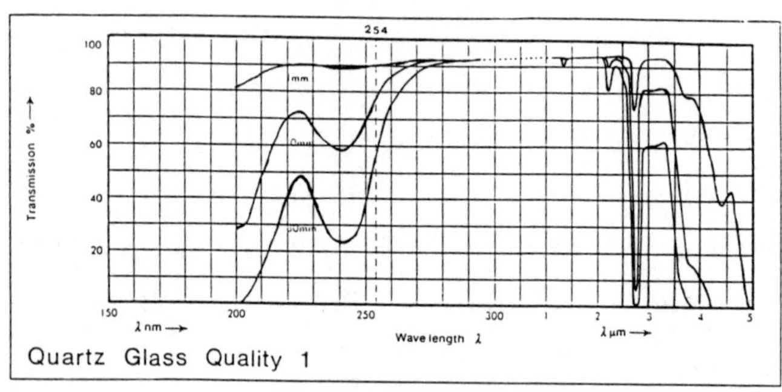

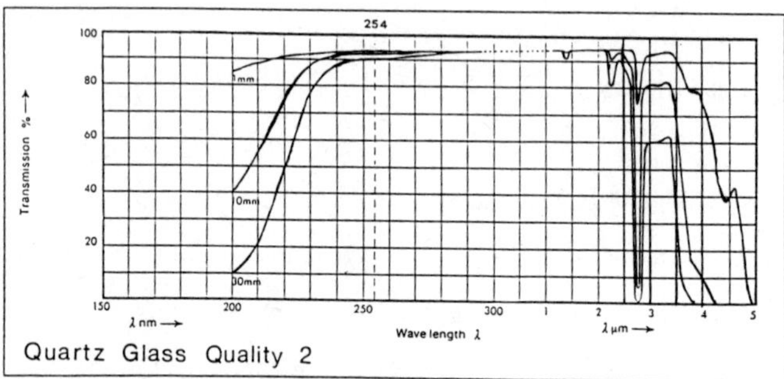

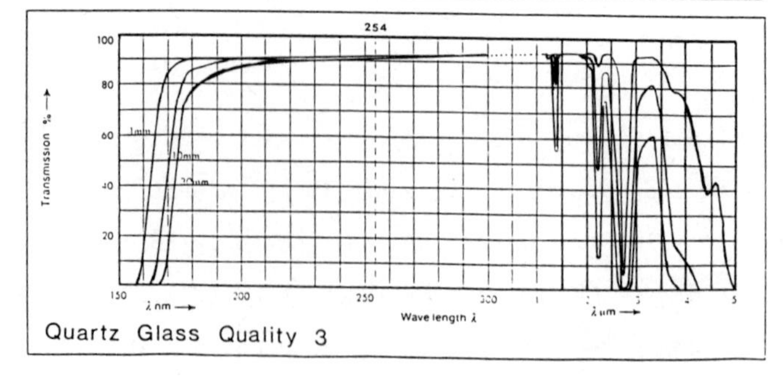

Figure 17

Economic considerations

Assuming that only about 10 % of the power input is utilized as photochemically effective light, nearly the total power input has to be removed as heat. The increased electrical cost must be calculated by multiplication of the power input of the lamp by a factor of 3. Already it has been mentioned that the molecular weight of the photoproduct must be taken into account (Fig.7). To avoid side reactions many photoreactions must be performed in dilute solutions. Then the cost resulting from solvent losses and recycling may become a limiting factor for scaling up a photoreaction. Surprisingly, this important fact is unknown to most photochemists.

Conclusions

The viability of scaling up a photoreaction increases with increasing molecular weight of the photoproduct, with increasing refinement of the starting material, and with increasing quantum yield, i.e. with decreasing irradiation time. Competing with conventional chemical procedures a quick realization of the project is very important, too. For this reason industrial photochemists tend to avoid time-consuming measurements and replace extensive computer calculations - wherever possible - by empirical methods. Thus, realistic information concerning the necessary irradiation time can be obtained by quantitative determination of the amount produced per unit of time, e.g. in a stepwise scale-up using the analytical equipment available in the company. Measurements and calculations have only to be performed with complex reactions, in order to find the optimum substrate consumption where the reaction should be stopped to keep the amount of by-products low. Another obstacle could be that none of the types of photoreactor shown in Fig.16 meets the requirements arising from high substrate concentrations. Then the alternative is to produce turbulences in the rapidly upwards streaming solution of the reactor example a by generating many small gas bubbles in it. This cannot be achieved by separate injection of gas and liquid (Fig. 18), but only by a symmetrical distribution of the structural element Fig.19 on the bottom of the photoreactor[15]. Here, gas and liquid are injected together by a system acting like little water jet pumps. Of course, in this case one has to consider that solvent vapor taken along with the gas has to be condensed when leaving the photoreactor.

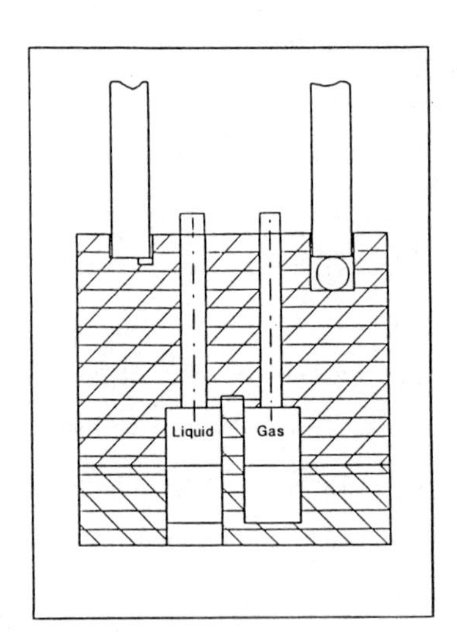

Figure 18

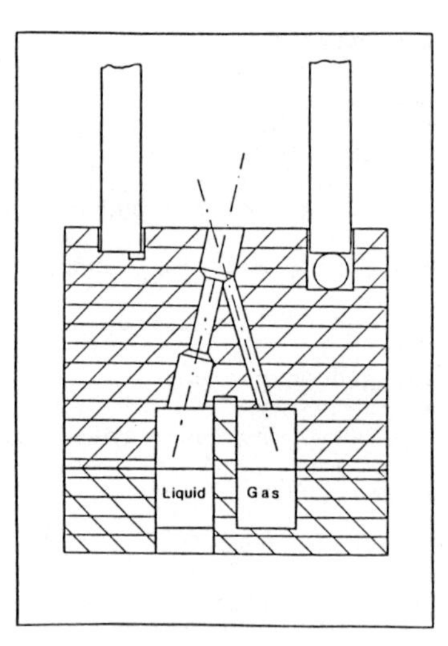

Figure 19

References

[1]EPA Newsletter 1985 (25), 14: Although widely used, "Einstein" is not a IUPAC-sanctioned term. It is sometimes defined as the energy of 1 mol of photons. This use is discouraged. Also,instead of expressing the concentration c in mol l^{-1}, mol dm^{-3} is recommended, and the term "radiant flux I" should be replaced by "radiant power P".

[2]O. Kling, E. Nikolaiski and H.L. Schläfer, Ber.Bunsenges.Phys. Chem., 1963, 67, 883.

[3]R. Srinivasan and T.D. Roberts (eds), "Organic Photochemical Syntheses", Wiley-Interscience, New York, 1971, Vol.1.

[4]K.H. Pfoertner, unpublished results, 1975.

[5]K.H. Pfoertner, J. Photochem., 1984, 25, 91.

[6]J.M. Fitzgerald (ed),"Analytical Photochemistry and Photochemical Analysis", Marcel Dekker, New York, 1971, p.171.

[7]J.F. Rabek, "Experimental Methods in Photochemistry and Photophysics", John Wiley, Chichester, 1982, Part 2.

[8]I. Amdur and G.G. Hames,"Chemical Kinetics: Principles and Selected Topics", McGraw-Hill, New York, 1966, p.14.

[9]K.J. Laidler, "Chemical Kinetics", McGraw-Hill, New York, 1955, p.21.

[10]K.H. Pfoertner and J.P. Weber, Helv. Chim. Acta, 1972, 55, 921.

[11]E. Müller (ed), Houben-Weyl, "Methoden der Organischen Chemie", Thieme Stuttgart, 1975, Vol.IV, Part 5a (Photochemie Part 1), p.9.

[12]K.H. Pfoertner and H.J. Hansen, Hoffmann-LaRoche and Co. Ltd., Swiss Patent Appl. 3637/83, July 1, 1983.

[13]K.H. Pfoertner, unpublished results, 1968.

[14]S.D. Cohen, M.V. Mijovic, G.A. Newman and E. Pitts, Chem. Ind. (London), 1967, 1079.

[15]K.H. Pfoertner in A.M. Braun, M.T. Maurette and E. Oliveros, "Technologie Photochimique", Presses Polytechniques Romandes, Lausanne, in press.

14
Aromatic Compounds: Substitution and Cyclisation

By A. Gilbert
CHEMISTRY DEPARTMENT, UNIVERSITY OF READING, WHITEKNIGHTS, PO BOX 224, READING, BERKS. RG6 2AD, UK

The literature concerning the substitution and cyclisation of aromatic compounds is enormous[1] and reflects the study of these processes from many different viewpoints. The present treatment of the subject matter is necessarily restricted to those topics which have attracted the greatest interest and which offer synthetic potential but the substitution of arenes by photo-generated radicals is outside the scope of this Chapter.

1. Substitution Processes

In marked contrast to the thermal reaction, there are very few examples of photo-induced electrophilic aromatic substitution and the process is largely restricted to the study of aryl hydrogen-deuterium exchanges,[2] and acid catalysed reactions of certain electrophilic ethylenes and acetylenes with benzene. Thus for example phenyl succinic anhydride is produced in low yield from irradiation of the benzene-maleic anhydride system in the presence of trifluoroacetic acid.[3] The reasons for the low efficiency and paucity of electrophilic photosubstitutions probably reflects the quenching of the electronically excited species by common electrophiles and the fact that $\pi\pi^*$ excited arenes would be expected to donate an electron to the electrophilic media: this would produce a radical cation of very little tendency for reaction with electrophiles.

On the other hand, nucleophilic photosubstitution of arenes is a commonly observed process and is the single largest area in this particular field of interest. The reaction arises from the $\pi\pi^*$ triplet state of the arene, and from an empirical classification of the numerous reported examples of this process the following four rules have been proposed in order to predict and rationalise the orientation of attack of the nucleophile onto the arene.[4]

(I) Electron withdrawing substituents on the arene (e.g. NO_2, COMe, CN) activate the meta position towards nucleophilic attack.

(II) Electron donating substituents on the arene (e.g. OMe, OH, alkyl) activate the ortho and para positions towards nucleophilic attack.

(III) Nucleophilic attack occurs at the α-positions of photo-excited polynuclear arenes and at the 2- and/or 4-positions of biphenyls.

(IV) The nature of the nucleophile affects the course of the reaction and merging resonance stabilisation is a feature for consideration in product formation.

Deductions of reaction orientation based on calculated charge densities of $\pi\pi^*$ excited arenes are generally consistent with experimental observations and hence provide support for rules I and II. The photonucleophilic reactions of 4-nitroveratrole, p-chloroanisole, and 1-methoxy-4-nitronaphthalene given in Scheme 1 illustrate the features of rules I, II, and IV respectively. The last example shows that factors more dominant than

Scheme 1

nitro group meta activation (rule I) can operate and the formation of (1), while being an example of rule III, is better rationalised in terms of rule IV. In (1) the combination of powerful electron donor and acceptor groups in a para relationship gives better resonance stabilisation than the product in which the nitro group is replaced by the NHMe group. Thus the stabilising electron donating (or withdrawing) effect that the incoming reagent will have as a substituent on the arene operates also during the product forming steps (merging resonance stabilisation). The chemical yields from photonucleophilic aromatic substitutions using CN^-, OH^- and amines are frequently good although the reactions are performed in dilute solution, but despite the potential of this route for the preparation of simple substituted arenes which may be difficult to obtain by conventional thermal means the process still lacks synthetic exploitation.

Replacement of halide in aromatic systems by enolate anions can also be accomplished with good yields and the process offers a convenient and versatile route to various compounds. Thus o-iodoanisole gives (2) and thence benzofurans in yields of 40-100% dependent on the R group in the enolate.[5] Similarly azaindoles (3) are accessible (23-100% yields) from 3-amino-2-chloropyridine,[6] and phenylacetone results from the irradiation of bromo or iodobenzene with the acetone enolate in 94 and 88% yields respectively.[7] This type of reaction also occurs intra-

R = alkyl

2

3

molecularly and thereby allows the synthesis of such ten-membered cyclic compounds as (4).[8] Another notable example of photoreplacement of halogen is the formation of (5) from o-dichlorobenzene and $(MeO)_3P$ at 60^oC.[9] The reaction requires long exposure times but can be carried out on a multimolar scale and gives ready access to o-$(H_2P)_2C_6H_4$, o-$(RHP)_2C_6H_4$, (where R=Me and Me_2CH) and o-$(Me_2P)_2C_6H_4$.

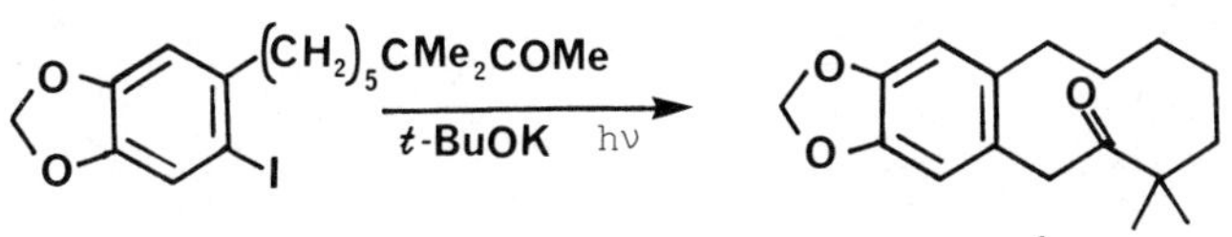

Intramolecular nucleophilic photosubstitution occurs on irradiation of 1-anilino-ω-(p-nitrophenoxy)alkenes (6) and in dilute solution gives good yields of N-(p-nitrophenyl)-ω-anilinoalkan-1-ols (7) (photo-Smiles reaction).[10] The efficiency of the process is dependent upon the length of the alkyl chain and generally product formation is accelerated in the presence of bases. This latter feature is interpreted in terms of the assistance rendered by the base in transforming the Meisenheimer complex (8) into the cyclised intermediate (9).

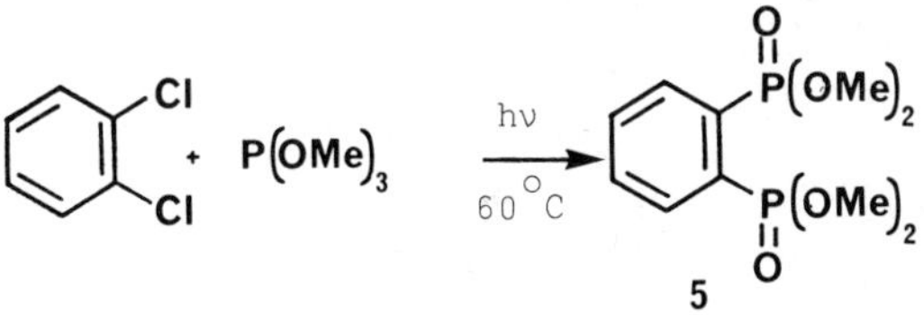

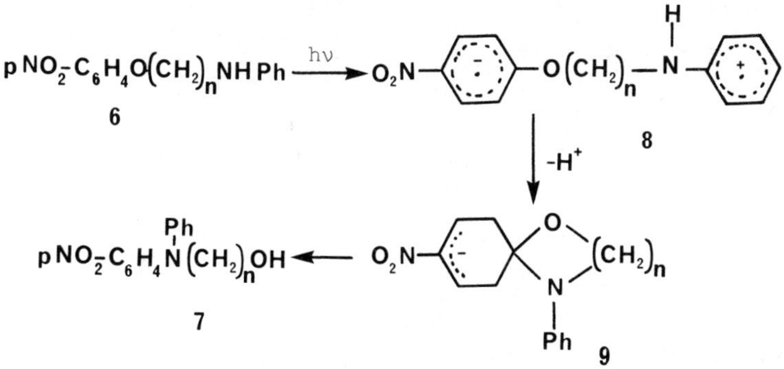

Photosubstitution in the aromatic rings of anthraquinones has attracted considerable attention and by this route a wide variety of derivatives of the quinone can be obtained. The parent molecule undergoes photohydroxylation in sulphuric acid with yields of the isomers dependent on the acid concentration.[11]

Scheme 3

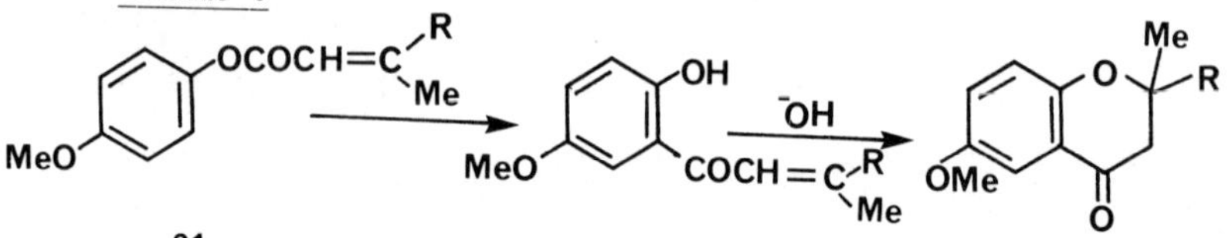

MeO OCOCH=C(R)Me (21) → MeO, OH, COCH=C(R)Me → $^{-}$OH → OMe, O, Me, R, O (22)

The photo-Fries reaction has also been used to modify the conformation and properties of polymers. The reactive chromophore (ArOCO-) may be part of the polymer chain or a pendant group: the former is exemplified by the photoconversion of the fluorene-based polyacrylates (23) to polymers (24) which have o-hydroxybenzophenone moieties.[18]

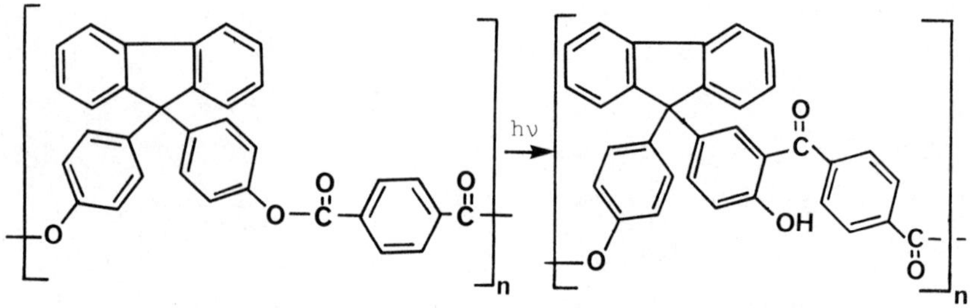

23 —hν→ 24

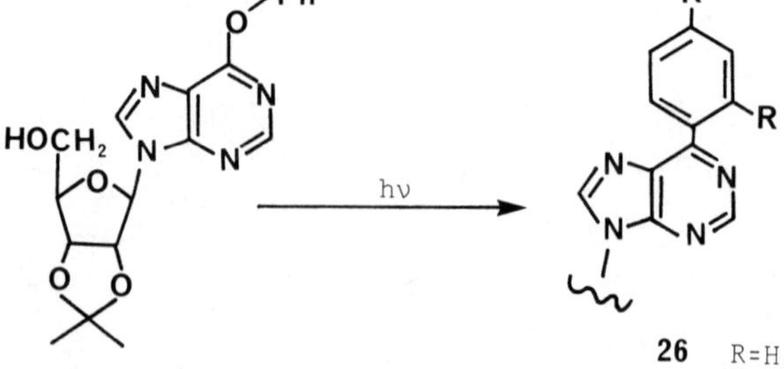

25 —hν→

26 R=H, R'=OH

27 R=OH, R'=H

In contrast to the photo-Fries, reported examples of the photo-Claisen reaction of aryl allyl ethers are rare and the process has received little attention as a synthetic procedure. With diaryl ethers, however, yields can be promising and, for example, irradiation of inosine (25) gives 42 and 28% respectively of the *ortho* and *para* rearrangement isomers (26) and (27).[19]

3. Intramolecular Cyclisation Processes

This process is very general for aromatic systems and occurs with a wide variety of different types of compounds. The process can arise by photo-oxidative cyclisation of two arene units (*i.e.* Ar-X-Ar systems) to give polynuclear arenes or by intramolecular attack of an ethylene, 1,3-diene, or acetylene moiety onto the aryl group (*e.g.* Ar-X-CH=CH_2 systems) to produce benzo compounds. In both classes of reaction various types of intervening unit (*e.g.* X=O, -CO-NR-) between the reacting chromophores have been used with little detrimental effect on the process. Irradiation of arene-arene and arene-ethylene systems which have appropriately positioned halogen substituents also leads to cyclisation products by the loss of halogen acid and some *o*-disubstituted arenes yield benzo compounds by photo-induced reactions between the substituents. Thus by these various intramolecular cyclisation routes, it has been possible to gain access into a number of different types of cyclic systems: this is illustrated below by examples from the above four classes of reaction.

3.1 Intramolecular Photocoupling of Arenes.

The reaction of this type which has received the greatest attention involves the photo-oxidative cyclisation of 1,2-diarylethylenes to give polynuclear arenes. This process has been outlined in the Chapter concerning rearrangements of alkenes but systems with units other than an ethylene moiety between the arenes also undergo oxidative cyclisation on irradiation. Thus diaryl ethers and diaryl amines respectively yield dibenzofurans and carbazoles. The latter reaction has been used to synthesise the previously unknown isomer (28) of the five possible indolocarbazoles by a two photon triplet state cyclisation of (29),[20] and this type of process with the dipyridyl thioether (30) leads to the parent compound of the novel heterocyclic system thieno[3,2-c, 4,5-c']-dipyridine (31).[21]

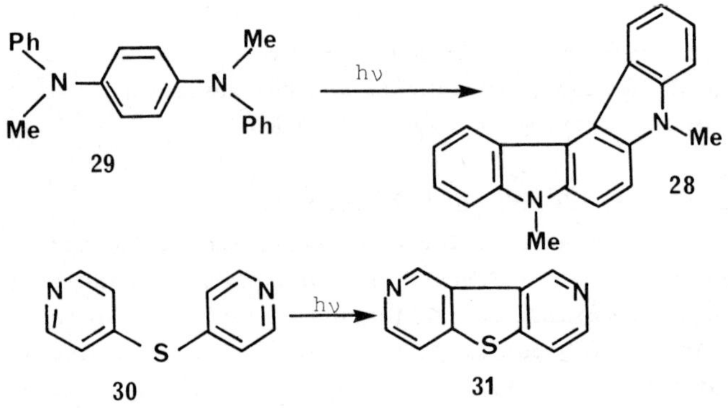

Unlike stilbenes, azobenzenes and benzylidene anilines do not undergo photo-oxidative cyclisation in neutral media since the lowest excited state has $n\pi^*$ character. Protonation or Lewis acid complexation of these compounds results in the lowest state being $\pi\pi^*$ and irradiation then results in benzo[c]cinnolines and azaphenanthrenes respectively. For example this procedure with phenylazopyridine produces (32) and (33),[22] and similarly diazaphenanthrenes are obtained from irradiation of benzylidenaminopyridine (34) or pyridalaniline (35).[23]

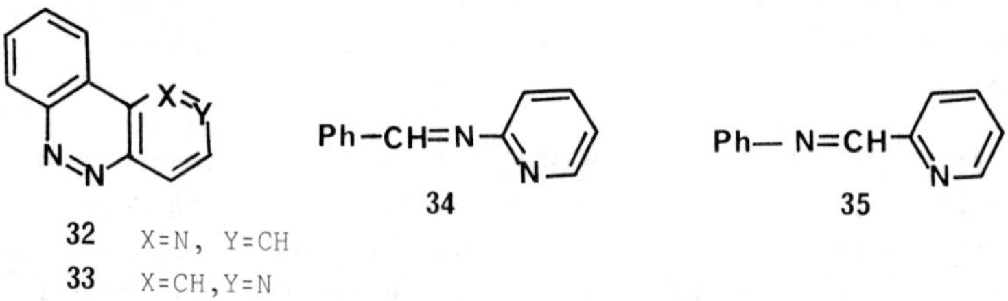

1,2-Diarylarene systems also undergo photocyclisation and polynuclear aromatic compounds result. The conversion of (36) to the novel condensed ring system (37),[24] and of the N-phenylpyridinium perchlorate (38) to (39) which occurs in 80% yield[25] illustrate the variety of compounds for which this particular type of cyclisation has been observed.

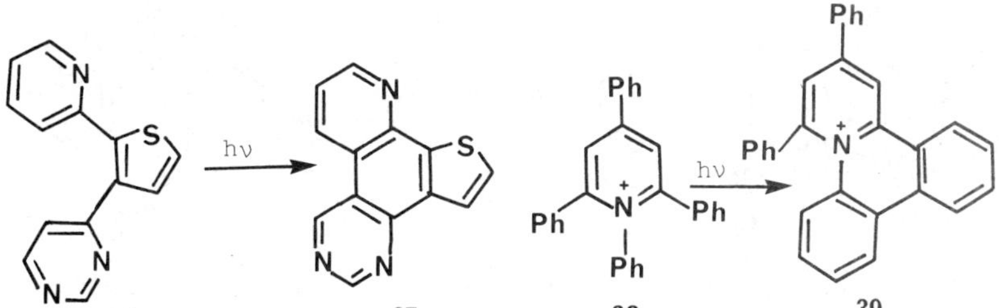

36 37 38 39

3.2 Photocyclisation of Arene-Ethylene Systems. There are numerous reported examples of this type of cyclisation involving a variety of different classes of compounds, and many instances of synthetic use of the process have been described. The reaction for compounds in which the intervening unit, X, between the arene and ethylene is a heteroatom having an unshared electron pair available for involvement in electrocyclisation is outlined in Scheme 4, and such processes and their synthetic uses have recently been reviewed.[26]

Scheme 4

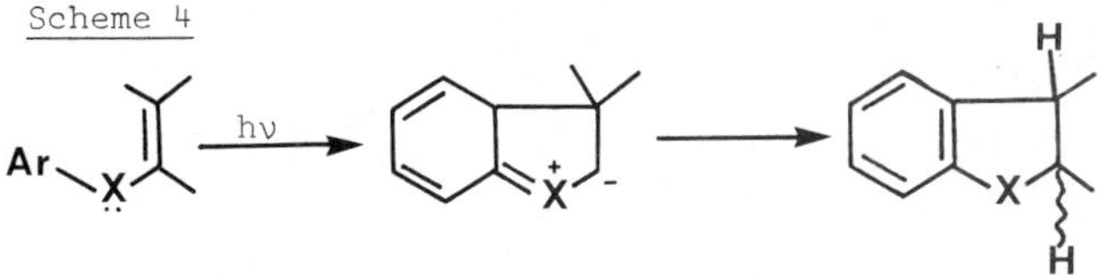

Yields of benzofurans, benzothiophens and indoles from respective cyclisation of aryl ethenyl ethers, thioethers, and N-arylenamines are generally good and for example (40) gives, on a 30g scale, a 95% yield of (41) which is an intermediate for morphine type

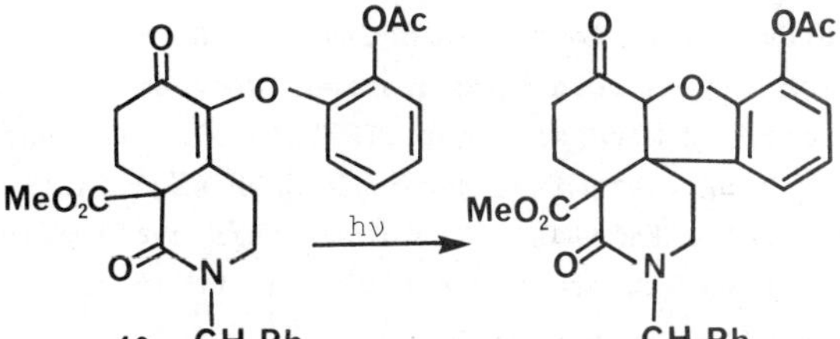

40 41

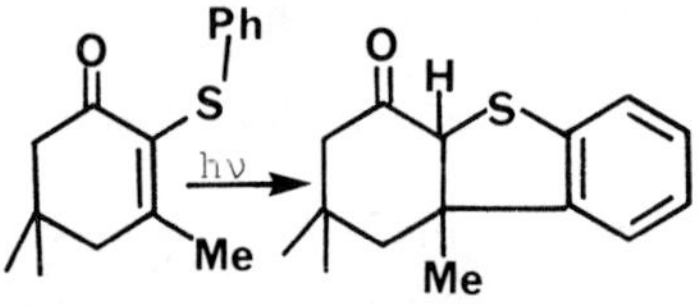

alkaloids, and similarly, irradiation of (42) and (43) leads to 91 and 90% respectively of the *cis* decalone derivatives (44)[28] and (45).[29]

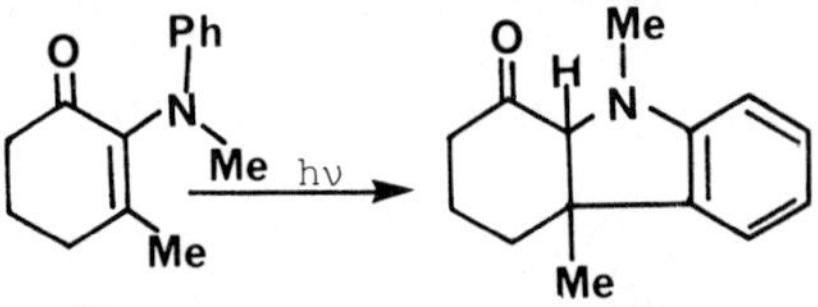

This type of cyclisation also readily occurs for aryl enones *N*-arylanilides, *N*-benzoylenamines, 1-arylbut-1,3-dienes, 1-arylbut-1-en-3-ynes and *N*-allyl-pyridinium and quinolinium perchlorates. The reaction with aryl enones has been studied with a variety of derivatives and the cyclisation of (46) is the key step in the convenient synthesis of 1,2,3,4-tetrahydro-1-methylfluorene-1-carboxylic acid (47).[30]

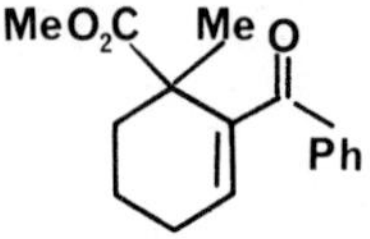

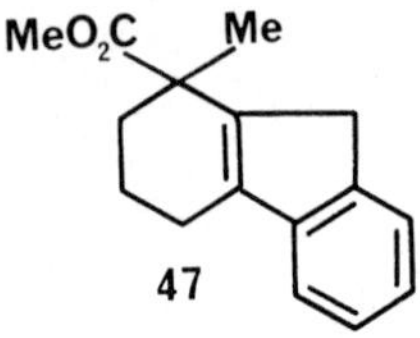

Of all the compounds studied under this classification, *N*-aryl-acrylamides and *N*-benzoylenamines are those most frequently used in synthetic pathways and their cyclisation reactions offer versatile routes to many naturally occurring alkaloids. Although the latter of the two arene-ethylene systems to be that favoured in synthetic procedures, irradiation of the *N*-naphthylacrylamides [*e.g.* (48)] conveniently yields benzoquinoline derivatives [e.g. (49)] which can be readily converted into compounds having the indoloquinoline nucleus of ergot alkaloids.[31] The photo-cyclisation of *N*-benzoylenamines has also been used in the synthesis of several alkaloids. Yields of this cyclisation are dependent on the substituents present but can be in excess of 70% for such conversions as (50) to (51).[32] Examples of the synthetic exploitation of this process include the formation of the key intermediate (52) for the synthesis of lycorane from irradiation of the N-benzoyl derivatives (53) of 1,2,3,3a,4,5-hexahydroindole-6-one,[33] and as a route to nauclefine (54) and isonauclefine (55) from (56).[34]

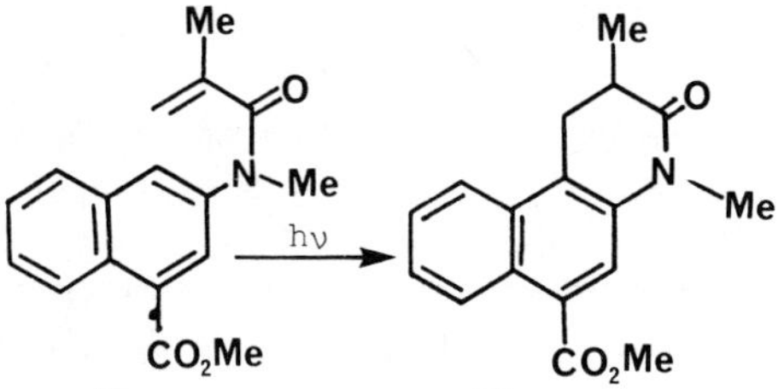

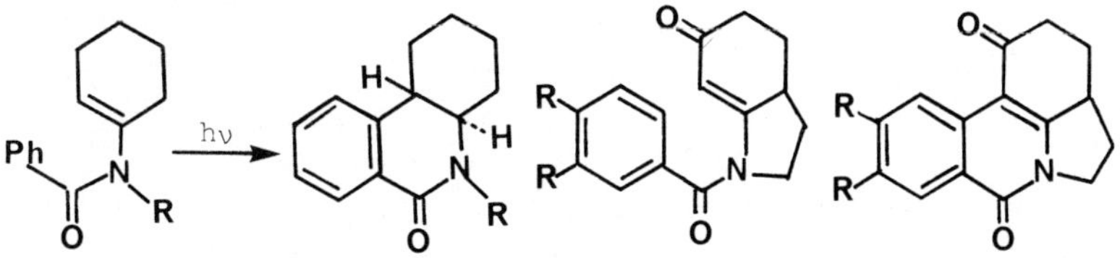

R=H or R-R= $-OCH_2O-$

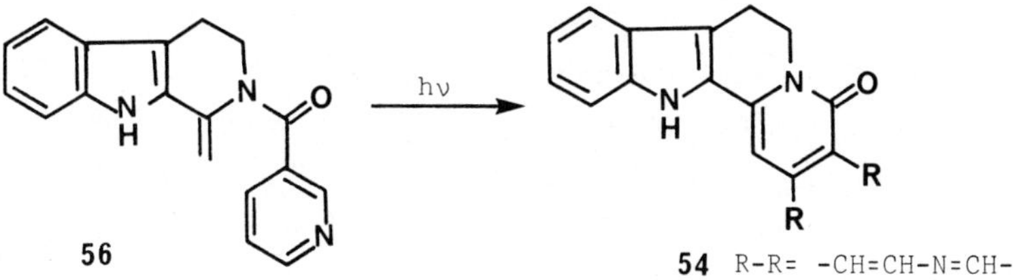

54 R-R= -CH=CH-N=CH-

55 R-R= -N=CH-CH=CH-

The first report of photocyclisation of 1-arylbuta-1,3-diene to yield naphthalene derivatives appeared in 1962,[35] and since this time the reaction has attracted interest both for its synthetic potential and, with certain derivatives, for its photochromic properties. The former of these two aspects is well illustrated by the photocyclisation of the styrylchromones (57) in the presence of iodine to give yields of the 12-benzo[a]xanthen-12-ones (58) in the region of 80%.[36] The photochromic interests in these systems originate from studies of the photochemistry of the overcrowded molecules bis arylmethylene succinic anhydrides and imides (59). Irradiation of such compounds results in

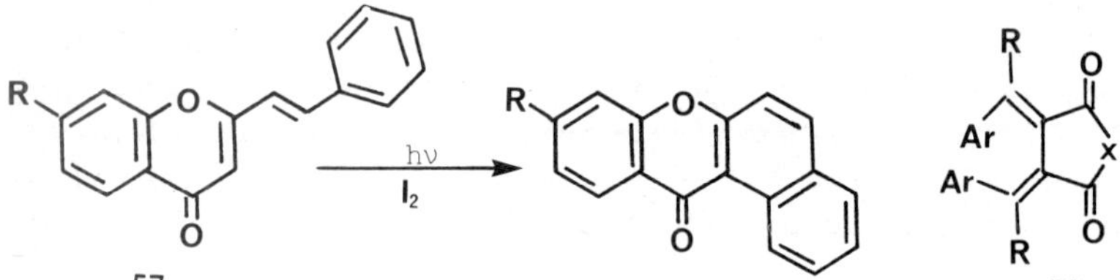

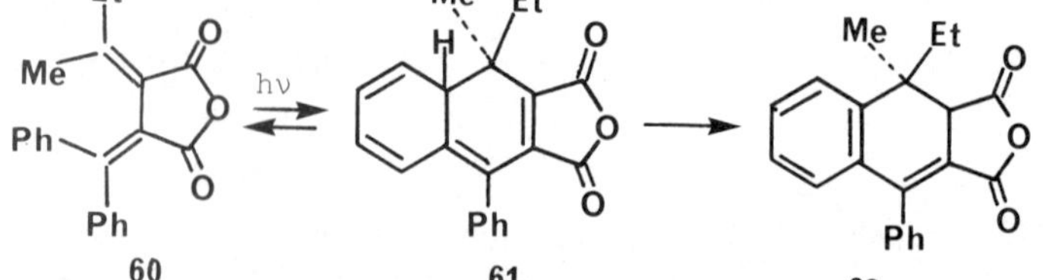

cyclisation and the formation of coloured 1,8a dihydronaphthalenes which may undergo a 1,5-H shift to the 1,2-dihydronaphthalene. For example (60) yields the red isomer (61) and thence (62).[37] The efficiencies of the reactions and the colours produced are markedly dependent on the type of arenes in (59) and the nature and position of substituents. In some cases where there is no hydrogen atom to participate in a 1,5-shift [e.g. the photoisomer (63) of (64)[38]], the coloured compound is thermally stable and simply undergoes reversible photocyclisation. The high quantum yields for photocoloration, the photodegradation resistance of the systems, and the display of these properties in a range of media, make compounds (59) attractive for commercial photochromic applications. This type of photocyclisation followed by a 1,7 H-shift to yield 1,4-dihydronaphthalene derivatives has been used towards the synthesis of β-apolignans and by such a process the trimethyl ether of β-apoplicatitoxin (65) has been obtained in 34% yield.[39]

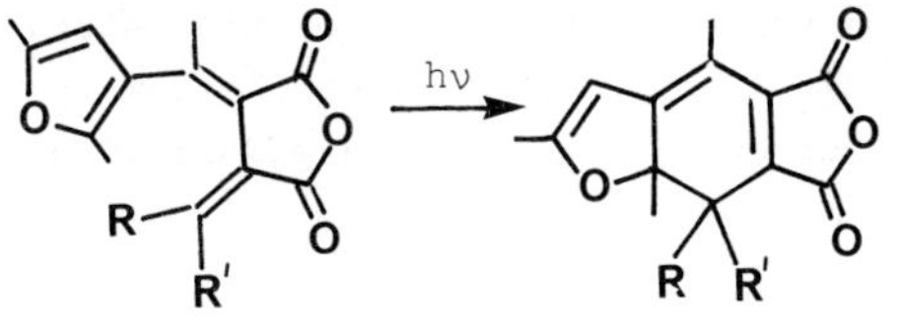

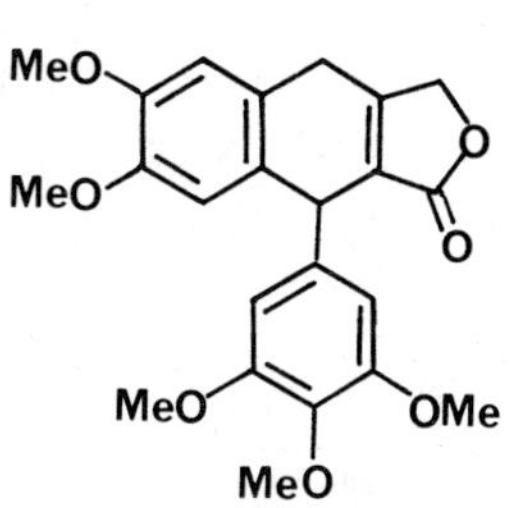

Formation of condensed arenes by photocyclisation of 1-arylbut-1-en-3-ynes does not in principle require a dehydrogenation step but for several systems it has been observed that the presence of oxygen or iodine accelerates the reaction and improves product yields. From some systems, however, oxygenated products may result. Thus while irradiation of (66) gives 1-phenylphenanthrene in 45% yield and even crowded arenes such as 4,5-diphenylphenanthrene are accessible by this route, the 9-phenanthryl compound (67) yields (68) under an oxygen atmosphere.[40]

Ar-CH=CH-C≡C-Ph

66 Ar = 1-naphthyl

67 Ar = 9-phenanthryl

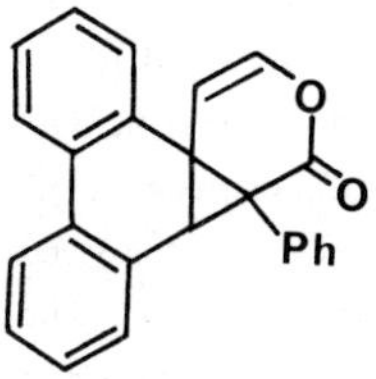

Cyclisation occurs on irradiation of N-allylpyridinium (69) and quinolinium (70) perchlorates in methanol solution.[41] The reaction pathway which is deduced to involve photo-induced intramolecular electron transfer and reaction of the radical-cation radical-anion species (71) with the solvent is outlined in

Scheme 5 for (69). For these systems, the crude reaction mixture can be catalytically hydrogenated and this gives synthetically useful yields of 90% at 54% conversion of perhydroimidolizidines.

Scheme 5

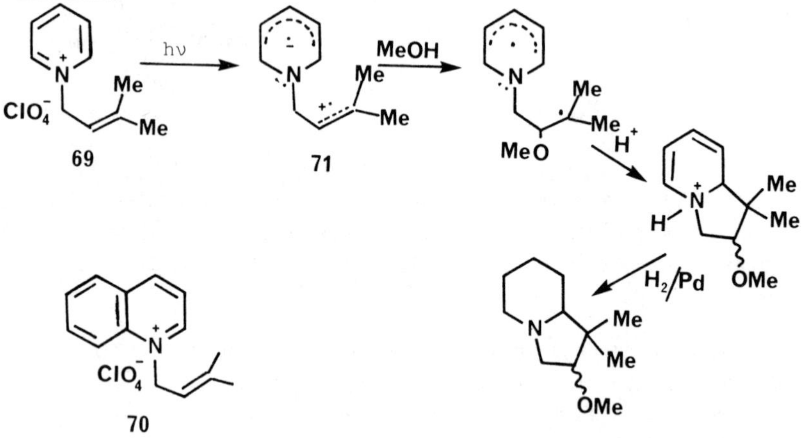

3.3 Cyclisation by Photo-Induced Elimination of Halogen Acid

This type of photocyclisation reaction is a versatile and well-used method for the synthesis of numerous polycyclic systems. The most commonly reported examples of this reaction involve intramolecular aryl-aryl coupling and the process which is complimentary to the photo-oxidative cyclisations described in Section 3.1 has been applied to the synthesis of a number of naturally occurring compounds. By this procedure phenanthridones (72) are obtained in 67-74% yield from cyclohexane solutions of 2-chlorobenzanilides (73),[42] and $(\pm)$-oliveridine (74) and its epimer are similarly produced from the two isomers (75) in aqueous hydrochloric acid.[43] Other examples of this widely applicable photoreaction include the formation of liriodenine (76) in 37% yield from (77),[44] and in view of the known biological activity of (76) it is not surprising that its synthesis has been achieved by an alternative route involving photo-induced cyclisation of (78) to (79) as the key step:[45] the latter procedure is also used in the synthesis of atherospermidine (80).[46]

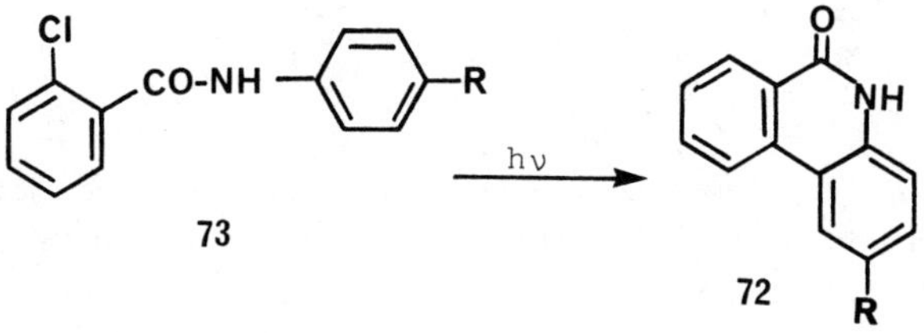

Cl
CO-NH
R
hν
O
NH
73
72
R

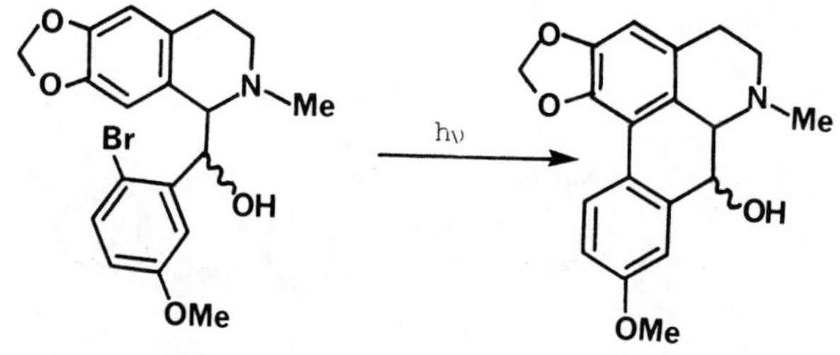

O
O
N
Me
Br
OH
OMe
75
hν
O
O
N
Me
OH
OMe
74

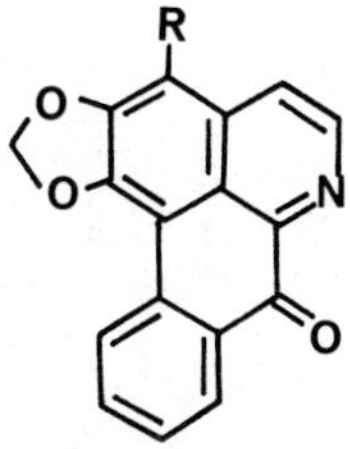

R
O
O
N
O
76 R=H
80 R=OMe

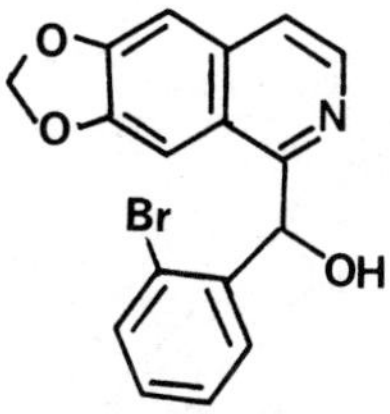

O
O
N
Br
OH
77

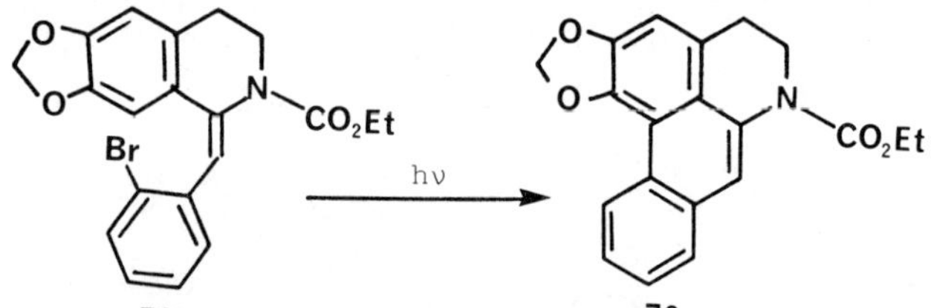

78 hν 79

Arene-ethylene coupling can also readily be achieved by photo-dehydrohalogenation and again this occurs with a variety of structural units. For example, irradiation of the iodocompound (81) in the presence of triethylamine gives 50% of the cyclised product (82) and this provides the key step in a novel synthesis of hexahydroapoerysopine dimethyl ether (83).[47] Similar photo-induced dehydrohalogenation-cyclisation is observed with the

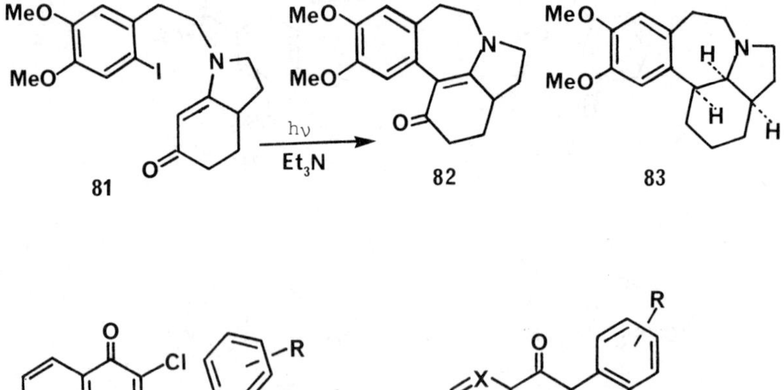

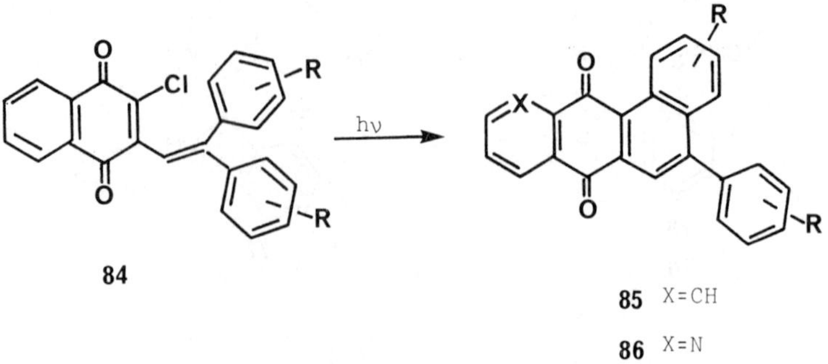

81 hν Et_3N 82 83

84 hν

85 X=CH

86 X=N

1,1-diaryl-4-chlorobutadiene residue in (84) which is obtained from irradiation of 2,3-dichloronaphtho-1,4-quinone and 1,1-diarylethylenes: this reaction offers a convenient 'one pot' synthesis of (85) and naphtho[2.1-g]quinoline-7,12-dione derivatives (86).[48]

The intramolecular photo-induced cyclisation between an aryl halide and an ester or thioamide unit provides access to various heterocyclic systems and can be a remarkably efficient process. This reaction is illustrated by the conversion of cinnamic acid derivatives (87) to 5-chlorocoumarin in yields approaching 100% at concentrations less than 10^{-3}M,[49] and by the novel synthesis of 4H-1,3-benzothiazin-4-ones (88) from (89) in 83-93% yield.[50] In the former system, amide derivatives of (87) give the corresponding imino analogues.

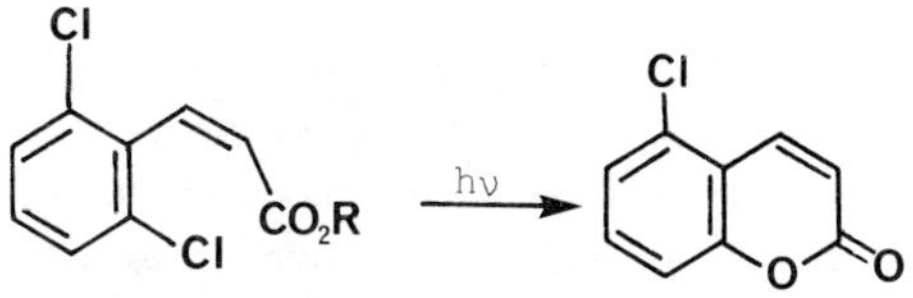

87 R = alkyl or aryl

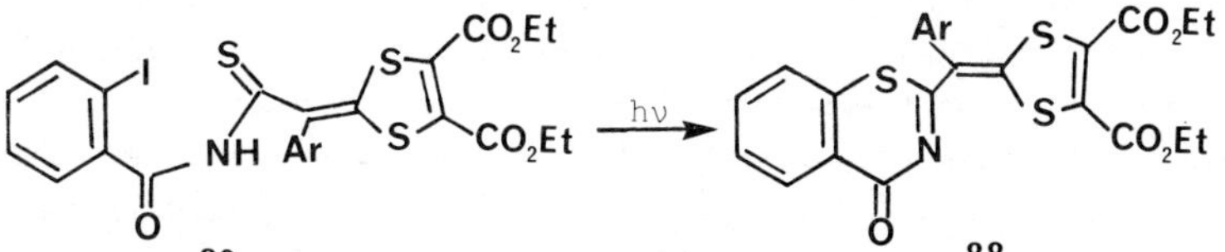

89 **88**

Chloroacetyl derivatives of pharmacodynamic amines are photoactive and undergo a variety of cyclisation reactions dependent on the type and position of substituent in the phenyl moiety. For example, the 2,5-dimethoxy compound (90) yields the benzazepin-2-ones (91) and (92) and the tricyclic enones (93) and (94) on irradiation:[51] these processes are also considered in a recent review.[52]

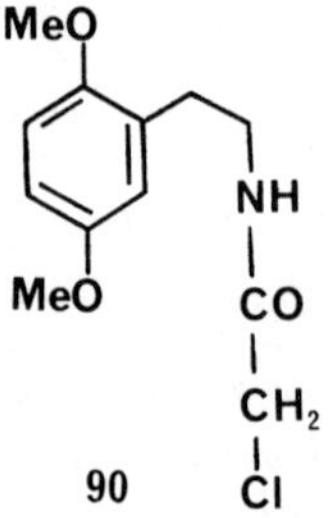

91 R=OMe, R'=H
92 R=H, R'=OMe

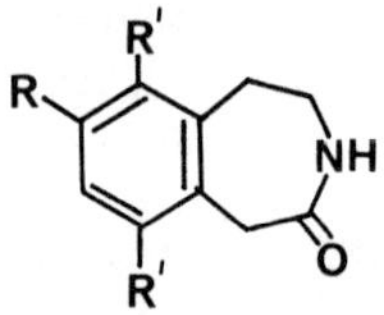

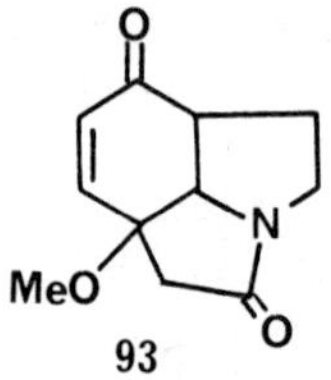

3.4 Cyclisation by Photoreaction of *ortho* Substituents. Although this type of reaction may not directly involve the aromatic ring, examples of the process are presented here since the cyclisation procedures are complimentary to those described above. Irradiation of *o*-allylanilines (95) in benzene solution provides an easy access to 2-methylindolines (96) in yields of 40-80%,[53] and the photo-induced reaction of the *ortho* substituents in (97) to give (98) in 85% has been used as an improved cyclisation step in the synthesis of 12-keto derivatives of tetracyclines.[54] Intramolecular photo-induced hydrogen abstraction in α-arylacetophenones (99) leads to 100% yields within experimental error, of the indane derivatives (100),[55] and by photodehydrohalogenation between the substituents in *o*-methoxyacetophenones (101) the benzofuran derivatives (102) are obtained.[56]

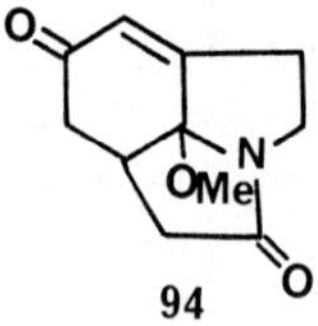

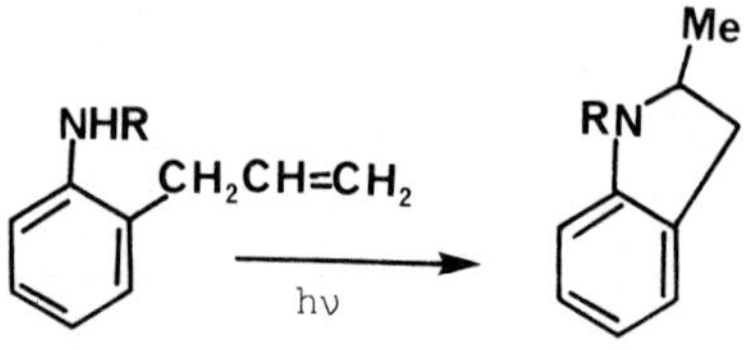

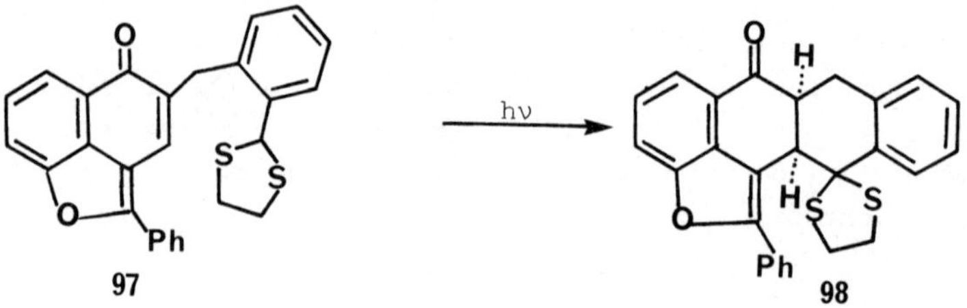

99 → **100**

101 → **102**

R's = H,Me,Ph

References

1. See references in A.Gilbert and J.D.Coyle, Photochemistry of Aromatic Compounds, "Photochemistry" ed. D.Bryce-Smith (Specialist Periodical Reports), The Royal Society of Chemistry, London, 1970-1985, vols. 1-16.

2. See for example H.Shizuka and S.Tobita, J.Amer.Chem.Soc., 1982, 104, 6919.

3. D.Bryce-Smith, A.Gilbert, N.Al-Jalal, R.R.Deshpande, J.Grzonka, M.A.Hems and P.Yianni, Z.Naturforsch, 1983, 38b, 1101.

4. E.Havinga and J.Cornelisse, Pure Appl.Chem., 1976, 47, 1.

5. R.Beugelmans and H.Ginsburg, J.Chem.Soc.Chem.Comm., 1980, 508.

6. R.Beugelmans, B.Boudet, and L.Quintero, Tetrahedron Letters, 1980, 21, 1943.

7. M.A.Fox, J.Younathan, and G.E.Fryxell, J.Org.Chem., 1983, 48, 3109.

8. M.F.Semmelhack and T.Bargar, J.Amer.Chem.Soc., 1980, 102, 7765.

9. E.P.Kyba, S.T.Liu, and R.L.Harris, Organometallics, 1983, 2, 1877.

10. See K.Yokoyama, J.Nakamura, K.Mutai, and S.Nagakura, Bull.Chem.Soc.Jpn., 1982, 55, 317; G.G.Wubbels and D.W.Celander, J.Amer.Chem.Soc., 1981, 103, 7669, and references therein.

11. O.P.Studzinskii and A.V.Eltsov, Zh.Org.Khim., 1982, 18,1904.

12. Y.V.Denisov and N.A.Pirogova, Zh.Org.Khim., 1982, 18, 2397.

13. K.Yoshida, T.Okugawa, E.Nagamatsu, Y.Yamashita, and M.Matsuoka, J.Chem.Soc.Perkin Trans.1, 1984, 529.

14. F.S.Tanaka, R.G.Wien, and B.L.Hoffer, Synth.Commun., 1983, 13, 951.

15. M.P.Moon, A.P.Komin, J.F.Wolfe, and G.F.Morris, J.Org.Chem., 1983, 48, 2392.

16. F.Farina, R.Martinez-Utrilla, and M.C.Paredes, Tetrahedron, 1982, 38, 1531.

17. J.Primo, R.Tormo, and M.A.Mirando, Heterocycles, 1982, 19, 1819.

18. J.Lo, S.N.Lee and E.M.Pearce, J.Appl.Polym.Sci., 1984, 29, 35.

19. T.Endo and J.Zemlicka, Nucleic Acids Symp.Ser., 1981, 9, 33.

20. H.Weller and K.H.Grellmann, J.Amer.Chem.Soc., 1983, 105,6268.

21. J.E.Rockley and L.A.Summers, Chem.Ind., 1979, 666.

22. J.W.Barton and R.B.Walker, Tetrahedron Letters, 1975, 569.

23. H.H.Perkampus and B.Behjati, J.Heterocyclic Chem., 1974, 11, 511.

24. A.Mitschker, U.Brandl, and T.Kauffmann, Tetrahedron Letters, 1974, 2343.

25. V.R.Tymyanskii, M.I.Knyazhanskii, Y.P.Andreichikov, G.E.Trukhan, and G.N.Gorofeenko, Zh.Org.Khim., 1976,12,1126.

26. A.G.Schultz, Acc.Chem.Res., 1983, 16, 210; A.G.Schultz and L.Motyka, Org.Photochem., 1983, 6, 1.

27. A.G.Schultz, R.D.Lucci, J.J.Napier, H.Kinoshita, R.Ravichandran, P.Shannon, and Y.K.Lee, J.Org.Chem., 1985, 50, 217.

28. A.G.Schultz, W.Y.Fu, R.D.Lucci, B.G.Kurr, K.M.Lo, and M.Boxer, J.Amer.Chem.Soc., 1978, 100, 2140.

29. A.G.Schultz and I.Ching Chiu, J.Chem.Soc.Chem.Comm., 1978, 29.

30. S.C.Roy and U.F.Ghatak, J.Chem.Res.(S), 1983, 138.

31. I.Ninomiya, C.Hashimoto, T.Kiguchi and T.Naito, J.Chem,Soc.Perkin I, 1983, 2967.

32. I.Ninomiya, T.Kiguchi, O.Yamamoto, and T.Naito, J.Chem.Soc.Perkin I, 1979, 1723.

33. H.Ida, S.Aoyagi, Y.Yuasa and C.Kibayashi, Heterocycles, 1977, 6, 1747.

34. M.Sainsbury and N.L.Uttley, J.Chem.Soc.Perkin I, 1977, 2109.

35. G.J.Fonken, Chem.Ind., 1962, 1327.

36. K.A.Kumar and G.Srimannarayana, Indian J.Chem.Sect.B, 1980, 19, 615.

37. R.J.Darcy, R.J.Hart, and H.G.Heller, J.Chem.Soc.Perkin I, 1978, 571.

38. R.J.Darcy, H.G.Heller, P.J.Strydom, and J.Whittall, J.Chem,Soc.Perkin I, 1981, 202.

39. T.Momose, T.Nakamura, and J.Kanai, Chem.Pharm.Bull., 1978, 26, 1592.

40. A.H.A.Tinnemans and W.H.Laarhoven, Tetrahedron Letters, 1973, 817.

41. U.C.Yoon, S.L.Quillen, P.S.Mariano, R.Swanson, J.L.Stavinoha and E.Bay, J.Am.Chem.Soc., 1983, 105, 1204.

42. J.Grimshaw and A.P. de Silva, J.Chem.Soc.Perkin Trans.2, 1982, 857.

43. S.V.Kessar, T.Mohammad and Y.P.Gupta, Indian J.Chem.Sect.B., 1983, 22, 321.

44. Y.P.Gupta, V.S.Yadav, and T.Mohammad, Indian J.Chem.Sect.B, 1983, 22, 429.

45. S.Nimgirawath and W.C.Taylor, Aust.J.Chem., 1983, 36, 1061.

46. S.Nimgirawath and W.C.Taylor, J.Sci.Soc.Thailand, 1983, 9, 73.

47. H.Iida, T.Takarai, and C.Kibayashi, J.Org.Chem., 1978, 43, 975.

48. K.Maruyama, S.Tai, and T.Otsuki, Heterocycles, 1983, 20, 1031.

49. R.Arad-Yellin, B.S.Green, and K.A.Muszkat, J.Org.Chem., 1983, 48, 2578.

50. T.Nishiwaki, E.Kawamura, N.Abe, Y.Sasaoka, and H.Kochi, Heterocycles, 1981, 16, 1203.

51. Y.Okuno and M.Kawamori, Tetrahedron Letters,1973, 3009; see also T.Iwakuma, H.Nakai, O.Yonemitsu, D.S.Jones, I.L.Karle, and B.Witkop, J.Amer.Chem.Soc., 1972, 94, 5136, and references therein.

52. R.J.Sundberg, Org.Chem., 1983, 6, 121.

53. U.Koch-Pomeranz, H.Schmid and H.J.Hansen, Helv.Chim.Acta, 1977, 60, 768.

54. D.H.R.Barton, M.T.Bielska, J.M.Cardoso, N.J.Cussans, and S.V.Ley, J.Chem.Soc.Perkin Trans.2, 1981, 1840.

55. M.A.Meador and P.J.Wagner, J.Amer.Chem.Soc., 1983, 105,4484.

56. Y.Izawa, Y.Watoh, and H.Tomioka, Chem.Lett., 1984, 33.

15
Alkaloids

By J.D. Coyle
CHEMISTRY DEPARTMENT, THE OPEN UNIVERSITY, MILTON KEYNES MK7 6AA, UK

The prime motivation in synthetic efforts directed towards alkaloids and related compounds used to be the completion of their characterization by total synthesis, and the production of larger quantities than could be extracted conveniently from natural sources. This still holds for some of the newer groups of alkaloids, but more commonly the nitrogen-heterocyclic systems found in "traditional" alkaloids are used as targets to evaluate new synthetic methods. In this brief review a selection of photochemical reactions is presented which exemplifies the application to alkaloid synthesis of photocyclization and related reaction of aromatic substrates. The major reaction types can be classified as six-electron electrocyclic ring-closure or as intramolecular substitution, and relevant mechanistic features are highlighted as appropriate.

Electrocyclic Reactions

The archetypal electrocyclic ring-closure in aromatic systems is the photochemical reaction of *cis*-stilbene to give 4a, 4b-dihydrophenanthrene (1). The process is stereospecific, and an "allowed" conrotatory mode of closure generates the *trans* isomer of the dihydrophenanthrene. Under oxidizing conditions, for example in the presence of air or with added iodine, the isolated product is phenanthrene. A very large number of reactions involving structurally related stilbenes are known,[2] and high chemical yields and quantum yields can be achieved in many cases.

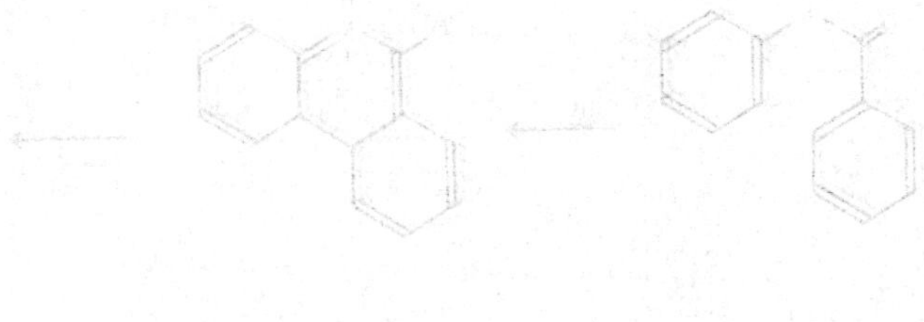

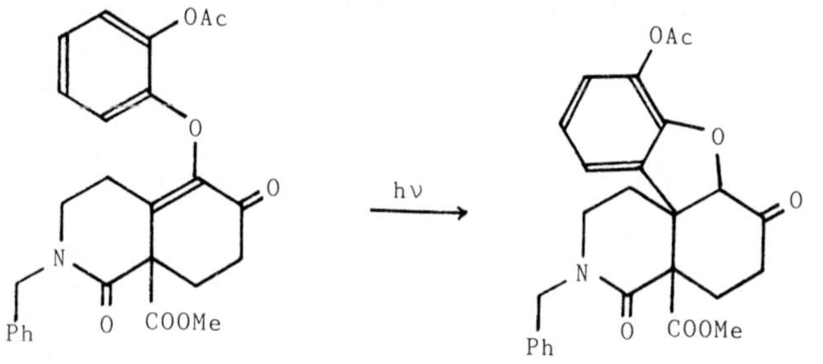

95%

(6)

Enamide Reactions

When one of the three participating double bonds of stilbene is replaced by an amide (-CO-NH-) unit, cyclization reactions remain feasible (7), and this broad class of enamide photocyclizations has been extensively developed and applied to the synthesis of alkaloids.[6] The most widely reported type of substrate is an N-aroyl derivative of an enamine. The reaction (8) used in a synthesis of the protoberberine framework exemplifies the operation of both oxidative (major) and non-oxidative processes,[7] though in this system both products could be used for subsequent transformation to xylopinine. Regioselectivity in respect of the position of cyclization is often very good, as seen in the reaction (9), employed to make γ-lycoran,[8] in which preferential attack occurs at the position para and meta (rather than ortho and meta) to the methylenedioxy substituent.

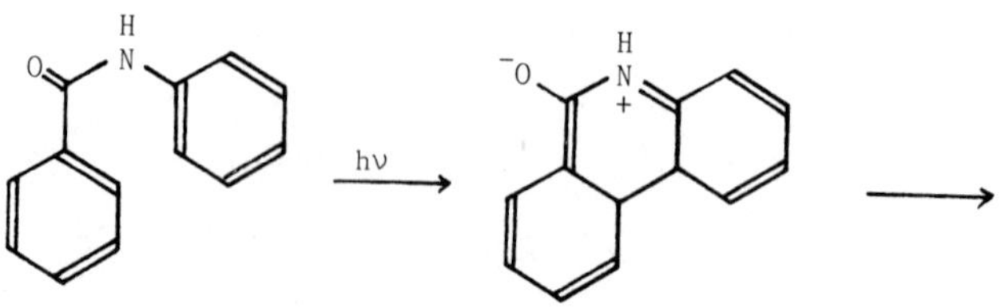

(7)

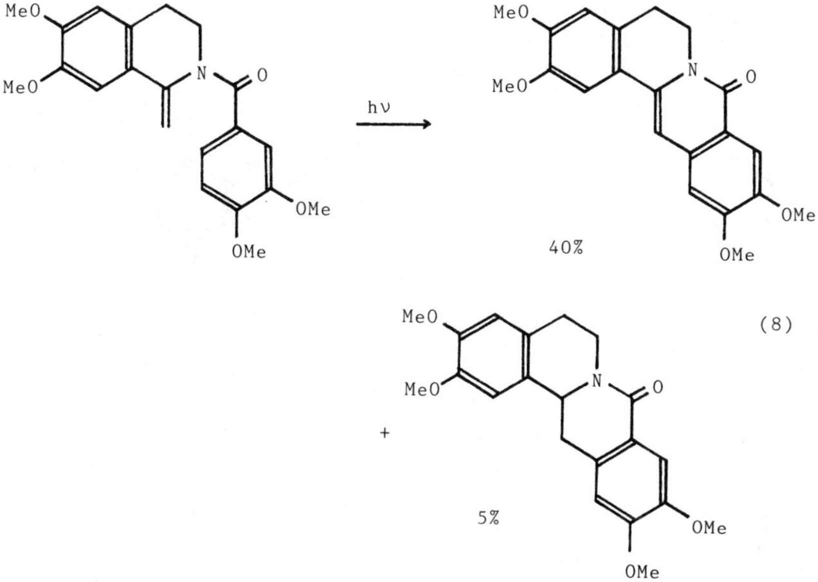

hν

40%

+

5%

(8)

hν

70%

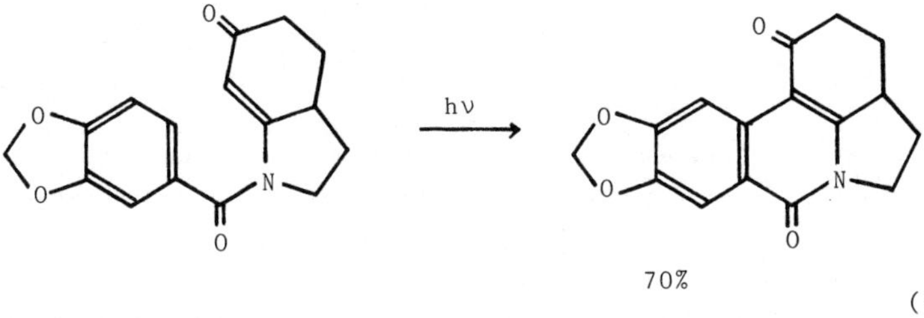

(9)

The use of an appropriate <u>ortho</u> methoxy group, which is eventually eliminated as methanol, allows a cyclization to be carried out that is effectively oxidative. This is seen in reaction (10) <u>en route</u> to a benzo[<u>c</u>]phenanthridine alkaloid, dihydronitidine.[9] The corresponding compound that lacks the <u>o</u>-methoxy group cyclizes to give mainly (51% isolated) the non-oxidative product. An extension to the scope of these reactions has been developed recently, whereby in the presence of sodium borohydride a reductive photocyclization takes place. This has been employed effectively in a reaction (11) that yields the ergot skeleton of lysergic acid.[10] Although three major stereoisomers are formed, the chiral centres at which they differ in configuration are no longer present in the target structure.

50% (10)

81%

lysergic acid

(11)

Although N-aroyl enamines are the most commonly encountered substrates in enamide photocyclizations, variations on the basic structure are possible. A different electronic pattern is found in N-arylamides of α,β-unsaturated carboxylic acids, and an example of such a compound is seen in a naphthyl methacrylamide (12) that forms the ergot skeleton on irradiation.[11] In these, as in other ring-closures, migration of an o-substituent at the position of closure can occur; because such a substituent can also have a considerable directing effect on the regioselectivity of reaction, it can be a useful tool in synthesis, for example in the reaction (13) leading to a reduced phenanthridine.[12]

hν

42%

(12)

hν

71%

(13)

N-Arylamides of aromatic carboxylic acids have not been as widely used, and effective cyclization of such benzanilides requires the presence of an ortho halogen substituent that can be eliminated in the overall process. This is apparent in a system (14) used to construct the lycorane framework for anhydrolycorine.[13]

hν

67% (14)

Another less common sequence of groups is C=C-C=C-N-C=O, which is present in the N-formyl substrate (15) that can be successfully converted to a protoberberine derivative capable of being reduced to xylopinine.[14] The final reaction in this section presents a puzzle, since the N-benzylenamine in (16) lacks the conjugation necessary for an electrocyclic reaction but nonetheless cyclizes efficiently;[15] a different mechanism must operate, probably initiated by attack of an excited aromatic ring on an alkene unit.

MeO MeO N O hν HI MeO MeO N+ I− OMe OMe 75% OMe OMe (15)

MeO N NH hν MeO N NH 80-85% (16)

Substitution Reactions

Aryl halides, particularly aryl iodides, can undergo efficient homolytic carbon-halogen bond fission on irradiation, and the aryl radical so formed is capable of attacking a ground-state aromatic molecule. Intramolecular reaction of this type is responsible for the ring-closure of a substrate with an o-iodostilbene unit (17), which generates an alkaloid of the aristolactam type.[16] Although this reaction could in principle involve an electrocyclic reaction followed by elimination of hydrogen halide, many similar reactions are known in which the aryl groups are not conjugated.[17]

MeO MeO I NH O hν → MeO MeO NH O 25%

cepharanone B

(17)

The excited states of aromatic compounds are much more susceptible to nucleophilic attack than are the ground states, and substitutions employing carbon nucleophiles, usually enolates, are particularly valuable in forming new carbon-carbon bonds to an aromatic ring.[18] An intramolecular example is seen in the reaction (18) employed to make cephalotaxine in good yield from a substrate for which various thermal methods of cyclization proved markedly inferior.[19]

O O I N O OMe hν → O O N O OMe 94%

cephalotaxine

(18)

Many aromatic substitution reactions take place by a mechanism involving initial electron transfer. Alkoxy-benzenes are especially prone to donate an electron from their excited state, and the application of this in compounds with a chloroacetamide group remotely attached to the aromatic ring offers a very useful source of medium-ring lactams.[20] An N-chloroacetyl 2-benzylpyrrolidine can be converted (19) to a compound containing three of the four rings necessary for the cephalotaxine skeleton (not counting the methylenedioxy substituent).[21] A more elaborate example is seen in the reaction (20) leading to an analogue of the iboga alkaloid catharanthine,[22] which was used subsequently in an approach to the di-indole alkaloid vinblastine.

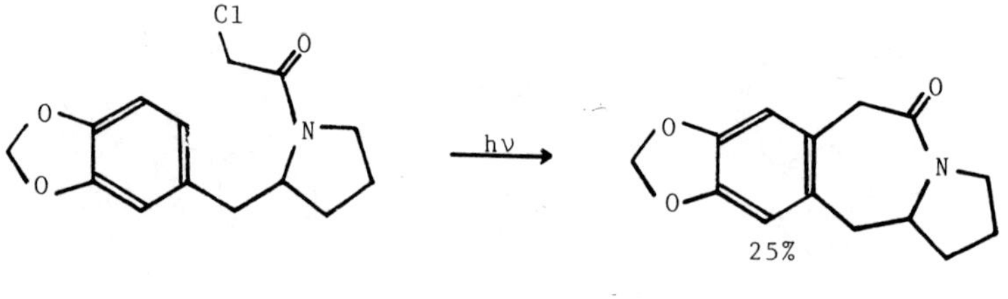

25%

(19)

45%

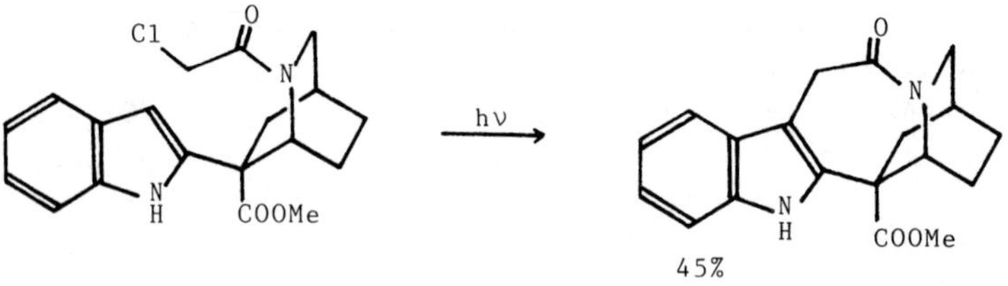

(20)

An excited aromatic compound is a good electron donor even if it does not carry alkoxy substituents, and electron transfer can occur from an unsubstituted phenyl ring to a more strongly accepting group; this is the basis for some of the photocyclization reactions of substituted iminium salts.[23] A simple N-benzyl iminium compound (21) gives a tricyclic nitrogen heterocycle in good yield.[24] A related process in which the aromatic ring is electron accepting is seen in the formation of an indolizidine product from an N-allyl pyridinium salt (22).[25] These types of reaction have been used more extensively for alkaloid synthesis in non-aromatic systems.

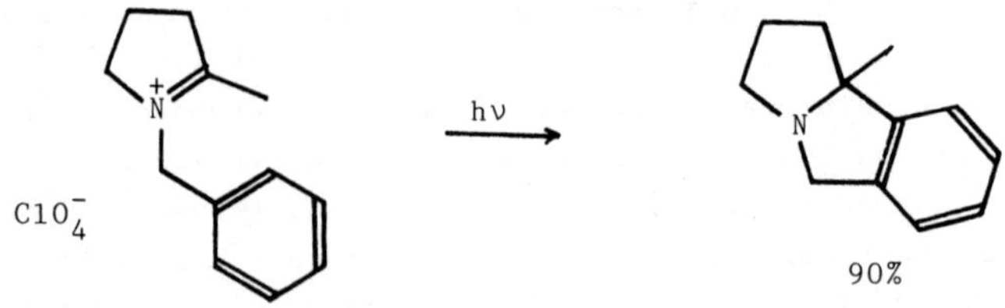

ClO_4^-

90%

(21)

ClO_4^- $\xrightarrow[MeOH]{h\nu}$ $\xrightarrow[PtO_2]{H_2}$ 60% OMe

(22)

Other reactions

The selection of reactions in this review has been restricted to processes in which an aromatic ring is involved as chromophore and takes part in the initial steps of the reaction. A range of other photochemical reactions can be adapted to the formation of alkaloid structures with aromatic groups, such as the cyclization of N-substituted phthalimides giving multicylcic products (23) whose treatment with acid yields an oxidized protoberberine derivative.[26]

MeO MeO O O N N OMe OMe $\xrightarrow{h\nu}$ MeO MeO O N HO N MeO OMe 76%

H^+

MeO MeO O N OH OMe OMe 59%

(23)

References

1. F.B. Mallory and C.W. Mallory, Organic Reactions, 1984, 30, 1.
2. T. Kametani and K. Fukumoto, Accounts Chem. Res., 1972, 5, 212.
3. J.B. Hendrickson and J.G. De Vries, J. Org. Chem., 1982, 47, 1148.
4. T. Onaka, Y. Kanda and M. Natsume, Tetrahedron Lett., 1974, 1179.
5. A.G. Schulz, R.D. Lucci, J.J. Napier, H. Kinoshita, R. Ravichandran, P. Shannon and Y.K. Kee, J. Org. Chem., 1985, 50, 217.
6. I. Ninomiya and T. Naito, in A. Brossi (ed), "The Alkaloids", Volume 22, Academic Press, 1983, p.189.
7. I. Ninomiya and T. Naito, Chem. Commun., 1973, 137.
8. H. Iida, S. Aoyagi and C. Kibayashi, Chem. Commun., 1974, 499.
9. H. Ishii, T. Ishida, M. Ueda and K. Harada, J. Chem. Soc., Perkin Trans. 1, 1975, 762.
10. I. Ninomiya, C. Hashimoto, T. Kiguchi and T. Naito, J. Chem. Soc., Perkin Trans. 1, 1985, 941.
11. I. Ninomiya, C. Hashimoto, T. Kiguchi and T. Naito, J. Chem. Soc., Perkin Trans. 1, 1983, 2967.
12. I. Ninomiya, T. Kiguchi, S. Yamauchi and T. Naito, J. Chem. Soc., Perkin Trans. 1, 1980, 197.
13. H. Hara, O. Hoshino and B. Umezawa, Tetrahedron Lett., 1972, 5031.
14. G.R. Lenz, J. Org. Chem., 1977, 42, 1117.
15. Atta-ur-Rachman and M. Ghazala, Heterocycles, 1981, 16, 261.
16. L. Castedo, E. Guitian, J.M. Saa and R. Suau, Heterocycles, 1982, 19, 279.
17. E.g. S.V. Kessar, Y.P. Gupta, V.S. Yadav, M. Narula and T. Mohammad, Tetrahedron Lett., 1980, 3307.
18. J.F. Bunnett, Accounts Chem. Res., 1978, 11, 413.
19. M.F. Semmelhack, B.P. Chong, R.D. Stauffer, T.D. Rogerson, A. Chong and L.D. Jones, J. Am. Chem. Soc;, 1975, 97, 2507.
20. R.J. Sundberg, in A. Padwa (ed), "Organic Photochemistry", Volume 6, Marcel Dekker, New York, 1983, chapter 2.
21. L.J. Dolby, S.J. Nelson and D. Senkovich, J. Org. Chem., 1972, 37, 3691.

22. R.J. Sundberg and D. Bloom, J. Org. Chem., 1980, 45, 3382.
23. P.S. Mariano, Tetrahedron, 1983, 39, 3845.
24. A.J.Y. Lan, S.L. Quillen, R.O. Heuckeroth and P.S. Mariano, J. Am. Chem. Soc., 1984, 106, 6439.
25. U.C. Yoon, S.L. Quillen, P.S. Mariano, R. Swanson, J.L. Stavinoha and E. Bay, J. Am. Chem. Soc., 1983, 105, 1204.
26. J.D. Coyle, L.R.B. Bryant, J.E. Cragg, J.F. Challiner and E.J. Haws, J. Chem. Soc., Perkin Trans. 1, 1985, 1177.

16
Photoinitiated Free-radical Chain Reactions

By J. Hutchison
THE ELECTRICITY COUNCIL RESEARCH CENTRE, CAPENHURST, CHESTER CHI 6ES, UK

Introduction

Most of the large scale industrial applications of "photochemistry in organic synthesis" involve the photoinitiation of free radical chain reactions[1] (the notable exception being photonitrosation of cyclohexane). The amplification of the primary photochemical (radical forming) process by a large number of successive thermal reactions can lead to very high quantum yields and consequently insignificant lamp power costs. Photoinitiation of short-chain reactions, requiring a relatively large number of initiating radicals, can also be economically favourable; more so than initiation of these reactions by thermal decomposition of peroxides or azo compounds when the costs of these materials are compared with the corresponding lamp power costs. A further advantage of photochemical initiation over thermal initiation is the ability to carry out the reactions at lower temperatures, which often results in higher selectivity with complicating side reactions occurring to a much lesser extent.

Halogens figure prominently in both academic studies and industrial applications of photoinitiated free radical chain reactions. As well as substitution and addition reactions involving the elemental halogens, participation by halogen atoms or halogen containing radicals occurs in photohydrohalogenation and dehydrohalogenation, alkyl polyhalide additions, photosulphochlorination, and in (non-chain) photonitrosation. After covering these areas, photoadditions to unsaturated systems of a variety of other compounds are dealt with. These compounds include hydrogen sulphide, amines, ketones, alcohols, ethers and carboxylic acid derivatives. Finally, the consequences of free radical chain kinetics and "spatial inhomogeneity" are considered, in relation to light intensity and photoinitiator concentration.

Photosubstitution reactions involving elemental halogens

Photohalogenation by the free radical chain subsitution mechanism is important as one of a limited number of ways of introducing functional groups into paraffins, and as a selective route to desired alkyl polyhalide isomers using starting materials which have already been halogenated, perhaps by a different method. For example gas phase photochlorination of 1,1-dichloroethane produces the commercially important 1,1,1-trichloroethane rather than the 1,1,2 isomer,the selectivity being higher than the liquid phase thermally initiated reaction[2].

$$CH_3CHCl_2 + Cl_2 \xrightarrow{h\nu} CH_3CCl_3 + HCl$$

As with all free radical chain reactions, the mechanism involves initiation, propagation and termination stages. Written in general terms, where X_2 is a halogen molecule and RH is, say, an alkane:

Initiation: $$X_2 \xrightarrow{h\nu} X\cdot + X\cdot \qquad (1)$$

Propagation: $$RH + X\cdot \longrightarrow R\cdot + HX \qquad (2)$$

$$R\cdot + X_2 \longrightarrow RX + X\cdot \qquad (3)$$

Termination:e.g. (two radicals react with each other)

$$R\cdot + R\cdot \longrightarrow R{-}R$$

$$R\cdot + X\cdot \longrightarrow RX$$

If propagation proceeds at a much higher rate than termination, then long chains and high quantum yields will be obtained for the overall reaction. In the photochlorination of paraffins, quantum yields of several thousand can be achieved, although in industrial plants the quantum yield may only be around 100, due to the presence of impurities such as oxygen which act as chain terminators[3]. Even so, a 400 watt medium pressure mercury lamp can bring about the reaction of 5 to 15 kg/h of chlorine[4]. Another example is the production of 27 tonnes per day of monochloroalkanes from a C_{11-14} n-paraffin cut, using a 7.5 kW mercury lamp[1] (150 kg/kWh). Monochloroalkanes are used to alkylate benzene in the production of alkylbenzene sulphonate detergents.

Photosubstitution reactions involving bromine and iodine are of much less practical importance than chlorination. (Fluorine reacts violently with hydrocarbons without any recourse to photoinitiation.) The lower reactivities of bromine and iodine are partly due to the reverse of reaction (2) in the above scheme playing a significant role, and in the case of iodine the reverse of reaction (3)[5]. Since chlorine absorbs, and is dissociated by, wavelengths covering the range 250 to 450 nm, a high proportion of the output from an undoped medium pressure mercury lamp may be utilised. Alternatively, phosphor coated low pressure lamps emitting within this range can be used. Reactor design must take into account the high exothermicity of these reactions.

Photochlorination of methane can be carried out in the gas phase[6] or in the liquid phase in carbon tetrachloride solution[3].

$$CH_4 \xrightarrow[Cl_2]{h\nu} CH_3Cl + CH_2Cl_2 + CHCl_3 + CCl_4 + HCl$$

Carbon tetrachloride tends to be the main product because the other chlorinated methanes are more readily chlorinated than methane itself.

Photochlorination of toluene is carried out using uncooled lamps, the heat from the lamps and the heat of reaction maintaining refluxing of the toluene[6]. The products result from substitution in the methyl group, rather than chlorination of the benzene ring.

$$C_6H_5CH_3 \xrightarrow[Cl_2]{h\nu} \underset{100}{C_6H_5CH_2Cl} + \underset{:\ 10\ :}{C_6H_5CHCl_2} + \underset{1}{C_6H_5CCl_3} + HCl$$

Benzyl chloride produced in this way is used to make phenylacetic acid, which is in turn used in the production of penicillin.

$$C_6H_5CH_2Cl \xrightarrow{CN^-} C_6H_5CH_2CN \longrightarrow C_6H_5CH_2COOH$$

The capacity in the United States for toluene photochlorination products is in the region of 85,000 tonnes per year.

Photochlorination of 1,1-difluoroethane[7] is used as a commercial route to vinylidine fluoride (monomer for Viton polymers).

$$CF_2HCH_3 \xrightarrow[Cl_2]{h\nu} CF_2ClCH_3 + HCl$$
$$>95\%$$

$$CF_2ClCH_3 \longrightarrow CF_2{=}CH_2 + HCl$$

As in the case of 1,1,1-trichloroethane production referred to earlier, the substitution proceeds via the more stable of the two possible intermediate radicals, $CH_3\dot{C}X_2$ being more stable than $\dot{C}H_2CHX_2$.

<u>Photoaddition reactions of elemental halogens</u>

Halogens undergo photoinitiated addition to a wide variety of unsaturated compounds by the following free radical chain mechanism.

Initiation: $X_2 \xrightarrow{h\nu} X\cdot + X\cdot$

Propagation: $X\cdot + \gt C{=}C\lt \longrightarrow X{-}\overset{|}{\underset{|}{C}}{-}\overset{|}{\underset{|}{C}}\cdot$

$X{-}\overset{|}{\underset{|}{C}}{-}\overset{|}{\underset{|}{C}}\cdot + X_2 \longrightarrow X{-}\overset{|}{\underset{|}{C}}{-}\overset{|}{\underset{|}{C}}{-}X + X\cdot$

Termination: radical-radical reactions.

The reactions are again retarded by the presence of oxygen, which readily adds to the intermediate radicals.

$$X{-}\overset{|}{\underset{|}{C}}{-}\overset{|}{\underset{|}{C}}\cdot + O_2 \longrightarrow X{-}\overset{|}{\underset{|}{C}}{-}\overset{|}{\underset{|}{C}}{-}O{-}O\cdot$$

The resulting peroxy radicals are unreactive towards the halogen molecule, and act mainly as chain terminators, <u>e.g</u>.

$$X{-}\overset{|}{\underset{|}{C}}{-}\overset{|}{\underset{|}{C}}{-}O{-}O\cdot + \cdot\overset{|}{\underset{|}{C}}{-}\overset{|}{\underset{|}{C}}{-}X \longrightarrow X{-}\overset{|}{\underset{|}{C}}{-}\overset{|}{\underset{|}{C}}{-}O{-}O{-}\overset{|}{\underset{|}{C}}{-}\overset{|}{\underset{|}{C}}{-}X$$

Addition of chlorine or bromine is usually more synthetically useful than addition of iodine. The reactivity sequence is $Cl\cdot > Br\cdot > I\cdot$ Chlorine additions tend to suffer less from complicating side reactions than bromine and iodine additions. Some typical examples[8] are given below.

$$Cl_2 + CH_2{=}CHCl \xrightarrow{h\nu} CH_2ClCHCl_2$$

$$Cl_2 + C_6H_5CH{=}CH{-}CH{=}CH{-}COOH \xrightarrow{h\nu} C_6H_5(CHCl)_4COOH$$

$$Cl_2 + CF_2{=}CH_2 \xrightarrow{h\nu} \underset{98\%}{CF_2ClCH_2Cl}$$

$$Cl_2 + CF_3CH{=}CH_2 \xrightarrow{h\nu} \underset{80\%}{CF_3CHClCH_2Cl}$$

$$Br_2 + CCl_3CH{=}CH_2 \xrightarrow{h\nu} CCl_3CHBrCH_2Br + CCl_2BrCHClCH_2Br$$

85% yield of mixture

In this last reaction, the second product results from isomerisation of the intermediate radical.

$$CCl_3\dot{C}HCH_2Br \longrightarrow \cdot CCl_2CHClCH_2Br$$

By far the most important industrial application of this type of reaction has been the addition of chlorine to benzene to produce hexachlorocyclohexane (benzene hexachloride). The photoinitiated process was developed industrially in the 1950's and was preferred to the thermally initiated process because the lower temperatures at which it could be carried out gave a higher yield (although still only 15%) of the desired gamma isomer, a versatile insecticide[1]. Quantum yields in the region of 2,500 were obtained using 40 watt fluorescent lamps as the light source.

Photohydrohalogenation

Free radical chain addition of hydrogen halides to unsaturated compounds is confined to hydrogen bromide as far as synthetic utility is concerned[8]. Hydrogen bromide absorbs, and is dissociated by, wavelengths shorter than 290 nm. Longer wavelengths can be used if the reactions are carried out in the presence of a photoinitiator such as acetone, absorption by acetone extending to 330nm. In the gas phase, dissociation of photoexcited acetone occurs, whereas in the liquid phase free radicals are produced by the photoexcited acetone performing a hydrogen abstraction reaction.

Gas phase: $(CH_3)_2C{=}O^* \longrightarrow \dot{C}H_3 + CH_3\dot{C}{=}O$

Liquid phase: $(CH_3)_2C{=}O^* + HBr \longrightarrow (CH_3)_2\dot{C}OH + Br\cdot$

The propagation steps for the addition of hydrogen bromide to a terminal olefin are

$$RCH{=}CH_2 + Br\cdot \longrightarrow R\dot{C}HCH_2Br$$

$$R\dot{C}HCH_2Br + HBr \longrightarrow RCH_2CH_2Br + Br\cdot$$

A terminal bromide is therefore produced because (i) it is a bromine atom which attacks the olefin and (ii) $R\dot{C}HCH_2Br$ is more stable than $RCHBr\dot{C}H_2$, since the alkyl group stabilises the free radical centre to a greater degree than hydrogen. This mode of addition (anti-Markownikoff) is in contrast to the ionic additions of hydrogen halides which give non-terminal (Markownikoff) products *via* initial proton attack on the terminal olefin. The almost exclusive formation of terminally substituted products by free radical chain addition to terminal olefins make this a particularly useful synthetic reaction *e.g.*[9]

$$CH_3CH{=}CH_2 + HBr \longrightarrow CH_3CH_2CH_2Br \quad (87\%)$$

$$C_2H_5CH{=}CH_2 + HBr \longrightarrow C_2H_5CH_2CH_2Br \quad (92\%)$$

$$n\text{-}C_3H_7CH{=}CH_2 + HBr \longrightarrow n\text{-}C_3H_7CH_2CH_2Br \quad (96\%)$$

The reaction is also applicable to terminal acetylenes[9].

$$CH_3C{\equiv}CH + HBr \xrightarrow[-60°C]{h\nu} cis\text{-}CH_3CH{=}CHBr \quad (88\%)$$

$$C_4H_9C{\equiv}CH + HBr \longrightarrow C_4H_9CH{=}CHBr \quad (73\%)$$

Some further examples[8]:

$$CH_2{=}CCl_2 + HBr \xrightarrow{h\nu} CH_2BrCHCl_2 \quad (62\%)$$

$$CF_3CH{=}CH_2 + HBr \xrightarrow{h\nu} CF_3CH_2CH_2Br \quad (90\%)$$

$$CH_2{=}CHCH_2CH_2CH{=}CH_2 + HBr \xrightarrow{h\nu} CH_2Br(CH_2)_4CH_2Br \quad (95\%)$$

Photodehydrohalogenation

Laser initiated dehydrochlorination has been proposed as a potentially economic method for the industrial production of vinyl chloride from 1,2-dichloroethane[10]. At elevated temperatures (350-500°C) the reverse of the above propagation sequence occurs.

$$Cl\cdot + CH_2ClCH_2Cl \longrightarrow CH_2Cl\dot{C}HCl + HCl$$

$$CH_2Cl\dot{C}HCl \longrightarrow CH_2{=}CHCl + Cl\cdot$$

In the thermally initiated reaction, used at present industrially, wall effects play an important part in the initiation and termination processes[11], and thereby complicate the overall kinetics. Laser light initiation of the reaction provides an ideal method for studying the reaction under wall-free conditions, and was found to give a cleaner reaction with a higher yield of vinyl chloride. Laser initiated dehydrochlorination could also be suitable for the industrial production of other olefin monomers such as vinyl fluoride, tetrafluoroethylene, chloroprene, propylene, vinylidine fluoride and vinylidine chloride.

Photoaddition of alkyl polyhalides

From a synthetic point of view, photoaddition of alkyl polyhalides to unsaturated systems can be considered as a further halogenation method. From a mechanistic point of view, exhaustive studies of these reactions, particularly of the polyhalomethanes, have provided valuable information relevant to the whole area of free radical chain addition.

In the addition of carbon tetrachloride to terminal olefins, a large excess of CCl_4 is required to avoid telomer formation, whereas in the case of carbon tetrabromide, only a slight excess is required, and the reaction proceeds at a much faster rate; so much so that CCl_4 can be used as the solvent for CBr_4 reactions[8].

Initiation: $\mathrm{CX_4 \xrightarrow{h\nu} \dot{C}X_3 + X\cdot}$

Propagation: $\mathrm{RCH{=}CH_2 + \dot{C}X_3 \longrightarrow R\dot{C}HCH_2CX_3}$

$$\mathrm{R\dot{C}HCH_2CX_3 + CX_4 \longrightarrow RCHXCH_2CX_3 + \dot{C}X_3}$$

Telomerisation:

$$\mathrm{R\dot{C}HCH_2CX_3 + RCH{=}CH_2 \longrightarrow \begin{array}{l} RCHCH_2CX_3 \\ \quad | \\ \ CH_2\underset{\bullet}{C}HR \end{array}}$$

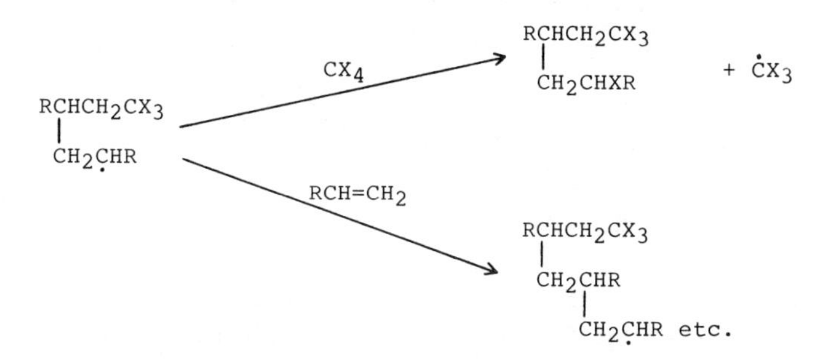

Termination: radical + radical $\longrightarrow$ non-radical products.

The observations are consistent with CBr_4 being much more reactive that CCl_4 in the propagation reaction with $\mathrm{R\dot{C}HCH_2CX_3}$.

Some examples of CCl_4 and CBr_4 photoaddition[8]:

$$\mathrm{CH_3CH{=}CH_2 + CCl_4 \xrightarrow{h\nu} CH_3CHClCH_2CCl_3} \qquad (55\%)$$

$$\mathrm{C_2H_5OCH{=}CH_2 + CCl_4 \xrightarrow{h\nu} C_2H_5OCHClCH_2CCl_3} \qquad (92\%)$$

$$\mathrm{C_6H_5CH{=}CH_2 + CBr_4 \xrightarrow{h\nu} C_6H_5CHBrCH_2CBr_3} \qquad (96\%)$$

$$\mathrm{C_6H_{13}CH{=}CH_2 + CBr_4 \xrightarrow{h\nu} C_6H_{13}CHBrCH_2CBr_3} \qquad (88\%)$$

Bromotrichloromethane is also much more reactive than CCl_4, the reactions proceeding by scission of the C-Br bond.

$$\mathrm{CH_3CH{=}CH_2 + BrCCl_3 \xrightarrow{h\nu} CH_3CHBrCH_2CCl_3} \qquad (62\%)$$

$$\mathrm{CH_3COOCH{=}CH_2 + BrCCl_3 \xrightarrow{h\nu} CH_3COOCHBrCH_2CCl_3} \qquad (90\%)$$

The reactivity of trifluoroiodomethane lies between $BrCCl_3$ and CCl_4:[12]

$$RCH{=}CH_2 + ICF_3 \xrightarrow{h\nu} RCHICH_2CF_3$$

$$CF_3CF{=}CF_2 + ICF_3 \xrightarrow{h\nu} CF_3CFICF_2CF_3$$

$$CF_3CH{=}CF_2 + ICF_3 \xrightarrow{h\nu} (CF_3)_2CHCF_2I$$

$$CH_3C{\equiv}CH + ICF_3 \xrightarrow{h\nu} CH_3CI{=}CHCF_3$$

The reactions of bromoform proceed by C-Br scission, in contrast to chloroform in which preferential C-H scission occurs:[8]

$$RCH{=}CH_2 + HCCl_3 \xrightarrow{h\nu} RCH_2CH_2CCl_3$$

$$RCH{=}CH_2 + BrCHBr_2 \xrightarrow{h\nu} RCHBrCH_2CHBr_2$$

Abstraction of an allylic hydrogen atom, rather than addition of the radical to the C=C double bond, is a further complication in polyhalomethane reactions and in free radical chain addition reactions generally. The allylic radical which is formed is resonance stabilised and therefore relatively unreactive. Instead of continuing the propagation of the reaction by abstracting an atom from the addend, allylic radicals tend to accumulate in the system and act mainly as chain terminators by reacting with other radicals[8].

$$RCH_2CH{=}CH_2 + \dot{C}X_3 \longrightarrow R\dot{C}HCH{=}CH_2 \leftrightarrow RCH{=}CH\dot{C}H_2 + HCX_3$$

<u>Photosulphochlorination and sulphoxidation</u>

The production of alkanesulphonyl chlorides by the photo-induced reaction of sulphur dioxide and chlorine with C_{10}-C_{20} paraffins predates photochlorination as an important industrial process[3].

Initiation: $Cl_2 \xrightarrow{h\nu} Cl\cdot + Cl\cdot$

Propagation: $RH + Cl\cdot \longrightarrow R\cdot + HCl$

$R\cdot + SO_2 \longrightarrow R\dot{S}O_2$

$R\dot{S}O_2 + Cl_2 \longrightarrow RSO_2Cl + Cl\cdot$

Termination: radical - radical reactions.

Reaction conditions are selected to favour the formation of monosulphonyl chlorides, however the position of substitution is virtually random[1]. Polysubstitution is undersirable, as is direct chlorination to give alkyl chlorides. The extent to which direct chlorination occurs increases with increasing temperature, and it becomes the main reaction above 100°C. Sulphochlorination is therefore carried out at 20-30°C, at which the Cl:S ratio in the product is approx. 1.2:1. Some alkyl chloride formation is considered to be inevitable owing to photodecomposition of the alkanesulphonyl chloride[3].

$$RSO_2Cl \xrightarrow{h\nu} RCl + SO_2$$

Quantum yields for sulphochlorination in the region of 35,000 have been obtained in the laboratory, coming down to around 2,000 for the industrial process. The fluorescent lamps used consume 30Wh per kg of sulphonyl chloride produced[1] (33 kg/kWh).

Hydrolysis of the product gives water soluble alkane sulphonates, which are used as emulsifiers for polymerisations. The sulphonyl chlorides themselves have been used as insecticides, tanning agents and high pressure lubricants.

Photosulphoxidation of n-alkanes is another industrial route to alkanesulphonic acids, which has the advantage of avoiding the use of chlorine. Peroxysulphonic acids are the primary products formed in a photoinduced free radical chain reaction involving n-alkane, sulphur dioxide and oxygen, the initiation step of which is ill defined.

$$\text{Photoinitiation} \longrightarrow R\cdot$$

$$R\cdot + SO_2 \longrightarrow R\dot{S}O_2$$

$$R\dot{S}O_2 + O_2 \longrightarrow RSO_2OO\cdot$$

$$RSO_2OO\cdot + RH \longrightarrow RSO_2OOH + R\cdot$$

The peroxysulphonic acids are continuously extracted into water (which prevents the formation of polysubstituted products) where further reaction with SO_2 produces the sulphonic acids.

$$RSO_2OOH + SO_2 + H_2O \longrightarrow RSO_3H + H_2SO_4$$

Quantum yields obtained for the reaction are very much lower than for sulphochlorination, being in the region of only 7 or 8[3]. Nevertheless, plants with a total capacity of 50,000 tonnes per year operate in France and Germany. The 40kW medium pressure lamps used, in immersion wells, consume 0.2 kWh per kg of sulphonate produced[1] (5 kg/kWh).

Photonitrosation

Another important industrial process in which the key step is a photo-induced free radical reaction is the production, in Japan, of 160,000 tonnes per year of caprolactam (Nylon 6 monomer) by the photonitrosation of cyclohexane[1]. This is not a chain reaction, and consequently proceeds with a quantum yield $\leqslant 1$. In practice, a quantum yield of 0.8 can be achieved.

$$\underset{\text{nitrosyl chloride}}{Cl{-}N{=}O} \xrightarrow{h\nu} Cl\cdot + \cdot NO$$

$$C_6H_{12} + Cl\cdot \longrightarrow C_6H_{11}\cdot + HCl$$

$$C_6H_{11}\cdot + \cdot NO \longrightarrow \underset{\text{nitrosocyclohexane}}{C_6H_{11}NO} \xrightarrow{H^+} \underset{\text{cyclohexanone oxime}}{C_6H_{10}{=}NOH}$$

HCl gas is fed in with the nitrosyl chloride so that the bis-hydrochloride of the cyclohexanone oxime will be formed. This separates to the bottom of the reactor as an insoluble oil and is continuously removed. In a separate stage of the process, the cyclohexanone oxime undergoes a Beckmann rearrangement to caprolactam.

NOH —H^+→ NH, O

Nitrosyl chloride absorbs over a wide range of ultraviolet and visible wavelengths. However, it was found that UV light, because it was very strongly absorbed, led to the formation of tarry deposits on the immersion well surfaces. Sodium nitrite is therefore added to the cooling water to filter out the UV, and thallium doped lamps emitting a strong line in the visible at 535nm are used as the light source. Each 60 kW lamp produces 24 kg/h of oxime (400 g/kWh).

A similar process has been developed in France to produce 8,000 tonnes per year of lauryl lactam (for Nylon 12) from cyclododecane.

The Barton reaction could be described as intramolecular photonitrosation, and certainly has been described as one of the most successful applications of photochemistry to the synthesis of organic molecules[13]. The reaction is exemplified by the photolysis of n-octyl nitrite[14].

ONO, C_4H_9 —hν→ •NO, O•, H, C_4H_9 → OH •NO, C_4H_9

OH, NOH, C_4H_9 ⇌ OH, NO, C_4H_9 → DIMER

Photocleavage of the nitrite produces nitric oxide and

an alkoxy radical. Intramolecular hydrogen abstraction by the alkoxy radical takes place via a six membered transition state (a hydrogen atom must be available in the 4 position) and the nitric oxide recombines at the new radical centre. Again, there is no chain sequence involved.

The synthetic utility of the reaction has been mainly in the area of steroid chemistry, for the functionalising of normally inactive sites, and it has been used commercially in the production of tritium labelled aldosterone, a medical diagnostic aid[1].

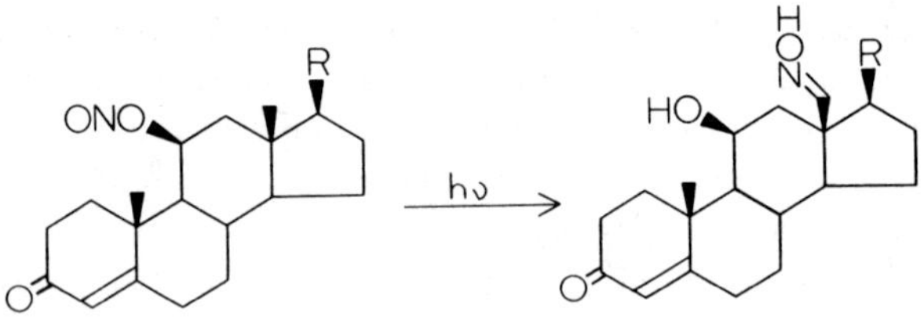

ONO … R —hν→ HO … HON … R

More photoinitiated free radical chain addition reactions

As well as the addition of halogens and halogen containing molecules to unsaturated systems, addition of a wide variety of other compounds can be brought about by the same photoinitiated free radical chain mechanism.

Photoinitiated addition of hydrogen sulphide to terminal olefins proceeds with very high quantum yields of anti-Markownikoff products, and is used industrially for the production of thiols (mercaptans). A temperature of less than 25°C is required, ruling out thermal initiation[2]. The reaction is complicated by the ability of the product thiols to take part in the chain reaction.

$$H_2S \xrightarrow[\text{acetone}]{h\nu} \cdot SH$$

$$RCH{=}CH_2 + \cdot SH \longrightarrow R\dot{C}HCH_2SH$$

$$R\dot{C}HCH_2SH + H_2S \longrightarrow \underset{\text{thiol}}{RCH_2CH_2SH} + \cdot SH$$

$$R\dot{C}HCH_2SH + R'SH \longrightarrow RCH_2CH_2SH + R'S\cdot$$

$$R'S\cdot + RCH{=}CH_2 \longrightarrow R\dot{C}HCH_2SR' \longrightarrow \text{thioether}$$

Thiol addition is itself a useful synthetic reaction, e.g.[8]

$$CH_3CH{=}CH_2 + n\text{-}C_3H_7SH \xrightarrow{h\nu} (n\text{-}C_3H_7)_2S \quad 96\%$$

Similarly with thioacids,

$$\text{cyclohexene} + C_6H_5C(O)SH \xrightarrow{h\nu} C_6H_{11}\text{-}S\text{-}C(O)\text{-}C_6H_5 \quad 89\%$$

Addition of polyfunctional thiols to polyenes has been developed as a commercial photopolymerisation system,[15] used for printing plates and "UV curable" inks and coatings.

Photoinitiated addition of phosphines and silanes can also be brought about, e.g.[8]

$$CF_2{=}CF_2 + PH_3 \xrightarrow{h\nu} CF_2HCF_2PH_2 \ (86\%)$$

$$CF_2{=}CF_2 + (CH_3)_2SiH_2 \xrightarrow{h\nu} CF_2HCF_2Si(CH_3)_2H \ (54\%)$$

The addition of aldehydes to olefins proceeds _via_ the formation of acyl radicals and produces ketones.

$$RCHO^* + RCHO \longrightarrow R\dot{C}HOH + R\dot{C}O$$

$$R\dot{C}O + R'CH{=}CH_2 \longrightarrow RCOCH_2\dot{C}HR'$$

$$RCOCH_2\dot{C}HR' + RCHO \longrightarrow RCOCH_2CH_2R' + R\dot{C}O$$

Ketone addition proceeds _via_ hydrogen abstraction from the carbon atom alpha to the carbonyl group, and is again self-initiated by the photoexcited state of the ketone reacting with a ground state ketone molecule, e.g.[16]

$$\text{cyclohexanone} + C_6H_{13}CH{=}CH_2 \xrightarrow{h\nu} \text{2-}(CH_2CH_2C_6H_{13})\text{cyclohexanone}$$

When, as in the above case, the addend is a sizeable molecule, the reacton may be termed "photoalkylation".

In the free radical chain addition reactions of amines, alcohols, ethers and acetals, hydrogen abstraction from the carbon atoms alpha to the N or O atoms occurs, e.g.[8,9]

piperidine + $CH_2{=}CHC_6H_{13}$ $\xrightarrow{h\nu}$ 2-($CH_2CH_2C_6H_{13}$)piperidine

$$CH_3(CH_2)_9CH{=}CH_2 + (CH_3)_2CHOH \longrightarrow CH_3(CH_2)_{11}C(CH_3)_2OH$$

51%

tetrahydrofuran + $RCH{=}CH_2$ $\xrightarrow[\text{acetone}]{h\nu}$ 2-(CH_2CH_2R)tetrahydrofuran

1,3-dioxolane + $RCH{=}CH_2$ $\xrightarrow[\text{acetone}]{h\nu}$ 2-(CH_2CH_2R)-1,3-dioxolane $\xrightarrow{\text{dil. HCl}}$ $RCH_2CH_2CHO + (CH_2OH)_2$

terminal aldehyde

The addition of formamide to olefins gives high yields of 1:1 adducts *via* the intermediacy of $\dot{C}ONH_2$ radicals, providing a useful synthetic method for the preparation of amides and the corresponding carboxylic acids[8]. No products are obtained arising from N-H bond scission.

$$C_n \text{ olefin} \longrightarrow C_{n+1} \text{ amide} \longrightarrow C_{n+1} \text{ acid}$$

With terminal olefins, the addition is very largely anti-Markownikoff, the yield of non-terminal adduct being around 5%, *e.g.*[17]

$$CH_3(CH_2)_7CH{=}CH_2 + HCONH_2 \xrightarrow[\text{acetone}]{h\nu} CH_3(CH_2)_9CONH_2 \quad (67\%)$$

"Photoamidation" is also applicable to a wide variety of non-terminal and cyclic olefins, and to α,β-unsaturated esters. With the latter, the orientation of $\dot{C}ONH_2$ addition is usually towards the β-carbon atom[8].

$$CH_3(CH_2)_nCH{=}CHCOOR + HCONH_2 \xrightarrow[\text{benzophenone}]{h\nu} CH_3(CH_2)_nCH(CONH_2)CH_2COOR$$

n = 4,5 or 6

Addition of N,N-dimethylformamide to olefins (thermally initiated[18]) produces a mixture of 1:1 adducts, hydrogen abstraction occurring from both possible sites.

$$RCH{=}CH_2 + HCON(CH_3)_2 \longrightarrow RCH_2CH_2CON(CH_3)_2 + HCON(CH_3)(CH_2CH_2CH_2R)$$

The photoinitiated addition of N-methylacetamide to olefins also gives a mixture of 1:1 adducts, the major product arising *via* hydrogen abstraction from the methyl group attached to the nitrogen atom[8].

$$CH_3CONHCH_3 + RCH{=}CH_2 \xrightarrow[\text{acetone}]{h\nu} \underset{\text{major}}{CH_3CONHCH_2CH_2CH_2R} + \underset{\text{minor}}{RCH_2CH_2CH_2CONHCH_3}$$

Similarly with lactams, *e.g.* pyrrolidone

pyrrolidone + $RCH{=}CH_2$ $\xrightarrow[\text{acetone}]{h\nu}$ 5-(CH_2CH_2R)-pyrrolidone (major) + 3-(CH_2CH_2R)-pyrrolidone (minor)

In contrast, lactone addition to olefins results in preferential alkylation at the carbon atom alpha to the carbonyl group.

$RCH{=}CH_2$ + butyrolactone $\xrightarrow[\text{acetone}]{h\nu}$ α-(CH_2CH_2R)-butyrolactone (major) + γ-(CH_2CH_2R)-butyrolactone (minor)

In several of the above acetone-photoinitiated reactions, it has been shown that isopropanol and 2-methylalkan-2-ols are also produced, providing evidence for the hydrogen abstraction initiation step. The free radical chain mechanism is further supported by the formation of telomers (referred to earlier) and dehydrodimers (e.g. oxamide in formamide reactions) as well as the anti-Markownikoff 1:1 adducts.

Consequences of radical chain kinetics and spatial inhomogeneity

A characteristic feature of most free radical chain reactions is the dependence of the propagation rate on the square root of the rate of initiation. This follows from kinetic treatment of the normal initiation, propagation and termination stages[19], and stems from the termination stage involving the reaction with each other of radicals which participate in the propagation stage.

In a photoinitiated free radical chain reaction, the rate of initiation varies throughout the reaction mixture (is spatially inhomogeneous), obviously being highest nearest the light source. For an unstirred system, the propagation rate in any small region will be proportional to the square root of the rate of light absorption (I_{abs}) by the photoinitiator within that region. When the reaction is brought about by a parallel beam of monochromatic light, and the photoinitiator is the only absorbing species, then by integrating Rate $\propto (I_{abs})^{\frac{1}{2}}$ over the depth of the reaction vessel, it can be shown that the overall propagation rate in the system will be proportional to

$$\left(\frac{Io}{\epsilon' c}\right)^{\frac{1}{2}} (1-e^{-\epsilon' cd/2})$$

where Io = incident light intensity (einstein $cm^{-2}h^{-1}$)

ϵ' = molar extinction coefficient of the photoinitiator at the irradiation wavelength, defined to the base e

c = photoinitiator concentration

d = depth of reaction vessel.

For a given photoinitiator concentration, the overall rate will thus be proportional to the square root of the incident light intensity, which means that the overall quantum yield observed for the reaction will be inversely proportional to the square root of the incident light intensity.

For a fixed incident light intensity, the dependence of the overall rate on the photoinitiator concentration is obviously not straightforward. At low concentrations, such that $\epsilon' cd$ is small, the following approximation holds

$$(1 - e^{-\epsilon' cd/2}) \doteqdot \epsilon' cd/2$$

$$(\text{from } \log_e(1 - y) = - y - y^2/2 - y^3/3 - - - - -)$$

which, substituted in the above expression, predicts that the overall rate will be proportional to the square root of the photoinitiator concentration. At high photoinitiator concentrations, i.e. $\epsilon' cd$ large,

$$(1 - e^{-\epsilon' cd/2}) = 1$$

and the overall rate becomes inversely proportional to the square root of the photoinitiator concentration. Between these two extremes, the overall rate passes through a maximum, as shown below.

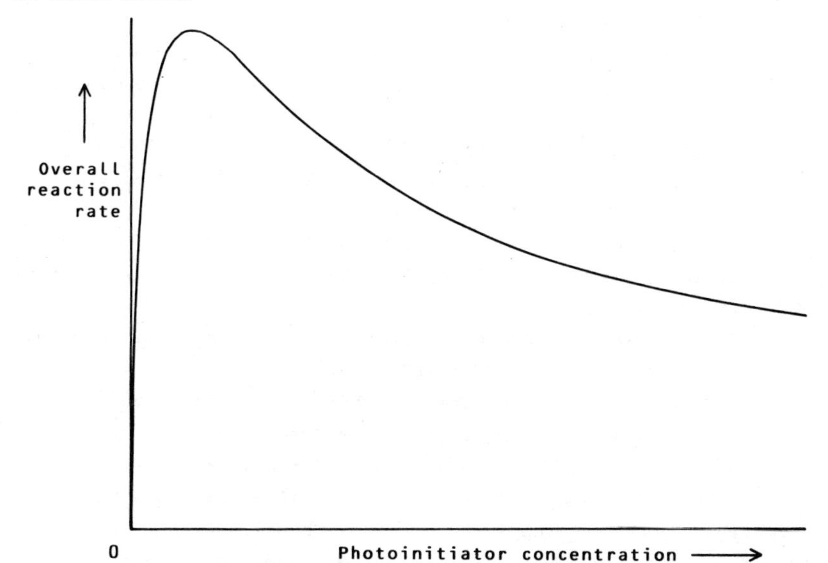

Differentiation of the expression with respect to photoinitiator concentration predicts that the maximum will occur when

$$1 + \epsilon' cd = e^{\epsilon' cd/2}$$

which corresponds to $\epsilon' cd = 2.513$, or an absorbance defined as normally, to the base 10, of 2.513/2.303 = 1.091. Such behaviour has been observed experimentally for the polymerisation of methyl methacrylate photoinitiated by benzoin methyl ether using the 366nm mercury line[20].

This consequence of free radical chain kinetics and spatial inhomogeneity should always be borne in mind before experimentally observed concentration or wavelength dependences are explained otherwise, for example in terms of inhibition, self - quenching, or a change in photochemical behaviour with excitation wavelength.[21] Theoretical treatments have been developed, for photoinitiated free radical polymerisations[22], in which components other than the photoinitiator also absorb the incident radiation, and stirring of the system is taken into account.

References

1. M. Fischer, Angew. Chem. Int. Ed. Engl., 1978, 17, 16.
2. A.D. Clements, Chem. Brit., 1980, 16, 464.
3. F. Asinger,"Paraffins Chemistry and Technology", Pergamon, 1968.
4. D.R. Arnold and P. de Mayo, Chem. Technol, 1971, 615.
5. M.L. Poutsma, "Free Radicals", Vol. 2, ed. J.K. Kochi, Wiley, New York, 1973, Chapter 15.
6. "Encyclopedia of Chemical Technology", 3rd edition, Wiley, New York, 1982, Vol. 17, pp 540-559.
7. U.S. Patent 1,142,181 (1969).
8. D. Elad, "Organic Photochemistry," Vol. 2, ed. O.L. Chapman, Dekker, New York, 1969, pp168-212.
9. G. Sosnovsky, "Free Radical Reactions in Preparative Organic Chemistry", Macmillan, New York, 1964.
10. Eur. Chem. News (Supplement), Dec.22 1980, 48.
11. J. Wolfram and M. Schneider, Proc. SPIE - Int. Soc. Opt. Eng., 1984, 458 (Appl. Lasers Ind. Chem.), 46.
12. R.N. Haszeldine and B.R. Steele, J. Chem. Soc., 1955, 3005.
13. W.M. Horspool, "Aspects of Organic Photochemistry", Academic Press, London, 1976, p254.

14. A. Schonberg, "Preparative Organic Photochemistry", Springer Verlag, New York, 1968, p 248.
15. G.E. Green, B.P. Stark and S.A. Zahir, J. Macromol. Sci., Rev. Macromol. Chem., 1981, C21, 187.
16. M.S. Kharash, J. Kuderna and W. Nudenberg, J. Org. Chem. 1953, 18, 1225.
17. D. Elad and J. Rokach, J. Org. Chem., 1964, 29, 1855.
18. L. Friedman and H. Schechter, Tetrahedron Letters, 1961, No. 7, 238.
19. C. Walling, "Free Radicals in Solution," Wiley, New York, 1963, p244.
20. E.A. Lissi and J. Garrido, J. Polym. Sci.: Polym. Lett. Ed., 1984, 22, 391.
21. E.A. Lissi and A. Zanocco, J. Polym. Sci.: Polym. Chem. Ed., 1983, 21, 2197.
22. A.R. Schultz and M.G. Joshi, J. Polym. Sci., Polym. Phys. Ed., 1984, 22, 1753.